Statistics for Health Management and Research

Mark Woodward

BSc, MSc, PhD, MIS

and

Lesley MA Francis

BA, BSc, MSc

GW00545581

EDWARD ARNOLD
A division of Hodder & Stoughton
LONDON   BALTIMORE   MELBOURNE   AUCKLAND

First published in Great Britain 1988 by
Edward Arnold, the educational, academic and medical
publishing division of Hodder & Stoughton Limited,
41 Bedford Square, London WC1B 3DQ

British Library Cataloguing Publication Data

Woodward, Mark
  Statistics for health management and
  research.
  1. Medical statistics
  I. Title   II. Francis, Lesley M.A.
  519.5′02461   RA409

  ISBN 0-340-42009-X

Typeset in 10/11pt Compugraphic Times
by Colset Private Limited, Singapore
Printed and bound in Great Britain for Edward Arnold, the educational, academic
and medical publishing division of Hodder and Stoughton Limited, 41 Bedford
Square, London WC1B 3DQ

Printed and bound by Richard Clay Ltd, Bungay, Suffolk

Contents

Acknowledgements

The authors thank the following for their permission to use data and other material: the British Medical Journal and the Journal of Epidemiology and Community Health and their Editors; the Central Statistical Office; the Office of Health Economics; the Office of Population Censuses and Surveys and Oxford Regional Health Authority. Data used in any table or figure where either the Department of Health and Social Security or the Office of Population Censuses and Surveys is given as the source have Crown Copyright. Data from Annual Abstract of Statistics, Monthly Digest of Statistics and Social Trends are reproduced by permission of the Controller of Her Majesty's Stationery Office.

Thanks to G.M. Clarke and D. Cooke for permission to use their table and to R. Mead and R.N. Curnow and publishers Chapman and Hall for permission to use their table and figure. Thanks also to Dr J. Bamford, Sir Austin Bradford-Hill, Dr G.A. Cook, Professor R.O. Cummins, Dr J.D. Edwards, Dr I.O. Ellis, Dr D. Field, Dr W. Gilchrist, J. Levin, Dr D.G. Neal and J. Sellwood who gave permission to use their data. The Birmingham Research Unit of the Royal College of General Practitioners was very helpful in providing information about the RCGP surveys of General Practitioners.

Finally, personal thanks to Jenny Edwards who produced the diagrams, Joan Knock who produced tables, librarians Joan Dyson and Enid Forsyth and to Dr P.N. Dixon, District Medical Officer, West Berkshire Health Authority, for his encouragement. Thank you to Lorna Turner and Muriel Morris for their many hours of typing, and in particular to Lorna for so often braving the ghost of Blandford Lodge.

Charvil, Berkshire
1987

M.W.
L.M.A.F.

To our sons, Robert and Philip

1
Introduction

1.1 Plan of the book

The motivation for this book was the need for a text dealing specifically with health statistics. It provides an introduction to the statistical methods most commonly used in health management and research, and illustrates them using examples drawn from these fields. There is, however, no claim that coverage is complete; the reader may find some subjects about which more detail is required, or even areas of interest that have been excluded due to their limited application or their technical difficulty. In most such cases, however, references are given to more complete expositions elsewhere.

Sections 1.2 and 1.3 in this first chapter introduce the role of statistics in health management and research, while Sections 1.4 and 1.5 cover topics which are necessary background to the rest of the book. Section 1.4 describes the structure of the National Health Service, and Section 1.5 gives guidelines for maintaining accuracy in numerical calculations.

Broadly speaking the remainder of this book divides into three parts. Chapters 2 and 10 are mainly concerned with planning statistical investigations and collecting and storing health data. Chapter 2 deals with general concepts whereas Chapter 10 is devoted to the special type of investigation known as a survey.

Chapters 3, 4 and 5 describe the basic methods of description and summarization of data, Chapters 3 and 4 dealing with general concepts which are applied to demographic data in Chapter 5.

The remaining chapters cover more complex methods of data analysis which allow particular results to be interpreted in a more general context. The link between the particular and the general is the theory of probability, which is the subject of Chapter 6. To escape the mathematical rigour of this theory, Chapter 6 frequently relies upon simple examples, such as gambling problems, to help illustrate the principles involved. Chapter 7 builds on the theory of the normal distribution, introduced in Chapter 6, to present the inferential methods of estimation and hypothesis testing. Chapter 8 deals with correlation and regression, that is, looking for relationships between particular items (*variables*) upon which data have been collected. Chapter 9 discusses the analysis of data in the form of tables.

This book makes no claim that it will turn the reader into a statistician. A basic knowledge of statistics is essential to a health manager or researcher, as

Sections 1.2 and 1.3 will show. This book does aim to provide this basic knowledge, but there will still be many situations in which it will be appropriate to seek the advice of a statistician.

One problem with any text on statistics for non-statisticians is the mathematical element of the subject. There is no viable alternative to the presentation of formulae in mathematical notation, but in this book the notation is explained and numerical examples are given for most of the formulae included. Statistical principles and methods are justified using rational argument, examples or by simple mathematics.

1.2 Statistics in health management

Statistics are of fundamental importance in the planning, administration, monitoring and evaluation of public health care programmes. Few would argue that health services can be run efficiently without 'the facts', yet in the past, many decisions about health policy have been made without adequate information, when health managers have relied upon their own observations or on anecdotal stories related in the press or elsewhere. In almost every problem facing health managers today, 'the facts' take the form of statistics, and may be anything from hospital elective admission list sizes to the relative costs and benefits of hospital and community care for a particular condition. Far too often these facts are not used to their full potential.

Why should this situation have arisen? Part of the reason is that many health managers have paid insufficient attention to the information requirements of their job, seeing statistics as being merely routine data collection demanded by a higher authority for accounting purposes. To such managers, statistics are a nuisance that can be left to a team of clerks (known by various titles) to get on with, completing official returns, compiling summaries and filing away the results for posterity.

Another reason why statistics have not always been properly used is that the statistics themselves have often been inadequate due to unreliability, lack of timeliness or the failure to link records. The reliability and timeliness problems have been caused by the sheer amount of data collected, too much to be able to check them properly or to process them quickly. Naturally the lack of use of data at a local level, as previously mentioned, has contributed to these problems because of unenthusiasm amongst data collectors and clerks, and because errors are not picked up as they might be if the data were routinely used.

Over the years, different record systems have evolved for community health, hospital inpatients, manpower, finance, etc. and data linking between systems has usually been impossible because the systems used different definitions for similar items and had no common access facilities. Even within the same system, linking of records may be very difficult. Linking different periods of hospital stays for a single patient was far from straightforward in the Hospital Activity Analysis system (see Section 2.4.3), as there was no requirement for a unique patient identifier. A secondary effect of this incompatibility of systems has been the inevitable duplication of effort.

Despite all these drawbacks, governments have promoted the idea of regular collection of extensive sets of health data in each locality because of the central role of health in daily life and a perfectly reasonable desire to quantify health and health care. In most health services in the world the result has been the collection and storage of vast amounts of data of considerable detail which are hardly ever used, despite the great value of health information.

In England recent events have made the outlook for statistics in health management much brighter. The first reorganization of the National Health Service (NHS) in 1974 created a hierarchical structure with each tier being responsible for the health care of a well-defined population. This produced a framework for the collection, analysis and dissemination of meaningful statistics to aid management. Although this framework did promote an improvement in statistics, for example through the need for information in the annual plans for the health authorities, the problems mentioned earlier certainly did not disappear. The real impetus for the serious use of statistics in health management came in the early 1980s through the Steering Group on Health Information headed by Edith Körner (see Section 2.4.2) and the coincidental development of cheap and powerful computers and software. The Körner Committee has tackled the problems of motivation in collecting and using statistics by giving district health authorities control over their own data. It has, furthermore, produced sets of common definitions for use in different health systems in order to facilitate linking, and has developed guidelines for the revision of routine data systems, to avoid much of the duplication that previously existed. The development of information technology has made possible the provision of accurate and timely information through speedy processing at, or near, the point of delivery of care, where errors can more easily be identified.

The adoption of the Körner Committee's recommendations and investment in new technology within the NHS has created an environment for the collection and dissemination of health data that is potentially useful to health management. To realize that potential, the data must be translated into *information*. The majority of this book is concerned with this process.

1.3 Statistics in health research

In health research the motivation to use statistics is generally strong, since most research requires data collection and subsequent description, summarization and possibly generalization from the particular cases observed. Examples include investigations into the relationship between smoking and lung cancer, an evaluation of the efficacy of a treatment for AIDS and a comparison of different policies for screening women for cervical cancer.

In recent years statistics have played an increasingly important role in health research due to an increased awareness amongst researchers of the resultant benefits, improved computing power and the demand for statistical analyses from editors of journals, the regulatory authorities of the pharmaceutical industry and others. However, the frequent misuse of statistics in medical journals, as described by Gore and Altman (1982), suggests that there is room for considerable improvement in the use of statistical methods by health researchers.

Health researchers are often hampered in their work by the shortcomings of the routine data systems mentioned in Section 1.2 but, by its very nature, research will often require special studies to be initiated to collect new data. Even when data on a subject of interest already exist, new data which are more detailed, have more items recorded or have wider coverage (or any combination of these), may need to be collected. A common fault is a lack of care in the initial data collection, which inevitably leads to poor quality research whatever the sophistication of the subsequent analyses. Decisions about what data to collect, how many subjects the data should cover and how the data will be recorded and stored all need particular attention. The collection process itself must then be monitored to ensure that high quality data are obtained. These points will be considered in more detail in Chapters 2 and 10.

The health researcher, then, needs to know how to decide which data are relevant to his problems, when special data collection is necessary, how to collect and store data when this is necessary and how to extract from the data the solution to his research problem. These are the issues that are discussed in this book.

The distinction between research and management made in this and the previous section is often blurred when statistical methods are applied to a particular problem. Frequently the same procedure will serve research and managerial functions. Research into road traffic accidents in a particular town could produce a cross-tabulation of geographical location against frequency and severity of accident that is later used by management in choosing locations for ambulance stations. Of course this process, of today's research aiding tomorrow's decision making, is often more generally true. It can also work the other way round, as when statistics on hospital bed use collected for operational and management purposes are used to study changes in morbidity against a background of changing environmental conditions. In some situations the health researcher and manager are one and the same person; community physicians certainly have this dual role. For these reasons it will not be appropriate to distinguish between managerial and research applications in this book; each technique introduced will have many applications and an example of one or more possible applications will usually be included.

1.4 The National Health Service

Some readers will not be familiar with the structure of the NHS and, as many of the examples given in this book relate to the NHS, it is necessary to give a brief description of it here.

There are, in fact, four separate services, for England, Wales, Scotland, and Northern Ireland. While their structures are different, the last three are essentially simplified versions of the first, so the description will be restricted to the NHS in England.

In the English National Health Service, Parliament sits at the top of the administrative hierarchy, followed by the Department of Health and Social Security (DHSS) which is the central government body responsible for both health and personal social security (pensions, unemployment benefits, etc.). The DHSS is headed by its Secretary of State who is a member of the Cabinet.

Within the DHSS there is a Health Services Supervisory Board, concerned with strategic planning, and an NHS Management Board.

Below the DHSS there are 14 regional health authorities (RHAs), each of which is responsible for health services within defined geographical boundaries. The populations served by RHAs vary in size roughly from 2 to 5 million people (see Table 8.3). Each RHA is divided into a number of district health authorities (DHAs), of which there are almost 200 in all. Regions are responsible for strategic planning, large scale capital developments, some operational services (such as the blood transfusion service), specialized management support services (such as computing) and are the employers of consultants and senior registrars. Districts manage the hospital and community services for people living within their boundaries, and employ all staff other than consultants and senior registrars. It does sometimes happen that people are treated outside the DHA, or even the RHA, in which they live (so-called 'cross-boundary flows') and these are allowed for when resources are allocated; the method of resource allocation in the NHS is described in Section 5.8. Each health authority is headed by a committee of responsible members of the community. Day-to-day management of each health authority, and also each 'unit' (such as a hospital), is the responsibility of a general manager supported by a management team of specialist professionals.

In parallel with the hospital and community services runs another line of administration. Directly under the DHSS sit the Family Practitioner Committees (FPCs), each of which is responsible for the general medical, dental, pharmaceutical and ophthalmic services provided within a defined local area. One task of a FPC is to maintain a list of people registered as patients with each general practitioner (GP) in the area. A GP, otherwise known as a family doctor, receives his income directly from his local FPC, this income being partially dependent upon the number of patients registered to him. However, the FPCs do not actually employ GPs; rather each GP acts as an independent contractor.

A few specialist bodies fall outside the general twofold administrative structure of the NHS. These include the special health authorities such as the NHS Training Authority and boards of governors of certain special hospitals, and the Public Health Laboratory Service. Local authorities (local government) also provide certain health, or health-related, services including environmental health, social work and homes for the elderly and the mentally infirm. Local authorities, DHAs and FPCs responsible for a common population liaise in Joint Consultative Committees. Each district has a further body, the Community Health Council, which is a public watchdog on health matters.

Beyond the services funded by central and local government there is a small private health sector. Services offered range from homes for the terminally ill, funded through donations, to large hospitals for acute illnesses run as commercial undertakings. Political influence in the 1980s has caused an expansion of the activities of the private sector, including the contracting-out of some ancillary services in hospitals, such as laundry and cleaning.

1.5 Accuracy in calculations

Most readers will be familiar with the concept of rounding, that is, limiting the number of digits in the result of a numerical calculation by writing the result to the nearest whole number, say, or to the first decimal place (e.g. 4/3 is 1.3 to one decimal place but 1 to the nearest whole number). What are not generally understood are the rules which should be applied when rounding: at what stages of complex calculations rounding should be employed, and to how many digits results should be rounded. These points are particularly important when intermediate results, as well as final ones, are to be presented, as happens when a table with a total is displayed.

Example 1.1: In 1986, a hospital had 5015 outpatient attendances, of which 1263 were general medicine, 1245 general surgery, 1635 ear, nose and throat, and 872 other specialities. The average number of outpatient attendances per day for the different specialities were thus (previous figures divided by 365): 3.46, 3.41, 4.48, and 2.39 respectively (to 2 decimal places). A table of these average daily attendances, each rounded to the nearest whole number, would be:

Average daily outpatient attendances

General medicine	3
General surgery	3
Ear, nose and throat	4
Others	2
Total	12

Here the total is obtained in the usual way, that is by adding up the component parts. In fact, there were 5015/365 = 13.74 (to 2 decimal places) attendances per day in total. Hence the total (12) of the component, rounded, figures is in considerable error.

In general when intermediate figures are added up, their errors (difference between the exact and the rounded value) also add up, so that the error in the total is the accumulation of the errors in the intermediate figures.

The problem of the accumulation of error can be avoided either by working with exact figures at the final stage (not using intermediate results) or by using intermediate results which have a greater degree of accuracy (at least one more digit) than that required in the final result. These methods have been used in this book. The only problem with them is that when intermediate and final results are all presented to the same degree of accuracy it can appear that a mistake has been made somewhere. This effect, an example of *rounding error*, is illustrated by Example 1.2.

Example 1.2: Using intermediate figures to one decimal place, the total number of attendances in Example 1.1 is 3.5 + 3.4 + 4.5 + 2.4 = 13.8, which is 14 to the nearest whole number. Alternatively, using exact figures, the total is 5015/365 which also is 14 to the nearest whole number. A more accurate presentation of the table would be:

Average daily outpatient attendances

General medicine	3
General surgery	3
Ear, nose and throat	4
Others	2
Total	14

It can be useful to indicate to the reader when rounding error has occurred. For instance a footnote could be added to this table which says, 'The total is not the sum of the component parts due to rounding error'.

One thing to remember is that the final result in any calculation cannot be more accurate than the numbers that have been used to produce it. Hence the final result should not be quoted to any higher degree of accuracy. For example, consider a regional health authority that is known to have a population of 2.3 million and 13 000 hospital beds (clearly both rounded figures). Now $13\,000/2.3 = 5652$ (to the nearest whole number), which implies that the region has 5.652 beds per thousand people. However, since both the population and bed figures were known only to two-digit accuracy, their ratio clearly cannot be known to the four-digit accuracy that 5.652 implies, and thus 5.652 has *spurious accuracy*. The answer 5.7 (with two digits) is preferred, although even the .7 here is not an exact result. Precise rules for the accumulation of errors during arithmetic operations are given in Chapter 7 of Thirkettle (1981).

Rounding to the nearest digit is an obvious process in almost every case (e.g., 56.2 becomes 56 and 56.7 becomes 57 to the nearest whole number). The exception is when the first digit to be lost is a 5. If more digits follow the 5, then the number should be rounded up (e.g., 56.51 is rounded to 57 since it is nearer to 57 than 56). If no non-zero digits follow the 5 there is an ambiguity (e.g., 56.5 and 56.50 are as near to 56 as they are to 57). This ambiguity is resolved by using some decision rule which ensures that half the time a number is rounded up and half the time rounded down. Tossing a coin is a simple mechanism for making this decision

2
The collection and management of health data

2.1 Introduction

Before discussing where health data come from, and how they should be managed, it is useful to try to define the concept of data. 'Data' is usually thought of as a technical term, implying lots of numbers stored in some incomprehensible way, but in fact is simply the plural of the word 'datum' meaning *given*. Data are those things that are given, in the technical sense, the things assumed to be in existence when a problem is to be solved.

Quite often, data do consist of a seemingly incomprehensible jumble of numbers, such as the output from an autoanalyser, or a list of lengths of stay in hospital wards, but the names of staff members which are held in the personnel department are data, as are the many different items contained in patients' records. What makes them all data is that they are the raw material from which information can be extracted by means of some kind of analysis. This analysis is not always statistical: the data in a patient's notes are analysed by a doctor in order to produce a diagnosis. Throughout the rest of this chapter, however, it will be assumed that the data under discussion are eventually to undergo some kind of statistical analysis.

Health data normally consist of alphanumeric characters, that is, numbers, letters and other characters such as plus signs and full stops. These may be natural language items such as names and addresses; they may be codes used either as a shorthand means of identification, like the World Health Organization (WHO) International Classification of Diseases (ICD) codes (see Section 2.3.1), or to obscure sensitive or confidential data, like the numeric codes used to identify patients at special diseases clinics. They may be dates, and finally they may be simple counts and measures, such as the number of new outpatients per year or distance travelled per day by health visitors. These data may either be collected specially, during the course of a study, or may be the by-product of operational systems. Examples of the latter are a patient master index, i.e. the index of all patients in a district, and elective admission lists. In general it is safe to say that every aspect of health care generates data in one form or another, and these are in turn collected by a very wide variety of personnel.

Once collected, data are stored in many different ways: sheets of paper in filing cabinets, libraries of patients' notes, computer databases, and so on. All of these methods are basically similar in that they store a certain amount of data relating to a specific topic in a known place; they differ only in the degree of organization of the data, and their accessibility. It is a simple matter to ask a personnel database system to list all employees on a certain grade between the ages of 20 and 30, but imagine trying to extract from patients' manually stored notes in a district general hospital a list of all those who had X-rays and blood samples taken in the last year!

The purpose of collecting data must, in the end, be the production of information, but unfortunately this does not always happen, either because the data are inappropriate or because they are stored in such a way as to make the extraction of information too difficult. Over the last few years, careful thought has been given in the NHS to the introduction of efficient methods for the collection of useful data. It is to be hoped that when such methods are in place in all sectors of the service, along with appropriate storage and access systems, it will be possible to obtain accurate and timely information about the demand for, and provision of, health care.

2.2 Deciding which data to collect

Anyone who has ever been involved in planning a research project involving statistical analysis of data would probably be able to list the steps in such an exercise.

(i)   Define the objectives.
(ii)   Design the study and specify which data are to be collected.
(iii)   Collect, organize and verify the data.
(iv)   Analyse the data.
(v)   Present the results and make recommendations.

This seems obvious but unfortunately the first two steps in this list, and particularly the first, are almost never given the time and effort to which their actual importance to the other three entitle them. This chapter will be concerned with steps (i) to (iii); later chapters will deal with (iv) and (v). The importance of being very clear about the aims of the project, and specifically the questions to be answered, cannot be overemphasized. These will dictate the data to be collected and analyses to be performed, and will have a major influence on the design of the study. Whether data are collected via a special study (see Section 2.5), or are readily available, say from a district information system (a general system for dealing with routine data in a district health authority), the two most important tools required in planning projects which involve the collection and analysis of data are clear, logical thought and common sense.

Once a question has been decided upon, the way in which the answer will be *measured* must be identified. In some cases this will be obvious. If the question were, "How much has the incidence of ischaemic heart disease changed over the last five years in this district?", the answer would be measured by the number of cases per unit population. A less obvious example is the question, "Is domiciliary physiotherapy more effective than hospital physiotherapy for some conditions?", where one possible measure of effectiveness might be the average number of treatments per patient. There are no hard and fast rules except that this issue of the measure of the answer must be resolved before steps (ii) to (v) can be embarked upon. Once again, this applies equally to research projects where data are specifically collected, and to the use of management information packages where the questions are answered at the touch of a button. In the latter case, steps (iii) to (v) are done automatically, so it is doubly important to ensure that the data used and the analyses performed are actually relevant to the questions asked. While an initial exploration of the data, using the techniques described in Chapter 3, is an invaluable way of getting a 'feel' for them, the temptation to do all possible analyses 'just in case something turns up' must be strenuously resisted!

Having decided which data to use, and collected them, there is still some preparatory work to be done before they can be analysed, and the amount of work required depends upon the collection method used. It is necessary to verify that the data are correct, as complete as possible and, if necessary, up-to-date. Computerized data collection can incorporate quite extensive checking for errors and omissions, but even so, can never be guaranteed infallible. Where such checking has not been incorporated, particularly where data collection is manual, extensive verification may be required before the data are ready to use. At least one health authority regularly reports men having hysterectomies; if number of hysterectomies were to be used in a study in this district, it would first be necessary to determine, for each such error, whether the sex or the operation code had been wrongly recorded or even whether a more fundamental error, such as a switch of columns, had been made. Even when using very sophisticated and highly regarded systems such as the DHSS Performance Indicators package (see Section 2.4.3), it is good practice to spot-check the data and analyses used from time to time (keeping in mind that if they disagree, either could be wrong).

2.3 Types of data

As mentioned in Section 2.1, data may take a variety of forms, but in general they are either *quantitative* or *qualitative*. Counts, such as the number of new outpatients per year, are an example of quantitative data, as are measures, like height and weight. Natural language items, such as names, codes and dates are all examples of qualitative data.

Qualitative data may in turn be subdivided into two types: *categorical* and *ordinal*. Categorical data are purely descriptive and imply no ordering of any kind. Sex, area of residence and ICD codes are all examples of categorical variables. Ordinal data are those which are not only descriptive, but which imply some kind of ordering, such as response to a drug recorded as either none, fair or good. Ranks are a special case of ordinal data, and arise whenever a group of data items are assigned order numbers according to some rule. The DHSS Performance Indicators package, for instance, ranks districts according to the value of any given indicator, such as cost per case.

The vast majority of routinely collected health data are in fact qualitative and are either analysed directly, for instance using contingency tables (see Chapter 9), or are used to subdivide quantitative data into meaningful groups. For example, the sex of hospital patients is always recorded and it is possible to use this directly by comparing the proportions of male and female patients admitted to hospital during a year or, in analysing lengths of stay, to compare lengths of stay of male and female patients.

Quantitative data may also be subdivided into two types: *discrete* and *continuous*. Most discrete data are counts, such as the number of patients present in a ward on a given day, but data such as lengths of stay of inpatients (which are always recorded to a whole number of days) are also discrete. It will be seen later (Chapter 4) that summary statistics derived from discrete data need not be discrete: mean length of stay is normally given to one decimal place, e.g. 6.8 days. Measures, like distance travelled or temperature, are the most common example of continuous data. In these cases the data could, in theory, be recorded to any number of decimal places (36°C, 36.1°C . . . 36.10592°C etc.), providing a sufficiently accurate measuring device was available.

Quantitative data lend themselves to most of the techniques described in this book, either directly, as in the calculation of averages, or indirectly, by converting to a related ordinal variable. If the ages of 50 patients were available, for instance, these could either be used to find the mean age of the 50 patients or, by forming age groups 0 to 14 years, 15 to 44 years, over 44 years, to split the patients into three age groups and perform analyses appropriate to qualitative data (although the process of grouping will inevitably lead to some loss of information).

Finally, data may be either *routinely* collected, in the way that all of the district database is collected (see Section 2.7), that is, as a matter of course and at regular intervals, or a *special study* may be undertaken to obtain them. Section 2.5 will describe the major types of special study encountered in the field of health and discuss the principles governing their use, while Section 2.4 will deal with sources of routinely collected data.

The purpose of the data collected routinely in the health service, whether in hospitals, in the community or in GP surgeries, is to produce information to support its clinical, administrative and financial management. Wherever possible, such data are collected as a by-product of operational systems or as summaries of data collected in this way. Bed-state statistics are examples of these so-called *secondary* data, that is, data collected for one purpose and used for another. In this case, ward admissions and discharges are recorded as a matter of course as part of the management of the patients themselves, but these may be compiled to produce hospital bed statistics (see Section 4.10). It is not always possible to take advantage of existing operational systems to collect routine data. Speech therapists have decided to record the time spent with each patient; these data are used, not for the care of the patient, but for the management of the speech therapists' time. Such specifically collected data, sometimes call *primary* data, are inevitably more expensive to collect. Secondary data, on the other hand, are easier and cheaper to collect, but may be much more difficult to verify.

2.3.1 Coding systems

When qualitative data are being collected, it is often useful to introduce codes to ensure that an unambiguous set of data items is defined. This is particularly the case where data are to be stored on computer: a computer is not, in general, intelligent enough to realize that though 'Peadiatrics' was entered, 'Paediatrics' was actually intended. Other reasons for coding were given in Section 2.1. Most systems use a series of 2 or 3 digit numbers for specialties, symbols M and F for male and female, and other obvious systems for coding qualitative data. The application of coding to questionnaire design is discussed in Section 10.4.1, but the points made in that section are, in fact, relevant to any data collection operation.

There are two standard coding systems in common use in the NHS, the WHO International Classification of Diseases (ICD), and the Office of Population Censuses and Surveys (OPCS) classification of operations. Another set of disease classifications, related to the ICD codes, has been devised by the Royal College of General Practitioners. Similarly, *OXMIS Problem Codes* (Oxford Community Health Project, 1978) are used by some pharmaceutical companies and GPs. Within the NHS, however, the policy is to use the ICD system; the advantages of the use of a single, internationally recognized coding system are obvious.

The ICD coding system is an attempt to allow all diagnoses to be represented by simple codes; these codes consist of three or four digits, or a letter followed by digits. The codes are divided into twelve chapters, each chapter dealing with a specific group of diagnoses. Codes are for individual diagnoses only, so that compound diagnoses are represented by two or more codes. For example, the diagnosis for a patient with both ischaemic heart disease and cerebral arteriosclerosis would be coded as 414.9,437 (ICD 9th Revision). The OPCS operation codes are similar to ICD codes in that an individual operation is represented by a numeric code of either three or four digits, and operations are grouped logically into related types or specialties. Code 894, for example, represents an operation on varicose veins, not elsewhere classified (nec), while adding a fourth digit makes the description more specific: 894.0 is the above operation with stripping (and ligation), while 894.1 is operation on varicose veins, nec, excision.

The Körner minimum dataset for inpatients (see Section 2.4.2) includes diagnoses and operations associated with each inpatient spell, and these are typically recorded as ICD codes and OPCS codes, respectively. ICD codes also appear in a number of OPCS publications, for example OPCS Monitor Series DH2, *Deaths by Cause*.

2.3.2 Security of data

Data often include material which in some way allows individuals to be identified and gives rise, therefore, to problems of confidentiality. Where these data are stored on electronic media, e.g. diskettes, they are subject to the regulations defined by the Data Protection Act (1984) which requires that appropriate security measures are taken, that data subjects have access to any data relevant to them, and that the use of the data is registered with the Data Protection Registrar. A good summary of the Act is given in the article 'Data Protection Act' (Institute of Statisticians, 1984). However such data are stored, the same level of security should be maintained so that files (whether manual or electronic) containing identifiable data should be stored securely, and access to them restricted. It is worth noting that diskettes and floppy disks are small and portable and therefore very easily stolen, so that it is just as important to lock these away as it is to lock away paper files. Within the NHS, districts and regions are charged with data protection in their authorities, and their information departments should be able to provide advice on security requirements and procedures.

2.4 Routine sources of health data

The majority of the data used in the NHS is routinely collected, whether within districts, by régions or the DHSS or by other bodies such as local authorities. Within districts, and in some cases across regions, the raw data themselves will be available for analysis, but in some cases, particularly where nationally collected or sensitive data are concerned, access may be restricted to published summaries.

Before listing the sources of routine data, it is necessary to describe how the data flow through the NHS; this will be covered in Section 2.4.1. Subsequent sections will outline the various sources.

2.4.1 Flows of data in the NHS

In general, routine data in the NHS flow upwards through the system. The particular arrangements within a district will depend upon the structure of the district information system, but will usually follow a hierarchical pattern, moving from wards, clinics and departments, via units and district departments, to district headquarters. Districts send regular summaries to regions, who pass these on to the DHSS, and it is regions who are charged with the responsibility of ensuring that all data submitted to the DHSS are correct and complete. Individual general practitioners make certain returns either to their Family Practitioner Committees (FPCs) or to the relevant district health authority, and the data move upwards from these points, either to region or directly to the DHSS. Bodies such as the Public Health Laboratory Service report upwards in a similar way. Data in the form of summary statistics also flow downwards from DHSS to region and from region to district, as well as outwards in the form of published tables and reports.

As an example, consider the routine collection and processing of bed statistics in the Oxford Regional Health Authority. Units collect daily ward listings which show the numbers of available and occupied beds at the previous midnight. These are aggregated at fixed intervals, usually monthly, and sent to district headquarters where they are summarized quarterly and annually, and various derived statistics such as average length of stay are produced. Copies of these statistics are sent to region (at least) annually where they are combined with returns from other districts before being submitted to the DHSS. The DHSS produce from these, and returns from other regions, annual figures for the whole of England and Wales, which are then distributed to regions. Summaries showing returns for each district in the region and for England and Wales are then prepared by the region and sent to districts. While the details may vary, bed statistics are collected in a similar way throughout the NHS.

In many regions and districts, this flow is not a manual process, but happens automatically as a by-product of computerized patient administration systems (see Section 2.6). Whatever the technical arrangements, however, districts and regions perform the same functions of collecting, verifying and disseminating data.

2.4.2 Data routinely collected in districts

In 1980, the Steering Group on Health Information, headed by Edith Körner, was established with the objective of developing guidelines for the development and implementation of information systems in the NHS. Such systems would replace the cumbersome, inefficient and fragmented methods of data collection in use, facilitate effective management of operational systems, and make available a wide range of data from which could be extracted information for monitoring existing services and planning future ones.

The result of the Steering Group's work was a series of six reports, along with associated publications (Körner, 1984, a,b,c,d,e,f,g), covering most aspects of the work of district health authorities. The six reports cover respectively hospital services, patient transport, manpower, paramedical and other services, services in the community, and finance. Each specifies a minimum set of data items to be collected, the *minimum dataset*, and gives a standard definition for each item. It is worth keeping in mind that data collected prior to the implementation of the reports may not comply with these definitions so that comparisons over time may be difficult.

Lists of items in the six minimum datasets are given in the relevant reports but, generally speaking, they aim to cover all basic aspects of patient care provided by a district on and off hospital sites, and the manpower and financial implications of these services. Districts themselves hold these data, and particular arrangements for storage and retrieval will vary from district to district. In addition to the raw data, districts compile regular summaries for their own use and for submission to region in the form of statutory returns, as mentioned in Section 2.4.1. Lists of these requirements, and samples of the forms are given in a series of DHSS circulars (references B1,B2,B3,B4,B5,B6 and B10). It should be stressed here that most districts collect a wider range of data relating to the six Körner report areas than are specified in the minimum datasets. This range will depend upon the particular services offered in the district and the arrangements for their management, as well as the personal interests of individuals involved. The most obvious example of this flexibility is in the collection of clinical details about births, which is deliberately left open for districts to decide. While many of these data items will be included in the district database (see Section 2.7), some will be collected by particular managers to satisfy their own needs and so will not normally be available on a district-wide basis. Certainly before undertaking any expensive data collection exercise, it is worth checking to determine whether the data are already being routinely collected somewhere in the district.

In addition to the data associated with the six Körner reports, others are routinely collected by a number of district-based computer systems, such as those which provide operational support for pharmacy, works and supplies departments. Child health computer systems contain data on immunization and vaccination of children, though these systems are sometimes run on a regional basis.

Finally, GPs are required by law to notify certain diseases to the District Medical Officer of Health (sometimes called Medical Officer of Environmental Health), responsible for the district in which the patient lives. These are generally infectious diseases such as measles and pertussis and some types of food poisoning. Summaries of the numbers of cases reported weekly are sent to the OPCS, who publish the national incidences in the Registrar General's Weekly Return for England and Wales (Series WR) and the notifications are held by district, either manually or on a computer, for retrospective analysis. Districts may also have data on certain services for which they pay GPs, such as rubella immunization.

While the foregoing is by no means an exhaustive list of the types of data collected by districts, it does give some idea of their variety and range. District information departments should be able to provide a comprehensive list of data sources and items associated with any district-wide system.

2.4.3 Other NHS sources

Data on incidence of cancers is collected by regional cancer registries (roughly corresponding to RHAs). Regions also collect data on regionally run services such as blood transfusion units. Usually, subsets of district-based data are held by regions for the purposes of regionwide planning and monitoring. In the past, Hospital Activity Analysis data (that is, records of inpatient events, excluding maternity and psychiatry) were held at regional level, and no equivalent data were held at district level. This has been replaced in many regions by the DIS/RIS computer system in which the Regional Information System (RIS) extracts inpatient data (now including maternity and psychiatry) from those held in the District Information Systems (DIS). These regionally held data are made available to districts, so they are able to perform regionwide analyses, such as interdistrict comparisons of particular services and calculation of cross-boundary flows within the region. Those regions not running the DIS/RIS system will have very similar arrangements. Regional information departments should be able to provide comprehensive lists of the data they collect and hold, and details of the facilities that exist for accessing them.

Public Health Laboratories, which are part of the Public Health Laboratory Service, hold all requests for tests which gave positive findings and their results for a statutory period of time. Summaries of their findings are sent to the central laboratory. These requests and results may be a valuable source of data in the investigation of outbreaks of infectious diseases such as legionnaires' disease, and may also be used to monitor the effectiveness of immunization programmes.

The statutory returns mentioned in Section 2.4.1 are used as input to the DHSS Performance Indicators (PI) package and the Inter-Authority Comparisons and Consultancy (IACC) package, which are available to all districts and regions. The PI package (set of computer programs and data) is updated yearly and distributed to all authorities, while the IACC package is available on request from the Health Services Management Centre at

Birmingham University. Both provide a wide range of information, and to some extent complement one another. The PI package has the facility to build up data from year to year, to allow analyses over time to be done.

Family Practitioner Committees have an enormous potential as a source of routine data on the population served by a health authority. GP surgery-based computers are proliferating, and are used in many practices to support immunization and screening programmes as well as for basic practice management. With the advent of large central FPC computer systems serving districts or groups of districts, it becomes feasible to investigate the possibility of linking population-based (FPC) and service-based (DHA) data, to provide information about the actual demand for and effectiveness of the services provided.

2.4.4 Sources of routine data outside the NHS

It would be impossible to list all the sources of routinely collected data which may be relevant to health issues. A few of the more commonly useful ones will be mentioned, but it should be borne in mind that most, if not all, public bodies keep records which may be of use in particular health projects.

The Royal College of General Practitioners (RCGP) receives reports weekly from a sample of GP practices; on average 40 practices, responsible for about 242 000 patients, respond each week. Each practice reports incidences of infectious diseases, acute respiratory disease and other events such as abortions, cot deaths and suicides. The rates of infectious and acute respiratory diseases are published weekly by the OPCS in the Monitor Series WR. From the returns, the Research Unit at Birmingham produces four-weekly summaries, as well as quarterly and annual reports which are sent to participating GP practices and the DHSS. While the practices included in the sample are not necessarily representative of practices across the country (the RCGP relies upon volunteers), they do provide a valuable source of data giving information on trends over time. Further details on the data collected and their use may be obtained from the RCGP, Birmingham Research Unit.

The DHSS receives direct notification of all abortions, and midwives notify OPCS of all congenital malformations. While, for reasons of confidentiality, the raw data may not be available, the OPCS publish periodic summaries of these returns in Monitor Series AB and WR respectively. In particular cases, it may be possible to arrange for *ad hoc*, non-standard analyses of the data to be performed.

Local authorities and county councils have data on social conditions, population movements, education and other areas which may be relevant to the management of health services. Voluntary organizations, such as the Family Planning Association, which provide services complementary to those provided by the NHS, should certainly not be overlooked.

2.4.5 Published data

Publications provide a rich and varied source of health data. They may be, for example, internal planning and policy documents, government reports, or papers dealing with specific research topics. Most of the data obtained from such sources will be in the form of summary statistics which will be used either to perform further analyses, or to verify or support locally produced analyses. One of their most common uses is to compare local trends with national ones, for example, to compare incidences of leukaemia in particular areas with the national incidence. As with other sources, it is impossible to list all the different types of publications which may yield data pertinent to health, but some of the most commonly used will be mentioned.

The first, and one of the most important, is the collection of planning and policy documents produced at all levels of the NHS. These may be district and regional strategic plans, departmental annual reports, or one-off papers supporting suggested changes in service provision. As there is no comprehensive system for collecting, storing and organizing all such documents in most districts, a certain amount of detective work may be required to unearth some of them. The effort involved in obtaining such documents, however, is often rewarded by the discovery that much of the background work required to initiate a project has already been done. In a study of waiting lists for hearing-aid fitting, for example, valuable data on staffing levels was obtained from an Audiology department's annual reports.

The three National Morbidity Surveys carried out by the RCGP, in collaboration with OPCS, have resulted in a series of very useful publications (RCGP *et al.*, 1958; 1982; and 1986). During each survey a number of GP practices submitted to the RCGP, for each patient seen, a record of each visit and associated diagnoses and referrals. Like the regular weekly survey, the analyses presented in the reports provide valuable information about trends in diagnoses and treatments over time. A good description of the surveys is given by Coulter (1987).

Another very important source of health data is the wide range of regular government statistical publications (see also Section 5.2). The Office of Population Censuses and Surveys is a particularly rich source. As already mentioned it publishes a range of *Monitors* at regular intervals on such topics as birth and fertility rates, congenital malformations, infectious disease notifications, population projections, cancer statistics and deaths by cause. Also available are analyses arising from the *Hospital Inpatient Enquiry* (DHSS) (a 1 in 10 sample of patients), the *General Household Survey* (OPCS) (an annual nationwide sample survey of households) and the regular censuses of population. The quarterly publication *Health Trends* (DHSS and the Welsh Office) and the annual publications *Regional Trends* and *Social Trends* (Central Statistical Office) contain tables and commentary on current health issues. The *Annual Abstract of Statistics* (Central Statistical Office) also includes tables of health data. Such useful government publications are mainly produced by the Government Statistical Service (which includes the Central Statistical Office and OPCS) or by the DHSS. A brief list of such sources is given in the booklet *Government Statistics: A Brief Guide to Sources*, produced annually by the Government Statistical Service. The more detailed *Guide to Official Statistics* (Central Statistical Office, 1986) gives a complete list. Another useful source is *A Guide to Health and Social Services Statistics* (1986), produced by the DHSS. A thorough bibliography of sources of medical statistics is given by Cowie (1986).

One source not mentioned in these bibliographies is the *Compendium of*

Health Statistics published by the Office of Health Economics, which combines data from FPCs and health authorities to produce interesting and useful analyses. Recent articles have discussed trends in prescription prices over time, and the distribution of grades of hospital medical staff by sex.

Finally there is the extensive literature dealing with topics in medical research and health management. Any district or regional information department should have access to many of the most useful journals such as the *British Medical Journal* and the *Health Services Journal*. They should also have, or be able to arrange, access to public databases for online searching about particular topics.

2.5 Special studies

Despite their considerable scope, routinely collected health data do not provide the appropriate information for all needs, and frequently a study must be undertaken to generate specific data. As an example, consider a district health authority that wishes to investigate the extent of mental illness in its managed population. Any records of contact with the health service, social services, police or other bodies are certain to indicate only part of the problem. Such records show the demand for psychiatric care that is met, though the records may also show something about unmet demand, such as the number awaiting the visit of a social worker. Almost certainly, no records will indicate the unexpressed need for psychiatric care, such as the person who experiences periods of depression but has not sought help. To obtain the necessary data, then, a special study would be required, perhaps a door-to-door *survey* where the relevant questions were asked by trained personnel.

Since special studies are relatively expensive in both time and money, it is a good idea to consider whether they are really necessary by first checking potential sources of routine data. Where a publication containing relevant data exists, the data included may be a selection from, or a summary of, the primary source data, and these primary data may be obtainable in a convenient form for analysis by the user (e.g., on a floppy disk). Even where the available data are not directly relevant, they may still be sufficient for some purposes; in the example above, it might be possible to use the results of a national survey on mental health to represent, pro rata, the situation in a particular district health authority. This would, of course, not be sufficient if a region wished to compare different districts or if the district were known to be in some way special.

Four major types of health study are possible: a survey, a cohort study, a case-control study and an intervention study. The essential features of each of these methods of data collection will be described in this section. Undoubtedly the most common of the methods is the survey and because of its importance, a thorough description is given in Chapter 10, after ideas of statistical analysis have been presented in earlier chapters. All the methods of data collection may require the selection of a sample of individuals to be studied, but the principles of sampling are also deferred to Chapter 10 where they are presented in the context of surveys.

More complete accounts of special studies are given by Lilienfield and Lilienfield (1980), and by Alderson (1983). Bauman (1980) also gives an interesting discussion of the underlying principles in his early chapters and Barker and Rose (1984) present an extensive collection of examples of medical studies.

2.5.1 Surveys

A survey is an investigation at a particular point in time of a pre-defined population, in which members of that population are either measured, observed or questioned. Surveys take two distinct forms: the *census*, in which every member of the population is included, and the *sample survey*, in which only a selection are included.

Surveys are particularly useful for describing an existing state of affairs, such as patients' opinions of hospital meals, the distribution of nurses' time spent on their various duties, the lengths of time spent waiting in a GP surgery and the levels of lead in the blood of children at a school situated close to a glass factory. They are less useful when the object of the study is to determine the cause of a given effect, because there is no control over the exposure of individuals to outside influences. Thus in the lead example, high lead levels could be due to factors unassociated with glass production, such as high road traffic densities near the children's homes. More particularly, there is no guarantee that accurate comparisons between those with and without the cause or effect will be possible.

Since a survey is carried out at a single point (or, at least, period) in time it is called *cross-sectional* to show that it cuts across the time continuum. This imposes two limitations. First, an essential ingredient of a cause and effect relationship is that the effect should be preceded by the cause. So to demonstrate such a relationship, the correct sequence of events must be established, which cannot always be accurately done in a survey. Second, a cross-sectional survey cannot directly measure changes over time, as would be needed to evaluate the changes in health status resulting from a new public health programme.

The limitation of a single cross-sectional study can be overcome by carrying out two (or more) such surveys at different times, sometimes called a *trend analysis*. When routinely collected data covering the period between successive surveys are available they can be used with the survey data to calculate rates of change. This happens in demography where population censuses are carried out every ten years and data on births and deaths are collected in the intervening period. Taken together these two sources of data can be used to calculate birth and death rates, as shown in Chapter 5. In a trend analysis, records of a single individual are not linked from survey to survey (unlike in a cohort study) and, indeed, there is no reason why the same individual should be included in each survey. Yet again this puts a serious limitation on any analysis of cause and effect.

Despite the drawbacks mentioned above, the survey is a very useful tool in health studies, particularly in those studies in which description is, at least in part, the object. A comprehensive account of survey methods for human populations is given by Moser and Kalton (1971).

2.5.2 Cohort studies

In a cohort study a group of people is selected at some initial point in time and is then followed up in the future. Since data collection proceeds forwards in time these are also called *longitudinal* or *prospective* studies. Normally the group chosen for study would have some factor which the study seeks to investigate, as when a group of healthy male workers at a nuclear power station is followed up to look for possible future ill health. Here the factor of interest is their occupation.

Since cohort studies progress through time they give information about the sequence of events that is essential in determining causality. For instance, if the employees develop testicular cancer after having worked at the power station for several years then this may be evidence in favour of the hypothesis that the occupation causes the cancer. In fact, although there is often one outcome of special interest (testicular cancer in the example), many outcomes can be explored in a cohort study (perhaps leukaemia and brain tumours as well as testicular cancer), provided that the necessary data are obtained.

Following a single cohort through time gives interesting information about the history of a group of people with a special factor but does not, by itself, say anything about the importance of that factor to the outcome observed. For instance, the incidence of testicular cancer in the five-year period from the start of the study may be high amongst nuclear power station employees, but perhaps not much higher than amongst men of a similar age who follow other occupations.

If the presence of this special factor contributes to the occurrence of a positive outcome then there should be an excess of positive outcomes when the factor is present. This is just common sense; it means simply that if working at a nuclear power station is one of the causes of testicular cancer, then more such cancers should be observed among men with this occupation than among those with other occupations. The question of how many more constitute reliable evidence is a crucial one, and is related to the 'relative risk' defined later in this section.

For this kind of comparison to be possible, the study must include a 'non-factor group', that is, people without the special factor, and they should, as far as possible, be similar to the factor group in every other way (that is, same age range, living conditions, medical history, etc.).

The two groups should be followed in parallel to ensure that any changes in the background environment over time affect them equally. For example, testicular cancer might be generally on the increase due to increased air pollution, changes in diet, etc. Sometimes the factor and non-factor groups are selected as two distinct cohorts, in other cases they are formed by splitting an initial cohort into the 'have factors' and 'have not factors'.

A classic example of a cohort study is the investigation into the effects of smoking carried out by Doll and Hill. Their cohort was the 59 600 men and women on the British Medical Register at October 1951. These doctors were initially questioned about their smoking habits and, as a consequence, divided into the factor (smoking) and non-factor (non-smoking) groups. The subsequent mortality of the two groups was then monitored through routine record sources. Results were published after 10 years (Doll and Hill, 1964) and 20 years (Doll and Peto, 1976). In each ten-year period there were, for individual age and sex groups, many more deaths amongst the smokers than the non-smokers, especially deaths due to specific causes such as lung cancer. This suggests that smoking has contributed to some of the deaths. The Doll and Hill cohort study, of which this is a much simplified account, has contributed valuable information to the study of the effects of smoking.

An extremely useful summary measure from a cohort study is the *risk* of an individual getting the outcome being studied. If, say, 100 nuclear power station workers were studied and 3 died from testicular cancer then their risk of such death would be 3/100. The risks are calculated for both the factor group and the non-factor group and the two compared using the ratio of the former to the latter, called the *relative risk* (see Section 9.2.6). If this is substantially above one, then there is evidence of a higher risk for people with the special factor. In many situations risk changes with age or with length of exposure to the factor. Such changes can be represented in a life table (see Section 5.7). When the factor and non-factor groups are dissimilar with respect to important variables such as age and sex, then the relative risk should be adjusted for such variables by, for example, the method of standardization (see Section 5.5.2).

The main disadvantage of cohort studies is that they are expensive and time-consuming; expensive because people have to be continually accounted for, and time-consuming because the accounting may well need to continue for many years. For this reason cohort studies are not recommended for the study of outcomes with a long latency. Another reason for the relative expense of cohort studies is that many subjects are needed to study rare outcomes. For example, if a disease has a typical incidence rate of 1 per 100 000 people each year then even with a cohort as large as 10 000 only one case of the disease can be expected during a ten-year follow-up. Case-control studies will be better in such circumstances.

Other disadvantages of cohort studies are that the process of being studied may, in itself, influence the results (another good reason for concurrently observing a non-factor group), and that almost inevitably, some subjects will be withdrawn during the study, due to emigration or for other reasons.

One special disadvantage with the nuclear power station example used here is the so-called 'healthy worker' effect. It can be argued that the very fact of being employed in the nuclear industry is an indication of above-average healthiness, as stringent medical tests may precede the offer of a job. This would lead to an underestimate of the relative risk of working at nuclear power stations.

A wide selection of cohort studies is described by Mednick and Baert (1981). Two variations on the basic theme described here are, first, retrospective cohort studies, where data are traced back in time to obtain cohort data that begin in the past and, second, comparison with national outcomes to replace the use of a true non-factor group. An example of the latter is where the death rate due to testicular cancer amongst the studied power station employees is compared with the national death rate for the same cause in the same period. Both these variations are only feasible when the data to be used are complete; if some deaths due to testicular cancer were, for some reason, regularly misreported on death certificates then the national rate would be an underestimate. If the outcome studied is morbidity rather than mortality, serious omissions are likely in whatever data are routinely available.

2.5.3 Case-control studies

Whereas in a cohort study individuals are observed prospectively to see whether or not they achieve some outcome, in a case-control study individuals who already have the outcome (the cases) and individuals who do not have the outcome (the controls) are studied *retrospectively* to see what factors could have led to the outcome. As an example, consider another study of smoking by Doll and Hill (1950). They took 709 cases with lung cancer from a number of hospitals and the same number of controls without lung cancer from the same hospitals and compared the prior smoking habits of the two groups (see Table 9.2). They found a much higher consumption of tobacco amongst the cases, suggesting a possible causal relationship between smoking and lung cancer (further details are given in Chapter 9). Notice that in a case-control study the movement is from effect to cause, not cause to effect as it would be in a cohort study.

Cases are normally found from hospital or other records, as in Doll and Hill's study. Controls are usually selected to be as similar to the cases as possible in all important ways except in relation to the factor being studied (smoking in the example). This is to prevent differences between cases and controls in other variables influencing the results. The best way of doing this is to match each case with a control (producing 'paired' data) or with many controls. Typical matching criteria are age and sex.

Care must be taken to ensure equal quality of data for both cases and controls; cases are more 'interesting' and hence there can be a temptation to research their life-styles etc. more thoroughly which can lead to biased results. Case-control studies often rely upon the memory of the subjects involved, for instance they may be asked whether they have visited a certain place, suspected as the source of an epidemic, in the past month (see Example 7.20). In such situations the memory of a case may be more reliable than that of a control.

Case-control studies may be used to investigate many factors concurrently. In a study of lung cancer, for instance, people with and without lung cancer can be compared for exposure to coal dust and any other factors which might contribute towards lung cancer, in addition to smoking habits. Unlike cohort studies, they cannot investigate more than one outcome, since cases and controls are selected for a solitary outcome.

In general, case-control studies are quicker to carry out and cheaper than cohort studies, and better suited to the study of rare outcomes or outcomes that take many years to be identified. They are, however, less convincing in the study of cause and effect because they do not show the sequence from suspected cause to suspected effect, unlike cohort studies. The other major disadvantage is that case-control data *cannot* be used to calculate the risk of a factor (such as smoking) leading to an outcome (such as lung cancer). Provided the outcome is rare, case-control data can, nevertheless, be used to find an approximate relative risk for those with and without the factor (see Section 9.2.6).

A clear and comprehensive account of the design, conduct and analysis of case-control studies is given by Schlesselman (1982).

2.5.4 Intervention studies

In an intervention study a treatment is applied to one group of individuals and a second treatment is applied to a second group of individuals; both groups are then observed prospectively and their outcomes compared. There are two methods of applying treatments, either to individuals separately or to an entire group as a whole.

Where a new pain-relief drug for use by patients with arthritis is being tested, each patient is assigned to one of two groups; individuals in one of the groups are given the new drug, while those in the other are given the existing, standard treatment. At some predetermined time after the drugs have been administered, the pain relief experienced by each patient is evaluated, and the extent of pain relief in the two groups is then compared. Such a study is called a *clinical trial*. An example of the second method, called a *community trial*, is the comparison of numbers of dental caries in two towns, where one town's water supply has been fluoridated and the other's supply has been left untreated.

The 'treatments' in an intervention study can take many forms. Undoubtedly pharmaceutical companies carry out the majority of clinical trials to evaluate new drugs, but the technique can also be used to compare one method of medical treatment with another (such as surgery versus medication, anaesthetic versus acupuncture, etc.), one method of care against another (such as long stays in hospital versus short, hospital versus home confinement for the mentally ill, etc.), to evaluate the effects of immunization against disease, and in many other applications. Community trials are most commonly used to evaluate preventive health care policies such as in the fluoridation example, and in screening and health education programmes (see Example 7.1).

As with case-control and cohort studies, intervention studies are designed to compare one group with another. Intervention studies represent an application, to human health problems, of the scientific principle of experimentation to determine cause and effect. Because there is control over who receives the suspected cause and when they receive it, results from such studies are generally thought to be the most reliable when studying cause and effect. Clearly the amount of control is limited in a community trial, particularly as people will enter and leave the community during the course of the study. For this reason clinical trials are preferred when they are possible.

Due to the prospective nature of data collection, intervention studies share some properties with cohort studies. These include the need to ensure that data collection for both groups runs concurrently to avoid bias due to changes in the background environment. Also the disadvantages of long duration, high cost and the possibility of withdrawals are common to both. Withdrawals from intervention studies should be looked at particularly carefully since they may, for example, be a result of some undesirable side-effect of a treatment.

Despite their theoretical appeal, intervention studies are often impractical.

For instance, a clinical trial to determine the effects of heroin would require a selected group to take the drug and experience the consequences. It is socially unacceptable and medically unethical to force human beings to take potentially harmful substances. Some people take this argument to the extreme and say that all intervention studies are unethical. The counter argument to this is that it is at least as unethical to introduce a new procedure, such as fluoridation or a new drug formulation, without prior scientific study of the possible benefits and side-effects.

As already indicated, clinical trials are preferable to community trials, so the remainder of this section will be devoted to clinical trials. A clinical trial begins with the drawing up of an agreed *protocol* in which the objectives and methods to be used are specified. This will serve as a reference manual for all those involved in carrying out the trial. As well as defining the methods of assessment, the protocol for a clinical trial should say which people can and cannot be entered into the trial. For example, only patients with certain specific symptoms might be allowed, or patients thought to be particularly vulnerable to undesirable side-effects might be excluded. This may, of course, restrict the generality of the results. Also, if only volunteers are entered into the trial and the volunteers are different from the general population (perhaps they are more health conscious and hence take more exercise and eat sensibly), the results will not necessarily be meaningful for the world at large.

However subjects are recruited, it is essential that allocation to treatment group occurs *after* entry to the trial, and subjects may *not* be rejected from the trial merely because they have been assigned a particular treatment. This is to prevent bias, such as might happen in a trial of a new drug against a proven standard drug, where a doctor might be reluctant to allocate the new, untried drug to the most seriously ill of his patients. Then, even if the new drug is really inferior, it may appear to give better results merely because the patients using it are already in a better state of health when the trial begins. For the same reason it is essential to allocate subjects to treatment groups at random, that is so that every subject has an equal chance of receiving either treatment. Randomness is often restricted to ensure that each treatment group is similar in all aspects (age, sex, severity of illness, etc.) except in treatment received. This can be achieved by stratification methods similar to those described in Section 10.9.3. When these principles of allocation are followed, the clinical trial is called a *randomized controlled trial*.

Sometimes it is possible to give both treatments to every subject, but at different times, such as in the treatment of a chronic illness. Half the subjects would then start on one treatment and the other half on the other treatment. Provided that the treatments leave no residual effect this ensures compatability of treatment groups. Such a trial is called a *cross-over trial*.

Clinical trials may also be affected by psychological bias. Although this can occur with any pair of treatments, it is a particular risk when one of the two treatments is a *placebo*, a fake treatment such as a sugar pill. The patient who knows that he is taking a potential cure receives a psychological boost, while the patient who knows he is taking a placebo does not. To avoid such problems it is advisable to ensure that the identity of the treatment he is receiving is concealed from the patient, in which case the trial is said to be *single blind*. On the other hand, the person assessing the patients' health during the trial (probably a doctor or nurse) may himself be prejudiced towards one or other of the treatments, and this may affect his evaluation. It is, therefore, better if the assessor also does not know which treatment any patient has received. When both patient and assessor are unaware of the treatment allocation, the trial is said to be *double blind*.

A comprehensive account of the principles and strategies for the design and analysis of clinical trials is given by Pocock (1983). Although less detailed, the material on clinical trials in Gore and Altman (1982) provides a clear account of the subject.

2.6 Management of data

Whatever the type of study undertaken, a set of data will be generated which will then require manipulation in some way in order to obtain answers to the questions originally posed. The basic principles governing valid statistical investigation apply whether the data are processed manually or by a sophisticated computer package. In either case the steps in any study carried out are those listed in Section 2.2, but the ease with which collection, analysis and summarization are performed will certainly depend upon the methods used to manage the data.

Manual systems of data management have been in use for thousands of years and have changed very little, except in the media used for storage: we now use sheets of paper in a filing cabinet, rather than stone tablets. Microfiche, which is used by some medical records departments to archive patients' notes, is a rather more sophisticated, but arguably still manual, form of data management. In recent years, the ready availability of microcomputers and 'user-friendly' software has meant that computers are used to store and process data in a wide variety of areas in the health service. Examples are the handling of patient appointments in clinics, running a hospital bed bureau and analysing data from an epidemiological study. On a larger scale, mini or mainframe computers are used, for example to run complete patient administration systems (systems which manage the medical records and appointments functions in a hospital), and to integrate activity, financial and manpower data arising from many different sources. Speed is the only real difference between manual and computerized systems of data management – it is in fact theoretically possible, though usually not practical, to echo using manual methods every activity a computer carries out when storing and processing data.

The most important thing that happens to data when they are stored manually is that they are indexed, so in a personnel department's filing cabinet, employees' records may be stored by surname within cost centre, with the surname written on the employee's folder and the cost centre, say, on the drawer. This is the simplest type of filing system, with no requirement to store the index separately. Most personnel departments, however, keep subsidiary lists, such as lists of employees by grade or by professional qualification, and these are secondary indices. Similarly, administration departments often have quite extensive numbering systems for files on different topics and maintain a list linking numbers and topics. This list is the primary index and once again there may be secondary indices, perhaps by source of document or by date.

Manual filing systems used for data collected by special studies are usually very simple. The whole project is given a name, which is written on a file folder or container of some sort, and this is in turn placed in a labelled divider, in a filing cabinet, or on a shelf in plain view.

All of the above examples have described the naming and storing of and access to whole files, but what about the details contained in the files, the data themselves? Where each file contains only a few components, this may not be a problem, but large files need some kind of internal organization if anything is ever to be found in them; in order to accomplish this, indexing is used once again. In administrative files, accessions are normally in strict date order, but, as anyone who has ever tried to extract specific pieces of data from patients' hospital notes will attest, few other filing systems in the health service are so well organized!

Using patients' manual medical records as an example, certain common features of the records within a file may be noted; here, all the medical records in a hospital constitute one huge file, and each set of notes a record within a file. Firstly, each record has an identifier, in this case probably the patient's district number, and this identifier will be the means of uniquely identifying any record, i.e. the set of patients' numbers constitute the index within the file. There may also be secondary identifiers, such as the patient's name or, more likely, name and date of birth taken together. Secondly, each record has roughly the same layout and the potential to contain the same amount of data. The record of a patient who has only once been to hospital, say to an outpatient clinic, will actually contain very few items, but could if necessary contain data on an unlimited number of visits to and stays in hospital. Also, the form in which the data are stored in the records is very much the same from patient to patient, with preprinted forms to be filled in wherever possible.

For most files, the records they contain follow this general format of a main identifier, one or more secondary identifiers and a predefined set of data items. While this is most obvious in a file containing just alphanumeric data, even files of correspondence may be seen to conform to this pattern. In this case, the date on the letter is the identifier and the body of the letter the data. For alphanumeric data files, this layout gives rise to a rectangular array called a *data sheet* in which the rows are the records and the columns are data items or variables; at least one of the variables is a record identifier.

All computer data management is based on these simple concepts of manual data management. Data relating to a single application or study are stored in a file which has a name. Database packages (see Section 2.6.1) allow screens to be formatted to look exactly like any questionnaire or data collection form, and data may be entered into the computer using these forms. In order to save space and make calculations more efficient, within the computer data are always stored in data sheet format (conceptually, at least – some systems have very complicated ways of physically storing data, but these are fortunately invisible to the user). Where a formatted screen has been used to enter the data, the same screen may be used to retrieve them; the effect is of the storage and retrieval of individual questionnaires or forms.

One of the greatest advantages of computer storage is the speed and flexibility of indexing it offers, particularly when a database package has been used. To return to an example from Section 2.1, in a hospital with a computerized patient administration system linked (by patient number) to the pathology and radiology departments, it would be a simple matter to count the number of patients who had both blood tests and X-rays in a year.

Rules for assigning file names vary from system to system, but wherever possible they should be simple and meaningful. It is much easier to remember the name of a datafile containing bed-state statistics if it is BEDSTAT than if it is AB32X. Any computer will have facilities for listing and deleting files, and most will give additional information about them such as their size and date of creation. Most larger systems offer the facility for restricting access to files by password or username (the name used to start up the computer session), or both, though this may not be available on some microcomputers.

Most computers offer some kind of peripheral storage device. Depending on the size and sophistication of the machine, this may vary from a floppy disk which holds a relatively small amount of data (say 500 000 characters), to large hard disks which can provide storage for 60 million characters or more. These are really no more than filing cabinets in disguise, as they perform the same function of holding files and allowing access to them. They are merely faster to use. Disks and other storage media always contain one special file called a directory, which is simply the index of files the medium contains, and this allows filenames to be listed and makes it possible to perform housekeeping jobs such as copying and deleting files.

The idea of data security was mentioned in Section 2.3.2. The other side of the security 'coin' is the practical problem of actually protecting data files from loss or corruption. This entails maintaining back-up copies of files (and keeping these secure), and documenting them thoroughly, so that a clear record is kept of all filenames, variable names and codes used. This obviously becomes more important where computers are used; while in general, computers are far more efficient than filing cabinets at data storage and retrieval, when a filename has been forgotten, it can be very difficult to track the file down on a disk.

2.6.1 Database packages

In the previous section, the term 'database' was used without definition. While a database package can be a very powerful tool, and may in some cases require the user to learn an extensive computer language, the idea of a database is really a very simple one. A database is just a set of data organized in some logical fashion. The power of a database package is in the speed of access to the data, and the facilities which it provides for data entry, indexing and output. These can be rather limited or very extensive but at the end of the day, and no matter how thick the manual, the database remains just a set of data stored, as mentioned in Section 2.6, in 'data sheet' format. Because of the ever-increasing variety of packages on the market it is not possible to describe particular packages in detail, but some of the features common to most packages will be covered.

(1) *Data input and editing*   All database packages have some means of getting data into the computer. This may be by direct typing in of data in data sheet format, by use of a screen, which will be described below, or by copying data from an existing file, called *batch entry*. Once the data have been stored, facilities are available for viewing and editing it, that is adding new records, deleting records, changing data items within records, and so on.

(2) *Screens*   Most database packages provide some means of constructing a form on the computer screen (visual display unit or VDU), which consists of a mixture of descriptive text and spaces where data appear. These forms are often called screens, or *masks*, and are used both to enter data and to display them after entry. An example of a possible computer screen is shown in Fig. 2.1. The data items to be entered are patient number, name, age and sex.

A. B. HEALTH AUTHORITY
PATIENT REGISTRATION

PATIENT NO. _____
NAME _____
AGE _____   SEX (M/F) _____

Fig. 2.1   Example of a database screen.

When used for data entry, the lines shown in the example would be replaced by data typed in by the user. When data were retrieved using the screen, the patient number, name, age and sex would appear in place of the lines. The whole record and individual items could then be edited if necessary.

It is important to understand that the database package works by extracting the data from the screen and storing it in data sheet format, so that the descriptive text need not be stored for every record (patient in the example).

(3) *Keys*   Each record is identified by a *key* or label of some kind, which is usually definable by the user, and is normally the first item in each record. It is from these keys that the primary index of the database is constructed. Some packages make provision for multiple keys, so that it is possible to look records up in a variety of ways. Creation of keys is usually called *indexing*. In a personnel database, for example, the key would usually be the personnel number, but the file might also be indexed on cost centre, grade and professional qualification.

(4) *Data manipulation*   It is often necessary to perform operations on data once they have been entered, either to verify them (e.g., by range checking) or to derive new variables from existing ones (e.g., age from current date and date of birth). Facilities for processing data in this way are available in all database packages.

(5) *Data selection, retrieval and analysis*   The purpose of using a database package is to store data for later display or analysis, and database packages

therefore have extensive facilities for retrieving and displaying data. Selection procedures allow subsets of the data satisfying particular conditions to be specified, such as 'Select all records for which number of operations is greater than zero, source of admission is booked, or from the elective admission list, length of stay is zero, and patient was discharged (i.e., had not died)'. This would produce the subset of a hospital's inpatient data relating to day surgery.

It is also possible with many packages to produce relatively simple one- or two-way tables, such as length of stay by speciality on admission, and many of the analyses received by managers in districts will be of this type. More complicated analyses are beyond the scope of many database packages and when these are to be done it is far more efficient to export data from the database package to something more appropriate, typically a statistics or graphics package. All this means is that data are copied from the database to a separate file, the physical structure of which allows the data to be read by other packages.

2.6.2 Statistics packages

A very wide range of statistics packages exists, though not all packages may be used on all computers. These vary from quite general packages, which offer good data management facilities as well as a range of statistical techniques, to very specialized packages performing a narrow range of analyses appropriate to some special problem. It is useful to have a fairly general package which offers data editing facilities and graphical output, as well as a range of different types of statistical procedures. A number of these are available and for the majority of basic operations, there is little to choose between them. While most of the techniques covered in this book do not demand the use of a computer, the tedium involved in the calculations is reduced, and their accuracy increased, if a statistical package is used. District and regional computer or information departments and local university or polytechnic statistics departments should be able to provide advice on the availability of such packages.

2.6.3 Other computer packages

Probably the most familiar software in use is the word processing package. While its primary purpose is the production of documents such as letters and reports, its powerful editing facilities make it a useful tool for data entry and editing. Even when a database package is available, it may be more efficient to use a word processing package to enter small, simple datasets. Word processing packages may also play a very useful direct role in some projects; for example, in a study of rubella immunization (by Cook *et al.*, 1987) names and addresses of non-immune women were extracted from the database and used as input to a word processing package for the automatic production of follow-up letters.

Where tabular output from a database or statistics package needs to be supplemented by graphical display, graphics packages may be used to produce attractive and effective output. Some database and statistical packages have associated graphical facilities, either as an integral part or as a supplementary enhancement. Exported data from any database or statistics package may be used as input to graphics packages.

Spreadsheet packages have grown out of accounting activities but have found application in other areas, particularly where the natural format of the data being handled is a set of two-way tables. These packages allow the manipulation of tables and parts of tables, and in particular provide for arithmetic operations on whole tables, for example, divide one table by another. This may be particularly useful for demographic and epidemiological applications where rates are to be presented as tables of age group by sex (see Chapter 5).

While all of these types of package have been described separately, many packages actually combine two or more of these functions, and some attempt to cover all five areas of expertise. At the time of writing, however, there is none available that is actually best in every respect and it is therefore probably advisable to use for each aspect of data management and analysis, that package available which performs best in that area, and to move data between packages as necessary. A typical scenario would be to collect and verify data using a database package, obtain standard reports and very simple summary statistics using the tabulation facilities of the database (or possibly a spreadsheet) package, and where more complicated analyses are required, export the data to a statistics package. A word processing package could then be used to create the final report.

2.7 The district database

The concept of minimum datasets introduced in Section 2.4.2 extends naturally to that of a *district database*. It is obvious that if data collection and storage are to be as efficient as possible, data should ideally be collected only once and should, subject to confidentiality constraints, be accessible to all parts of the districts where they might be required. The district database can therefore be seen as being built up from subsets of data collected at different locations and in different ways. The data are not necessarily held in the same physical location, but ideally every person needing any of the data and having right of access to them is able to locate them simply and quickly.

The basis of a district database consists of activity, manpower and financial data but it may, and should wherever possible, be extended to include works, supplies, pharmacy and any other relevant data collected in the district. The art in constructing such a database is to achieve the right balance between sufficient repetition of identifying items to allow cross-linking, and minimization of the amount of data collected.

As a simple example, assume that a district maintains four databases: (1) hospital activity, (2) manpower, (3) supplies and (4) community nursing activity, and that while these are distinct, and may even be held in different locations, they are subject to common rules concerning frequency of data collection, identifying codes and level of detail. Also, computer links exist that allow any of the data to be accessed by any manager who may need them. While each dataset contains a different type of information there are common items. For example, database (1) and (4) have the same way of identifying patients seen (e.g., using a district patient number) while databases (1) and (2) refer to wards in the same way (using an agreed set of ward codes). The cost of treating particular groups of patients could be established, albeit in this example in a very simplistic and inadequate way, by selecting the appropriate items from each of the four datasets to say what services were used in hospital, how many whole-time equivalents of staff were required, what supplies were used in treating these patients, and what additional nursing time was required for their care in the community.

2.7.1 Data modelling

In the example of the last section, one particular view of the district database, the 'patient cost' view, was obtained by carefully selecting from it the particular set of items relevant to patient cost. Different views of the database could just as easily be obtained by selecting different sets of data. A 'nursing activity' view could be obtained by selecting some of the data items from sets (1) and (2), while a 'supplies inventory' view might be given by selected items from (1) and (3). The designation of these different views from one comprehensive dataset has been given the name of *data modelling*.

The concept of data modelling assumes that users of the information system are provided with tools for extracting from the district database only those items of data applicable to their particular application, so that to each user, the database appears to have been designed for his purposes alone and to contain no extraneous data. Database screens play a vital role in this process, by displaying selected data in a way tailored to be meaningful to a particular group of users. This is an extremely simple concept, but relies upon efficient design of the database, and to be at all practical, requires a district to have a comprehensive computer system linking all sites contributing to, or using, the district database. It must also be supported by a security system, for example, passwords controlling access to various subsets of data items and screens.

2.7.2 Producing information from a district database

The Körner reports specify in great detail what data must be collected, but say very little about how they may be used, apart from guidelines for statistical returns. The potential for the production of information from a comprehensive district database is obviously vast and care must be taken that this wealth of information is treated with the same caution and respect as would be due to data derived from any other source. This implies that before any information is extracted from the data, it must be verified that they are correct and up-to-date (methods for verifying the data should be an automatic part of the data collection procedures, but they can never take the place of common sense), and that the analyses performed are meaningful and relevant to the problem at hand. This book aims to provide methods for dealing with these issues.

When a comprehensive and well-designed integrated information system of the type described here exists, it has several advantages over the many data collection and record keeping systems which it replaces. Because it is much easier to verify data as and after they are collected, they are almost certainly more accurate; computerized systems obviate the need for time-consuming compilation of statistical returns and analyses can be performed as soon as data have been collected, and so are much more timely.

Because of the Körner standard definitions, there is a very high degree of compatibility between different types of data, and among different districts. It is therefore much easier to perform analyses using data from different sources. Finally, and perhaps most importantly, data are accessible, and not buried in the back of a filing cabinet, or in someone's head!

3
Descriptive statistics

3.1 Introduction

A set of data in its raw form, after collection and storage, forms the basic material for understanding a health system or population. Yet raw data are typically muddled, cumbersome and uninteresting, and need to be ordered, reduced and presented in some attractive way so that real information may be extracted from them.

Ordering and reduction of data can be achieved by the use of tables, pictures or summary statistics. They perform a basic function in producing health information and are also essential first steps in any more complex analysis of the data. In this chapter, tables and pictures will be considered, leaving a discussion of summary statistics to Chapter 4.

3.2 Basic principles of tabulation

A common way of presenting numerical health information is in the form of a table. If properly constructed, a table can make the information easier to assimilate by showing at a glance many of the properties of the data, such as patterns or groupings. Even when the information is not numeric, it may be beneficial to adopt a tabular presentation, although such situations will not be explored here.

In general there are three decisions that have to be made when designing a table:

  (i) the layout of the rows and columns;
 (ii) the content of the cells created by the rows and columns;
(iii) the annotation.

If the data are to be tabulated using a computer, the degree of choice when making these decisions may be less than if manual methods were used. The restrictions depend upon the actual software used. There is still, however, a need for human effort in table design; the real value of computerization is that tedious tasks, such as ordering numbers and creating totals, are done automatically.

Clearly the most appropriate format for a table will depend upon the data, and which aspects of the data it is important to portray. Some general rules can, however, be stated:

(i)   a concise, but informative, title should be given, which should include the date, place or whatever else is common to all the entries in the table;

(ii) a heading should be provided for the rows, which gives a brief description of the variable which changes in value from row to row;

(iii) similarly a heading should be provided for the columns;

(iv) each individual row and column should have a heading, these being descriptions of the values taken by the variable concerned (its *levels*);

(v)   the units of measurement should be given for all entries in the table, either in the title or in the row/column headings;

(vi) notes should be used to give, where appropriate,
   (a) the source of the data,
   (b) detailed specifications,
   (c) qualifications,
   (d) definitions of terms;

(vii) thicker ruling should be used to separate off distinct parts of the table;

(viii) figures should be given only to the degree of accuracy that is appropriate for the presentation, and after rounding, superfluous trailing zeros should be removed and the units of measurement altered accordingly.

Table 3.1 illustrates most of these points. The only other general point to make about table design is that it is a good idea to produce a sketch version of the table before beginning on the final version, whether this final version is to be produced by hand or on computer. At this stage, enhancements such as totals or extra annotation can be added and the overall layout altered without a great deal of effort.

Table 3.1 Accidental deaths by place, Great Britain. Note that this table excludes deaths undetermined whether accidentally or purposely inflicted, misadventures during medical care, abnormal reactions, late complications and late effects of accidental injury. Figures for Scotland include late effects. (Data from *Social Trends* (1985), Central Statistical Office.)

| Place of death | Year | | | | |
	1971	1976	1980	1982	1983
Railway	212	135	127	106	119
Road	7 970	6 956	6 653	6 124	5 869
Other transport system	219	215	165	177	150
Total transport	8 401	7 306	6 945	6 407	6 138
Work[1]	860	712	530	457	443
Home[2]	6 917	6 250	6 009	5 468	5 514
Other	3 068	2 831	2 516	2 781	2 459
Total	19 246	17 099	16 000	15 113	14 554

1 Includes a small amount of double counting; these are cases where the accident occurred while the deceased was working at home.

2 'Home' includes residential accommodation. Figures for Scotland comprise accidents occurring on farms and in forests, in mines and quarries and in individual places and premises.

3.3 Frequency distributions

When data on a single variable are to be ordered and reduced, one of the simplest and most effective methods is to prepare a table showing each of the distinct values in the data, in ascending order, together with counts of the number of times each value occurs. Such counts are known as *frequencies*.

The variable recorded in the table can either be quantitative (such as weight or age) or qualitative (such as sex or area of residence). When the variable is quantitative the table is known as a *frequency distribution*.

Example 3.1: Using sample data from a hospital ward, the frequency distribution shown in Table 3.2 was created.

Table 3.2 Length of stay for a sample of inpatients. (Data from Appendix 1.)

Length of stay/days	Frequency
0	4
1	6
2	4
3	7
4	8
5	2
6	1
7	2
8	1
11	1
12	1
13	1
15	1
18	1
19	2
22	2
24	5
30	3
31	1
44	1
54	1
65	1
Total	56

Similar tables can be created for the other variables in Appendix 1. Each variable can then be considered by itself in a much clearer fashion than from the raw data. As a second example, Table 3.3 shows the distribution of the number of operations per patient.

3.3.1 Grouped frequency distributions

When the values taken by a variable are almost all different, frequencies can be recorded for groups of values. This is a very useful method for condensing such data. It is also a useful method when values do tend to repeat but the number of distinct values is large.

Table 3.3   Number of operations for a sample of inpatients. (Data from Appendix 1.)

Number of operations	Frequency
0	47
1	8
2	0
3	1
Total	56

Example 3.2:   Although Table 3.2 has illustrated the length of stay data in a compacted format, it is still rather cumbersome and probably gives more detail than is required for administrative purposes. Table 3.4 shows the same data in a more compact, grouped form.

Table 3.4   Length of stay for a sample of inpatients (from Table 3.2).

Length of stay/days	Frequency
0	4
1–2	10
3–5	17
6–15	8
16–25	10
26–35	4
over 35	3
Total	56

Each group, defined by a lower and upper value (or just one value, where these are the same), is called a *class*, and the lower and upper values, the *class limits*. The interval between the limits is known as the *class interval*. Sometimes the final class is left open-ended, as in Table 3.4. This is usually done so as not to emphasize the largest values when these are relatively few and unimportant. Less often a similar thing is done for the first class of the table. Open-ended intervals lead to problems, however, when the information is to be represented pictorially (see Section 3.11) or summary statistics are to be calculated (see Chapter 4).

The class intervals should always be listed in ascending order, for clarity, and should never overlap. A common mistake is to label classes so that the adjacent limits are equal, for example 0–5, 5–10, 10–15, 15–20, etc. Such labelling should be avoided since it is not clear where a value exactly at a class limit should be recorded. For example, should 5 be recorded as a contributor to the 0–5 class or the 5–10 class (certainly not both)? On the other hand, it is just as bad to use limits with gaps when the variable recorded is continuous. For example 0–4, 5–9, 10–14, 15–19, etc. would leave no obvious place to record a number such as 4.8. This is only permissible when the data are discrete, such as length of stay which is always measured in whole days.

One possible solution to this problem is to give the recording convention as a footnote, but a neater solution is to use unambiguous labels on the classes themselves, as illustrated by Table 3.5.

Table 3.5 Ages of a sample of inpatients. (Data from Appendix 1.)

Age/years	Frequency
0 to less than 30	4
30 to less than 40	8
40 to less than 50	3
50 to less than 60	7
60 to less than 70	6
70 to less than 80	13
80 to less than 90	11
90 to less than 100	4
Total	56

In fact, although Table 3.5 gives a perfectly correct example of class limits for continuous variables, a convention has arisen that when age is tabulated it is recorded as 'age last birthday'. This is a *discrete* variable, since everyone is a whole number of years old on his birthday, and it is then permissible to label as 0–4, 5–9, etc. See Table 5.1 for an example.

In general, when designing a grouped frequency distribution two things have to be decided:

(i) the number of classes;
(ii) the size of each class interval.

These considerations are not independent since, for example, it is not possible to have many large classes when all the data values are very similar. Decisions should always be made with the reader in mind. Normally there will be no opportunity to look at the original data and so a complete, as well as informative, display will be needed. In most cases a reasonable balance can be achieved by using between 5 and 8 classes.

There is some advantage to be gained from keeping all class intervals equal in length, since this makes comparison of the frequencies more meaningful. However, when data are compacted at one end yet sparse at the other, as with the lengths of stay illustrated by Table 3.2, it is most sensible to group in the same fashion, as in Tables 3.4 and 3.5. Sometimes the application of the information may itself influence the choice of class intervals; for example, if lengths of stay were normally talked about in units of weeks, Table 3.4 would be better with classes 0, 1–7, 8–14, 15–21, etc.

Most statistical and database packages will create grouped frequency distributions automatically, using some sensible rules to decide upon the class limits. Some packages will also allow the user to specify class limits if some special grouping is desired.

3.3.2 Rescaled frequencies

Frequency distributions are usually prepared to enable comparisons to be made regarding the numbers of occurrences in the different classes. Such comparisons are easier to make if the frequencies are rescaled by making their sum equal one, yet retaining the same relative magnitudes. This is achieved by dividing each frequency by the overall sum of the frequencies. The result is

Table 3.6   Ages of a sample of inpatients (from Table 3.5).

Age/years	Relative frequency	Percentage frequency
0 to less than 30	0.07	7%
30 to less than 40	0.14	14%
40 to less than 50	0.05	5%
50 to less than 60	0.13	13%
60 to less than 70	0.11	11%
70 to less than 80	0.23	23%
80 to less than 90	0.20	20%
90 to less than 100	0.07	7%
Total	1	100%

called a *relative frequency distribution*. An alternative is to express each frequency as a percentage of the overall sum, achieved by multiplying the relative frequencies by 100. This is called a *percentage frequency distribution*. Examples of both of these distribution are given in Table 3.6.

There is one disadvantage with these rescaled frequency distributions. To enhance presentation it is advisable to round the rescaled frequencies, as in Table 3.6. However, if the person receiving this presentation wishes to carry out further analyses on the data, such as calculating a mean (see Section 4.3.1), he will not get as accurate a result using the rounded figures as he would from the basic frequency distribution. This may be avoided by presenting frequencies in both their original and rescaled forms as different columns of the same table (see Table 3.8 for an example).

3.3.3 Cumulative frequencies

Another way of expressing a frequency distribution is to accumulate frequencies as one moves down the table and so form a *cumulative frequency distribution*. This display enables the reader to see the frequency of occurrence of all values less than a certain amount, as illustrated by Table 3.7.

Table 3.7   Ages of a sample of inpatients (from Table 3.5).

Age/years	Cumulative frequency
Less than 30	4
Less than 40	12
Less than 50	15
Less than 60	22
Less than 70	28
Less than 80	41
Less than 90	52
Less than 100	56

Presentation can sometimes be improved by accumulating relative rather than actual frequencies to form a *cumulative relative frequency distribution*, or similarly a *cumulative percentage frequency distribution* could be formed

Table 3.8 Ages of a sample of inpatients (from Table 3.6).

Age/years	Cumulative frequency	Cumulative percentage frequency
Less than 30	4	7%
Less than 40	12	21%
Less than 50	15	26%
Less than 60	22	39%
Less than 70	28	50%
Less than 80	41	73%
Less than 90	52	93%
Less than 100	56	100%

in the obvious way. Another variation on the basic idea is to accumulate from the bottom upwards rather than the top downwards, although this is not commonly done and will not be discussed further. Table 3.8 shows a cumulative percentage frequency distribution for the age data, and also shows the cumulative frequencies themselves as a separate column.

3.4 Cross-classifications

When data have been recorded on two or more variables, there is a much wider choice of table design. The most basic design is where separate tables for each different variable are simply placed alongside one another. This has limited use since it says nothing about the relationship between the two variables.

An alternative is to cross-classify the variables so that one variable labels the rows and the other the columns. Then the distribution of values of one variable can be seen at every different value of the second variable, and hence these distributions can be compared to look for relationships between the variables (see Chapter 9). When, as would usually be the case, the entries in the cross-classification are frequencies the resulting table is called a *contingency table*. The variables presented in a contingency table can be discrete, grouped (continuous or discrete) or qualitative.

Example 3.3:   There is a considerable difference between the ages of patients who enter the hospital ward surveyed for Appendix 1 and there is interest in how age influences length of stay. Table 3.9 presents age and length of stay information.

Table 3.9 Age and length of stay for a sample of inpatients. (Data from Appendix 1.)

Length of stay/days	Number of patients	Age/yrs (last birthday)	Number of patients
0–2	14	0–29	4
3–5	17	30–49	11
6–15	8	50–69	13
16–25	10	70–79	13
26 and over	7	80 and over	15
Total	56	Total	56

Table 3.10   Age by length of stay for a sample of inpatients. (Data from Appendix 1.)

Age/years	Length of stay/days					Total
	0–2	3–5	6–15	16–25	26 and over	
0–29	2	2	0	0	0	4
30–49	6	3	2	0	0	11
50–69	3	4	1	1	4	13
70–79	1	5	3	2	2	13
80 and over	2	3	2	7	1	15
Total	14	17	8	10	7	56

This table certainly confirms that ages (and lengths of stay) are widely spread, but essentially this is merely a combination of two separate frequency distributions which fails to indicate the association between the variables in any way. On the other hand, Table 3.10 does indicate the association; the patients who stay longer tend to be older as might be expected.

Notice that the two individual frequency tables given as Table 3.9 are simply the row totals and column totals in Table 3.10. These are known as the *marginal tables*.

Table 3.11   Sex by age by length of stay for a sample of inpatients. (Data from Appendix 1.)

Male

Age/years	Length of stay/days			Total
	Below 10	10–19	20 and over	
Below 40	1	0	0	1
40–59	0	0	0	0
60 and over	2	1	1	4
Total	3	1	1	5

Female

Age/years	Length of stay/days			Total
	Below 10	10–19	20 and over	
Below 40	11	0	0	11
40–59	9	0	1	10
60 and over	19	10	1	30
Total	39	10	2	51

Persons (total)

Age/years	Length of stay/days			Total
	Below 10	10–19	20 and over	
Below 40	12	0	0	12
40–59	9	0	1	10
60 and over	21	11	2	34
Total	42	11	3	56

Table 3.12 Sex by age by length of stay for a sample of inpatients. (Data from Appendix 1.)

Age/years	Sex	Length of stay/days			Total
		Below 10	10–19	20 and over	
Below 40	Male	1	0	0	1
	Female	11	0	0	11
	Total	12	0	0	12
40–59	Male	0	0	0	0
	Female	9	0	1	10
	Total	9	0	1	10
60 and over	Male	2	1	1	4
	Female	19	10	1	30
	Total	21	11	2	34
Total	Male	3	1	1	5
	Female	39	10	2	51
	Total	42	11	3	56

When there are three variables recorded in the data the cross-classification idea can be generalized in one of two ways. Method one is to produce a separate two-way table of variable 1 against variable 2 for each value (or 'level') of variable 3, as shown by Table 3.11. Method two is to subdivide the rows (or columns) to make them cross-classify two of the three variables, as illustrated by Table 3.12. Method two takes up less space and hence tends to be more popular.

It is possible to present tables of cross-classification by more than three variables but such presentations are usually very difficult to understand, and are not recommended.

3.5 Statistics in tables

In order to lead the reader to the most important features of the data it is frequently useful to include summary statistics within a table. That is, the cells of the table contain the results of calculations instead of, or perhaps in addition to, the simple frequencies. Many statistical computer packages will allow tables to be constructed in this way, sometimes with many different summary statistics in each cell. Spreadsheet packages are also very suitable for constructing presentations of this type.

Table 3.13 Length of stay by sex for a sample of inpatients. (Data from Appendix 1.)

Sex	Length of stay/days			Total
	Below 10	10–19	20 and over	
Male	60%	20%	20%	100%
Female	76%	20%	4%	100%
Total	75%	20%	5%	100%

Table 3.14   Age by sex showing average length of stay (in days) for a sample of inpatients. (Data from Appendix 1.)

Sex	Age/years					Total
	Below 20	20–39	40–59	60–79	80 and over	
Male	1.0	—	—	29.8	—	24.0
Female	2.0	2.2	9.2	13.9	14.4	10.6
Total	1.7	2.2	9.2	17.2	14.4	11.8

A thorough discussion of summary statistics is deferred until Chapter 4, but for now it will be sufficient to give two simple examples. Table 3.13 shows the use of percentages, already mentioned in the context of frequency distributions, but here used in a cross-classification. Notice how the 100%s are written in the marginal row total column to signify to the reader that the percentages are row and not column percentages. Table 3.14 gives an example of the use of averages (see Section 4.3), and shows that for these data older people tend to stay longer.

3.6 Statistical pictures

Although tables certainly make data more easily understandable than they are in raw form, they may still require a considerable amount of close inspection before even the most basic aspects are discerned. Tables are also drab in appearance and the presence of many tables in a report tends to have a numbing effect, losing the reader's attention. Pictures provide a solution to these problems, since the human brain is more tolerant of visual presentations than it is of numerical ones, and can assimilate information more rapidly and retain it far longer when pictures are used.

There are, however, disadvantages with pictures as alternatives to tables. They are less precise and consequently not suitable as a starting point for further analysis of the data, or as a means of preserving an historical record. They are also more time-consuming to produce, even on a computer, and as the quality of production increases the differential increases considerably

Pictures which illustrate data are referred to as statistical pictures. Due to the differing nature of statistical pictures there are few conventions, or 'rules', that are appropriate in all cases. Nevertheless there are enough general rules for it to be worth listing them here.

  (i) A concise, but informative, title should be given.
 (ii) Notes should be used as for tables.
(iii) Axes should be labelled with both the subject and units of measurement.
(iv) The scale should be chosen to fit the paper space, that is the minimum and maximum values to be displayed should be found and the scale fixed by dividing their difference by the size of the paper space (rounded to a convenient number).
 (v) When different pictures are superimposed, the separate parts should be distinguished by using labels, colours or dotted lines, with a key if necessary.
(vi) The picture should not distort the facts (examples follow later).

In addition, just as for tables, it is worthwhile to sketch a trial drawing, where the scale and general appearance can be checked, before embarking on the final version. A detailed account of the principles underlying statistical pictures, and a beautiful collection of examples, is given by Tufte (1983).

The art of computer graphics is one of the fastest developing aspects of software. Most statistical packages produce pictures, sometimes in rough form as combinations of dots on a VDU or printer, and sometimes in high resolution on a special plotting device. Both have their uses, the first as a means for a rough or initial examination of data and the second as a formal presentation, such as in a health district's annual plan. In the latter context, software which enables pictures to be incorporated within word-processed text is an advantage.

The various types of statistical pictures are described in the sections which follow.

3.7 Graphs

The word 'graph' can be used for many different kinds of pictures, but here it will be taken to mean any picture where individual data values are represented as marks on a two-dimensional grid. The values of one quantity will label the vertical axis, and the values of a second quantity the horizontal. The marks may be joined by lines. In this case it is appropriate to use straight lines if the data are discrete and to draw a smooth line approximately through the points if the data are continuous. This is because in the former instance, but not the latter, points in between the observed data are meaningless.

Graphs should be contrasted with the other form of statistical picture in which data values are represented by shapes (see Sections 3.8–3.10).

When the object of a graph is to provide a means of comparison of relative size it is much better to provide a zero on the vertical scale, since otherwise

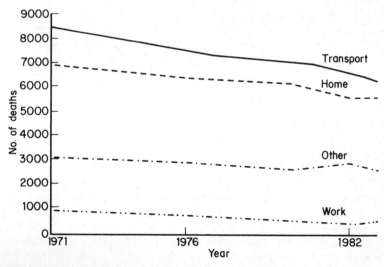

Fig. 3.1   Accidental deaths, Great Britain. (Data from Table 3.1.)

any differences may appear to be much larger than they really are.

When different graphs are to be compared an effective method is to put them on the same grid. This is most meaningful when the scales of measurement are the same, as in Fig. 3.1.

3.7.1 Layer graphs

When several lines on the graph are really components of some overall total, as was the case in Fig. 3.1, an alternative way of combining the separate graphs is to layer one above the other, so that the total accumulates upwards from line to line. The result is called a layer graph, as illustrated by Fig. 3.2.

Notice that Fig. 3.1 presents the changes in the number of accidental deaths in each category in an effective way, but does not bring the reader's attention to yearly changes overall. Figure 3.2 shows overall changes (by following the top line) and also shows how the relative importance of the different causes of accidental death changes from year to year. It does not, however, make very clear how the actual number of deaths due to any particular cause has changed. The choice of display depends entirely upon what aspects of the data one wishes to portray.

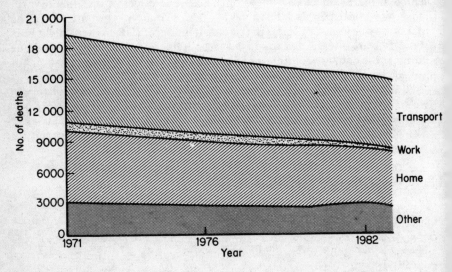

Fig. 3.2   A layer graph of accidental deaths, Great Britain. (Data from Table 3.1.)

3.8 Bar charts

Probably the simplest and most effective means of illustrating discrete or qualitative data is the bar chart. In this statistical picture, physical quantities are represented by the *lengths* of rectangles or 'bars'; that is the lengths are drawn in the same proportions as the physical quantities. Since the widths of the bars have no interpretation these are all the same. To emphasize the discrete or qualitative nature of the data, the bars are separated by gaps. The

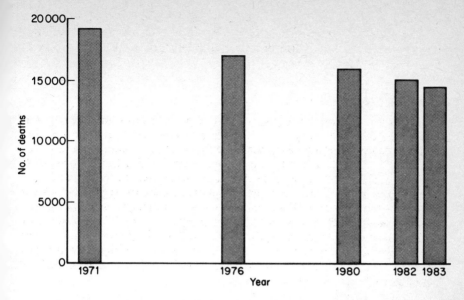

Fig. 3.3 A bar chart of accidental deaths, Great Britain. (Data from Table 3.1.)

bars can be drawn either vertically or horizontally, and in a report containing an extensive collection of bar charts it is a good idea to mix the types to provide variety. Since bar charts are essentially for comparative analyses it is essential to provide a true zero for the physical quantity, since otherwise the bars will be truncated and the lengths will not be in the correct proportions.

The simplest type of bar chart is that given in Fig. 3.3. Notice that the bars are spaced on the time axis according to the scale of this axis. If the bars were simply equally spaced, the presentation would be misleading.

3.8.1 Compound bar charts

Very often the physical quantity represented by a bar is made up of different parts, and it is then sensible to present these parts separately. One way of doing this is to have separate bars for each part, resulting in a compound bar chart such as Fig. 3.4. For ease of interpretation the bars within each group appear in the same order. Because the compound bar chart of the full set of data is too wide to fit comfortably across the page, only three years are illustrated by Fig. 3.4.

Figure 3.4 is excellent when comparisons of causes of death within years or actual number of deaths by cause are of interest. It is not so good for showing the overall total number of accidental deaths per year.

3.8.2 Component bar charts

An alternative to the compound is the component bar chart in which the bar which represents overall size is divided up into sections, the length of each section being in proportion to the parts they represent. Figure 3.5 gives an

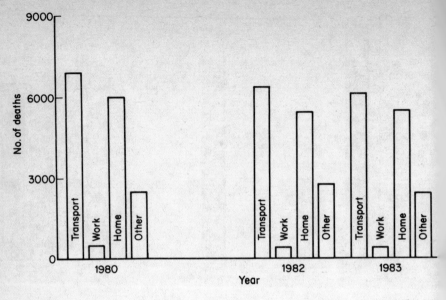

Fig. 3.4 A compound bar chart of accidental deaths, Great Britain. (Data from Table 3.1.)

example which will be seen to be comparable with the layer graph. Notice that here a key is provided as an alternative to annotating the bars themselves.

Figure 3.5 is useful for illustrating year-to-year changes in the overall number of accidental deaths and the proportion of deaths from each cause. It is not easy to see the actual number of deaths from the different causes because the starting point (bottom) of each section is not at zero, except in the case of the lowest section.

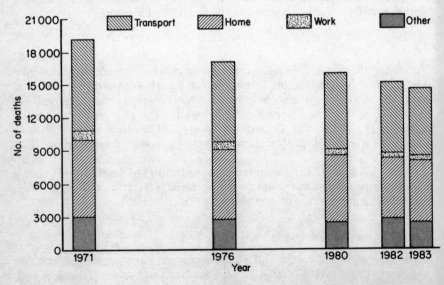

Fig. 3.5 A component bar chart of accidental deaths, Great Britain. (Data from Table 3.1.)

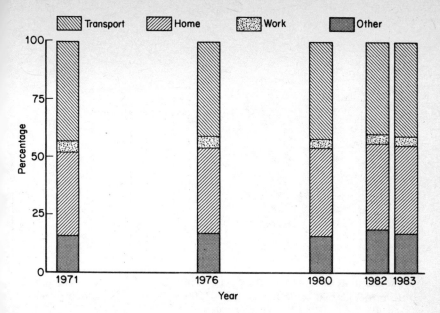

Fig. 3.6 A percentage component bar chart of accidental deaths, Great Britain. (Data from Table 3.1.)

3.8.3 Percentage component bar charts

Neither the compound nor the component bar chart is particularly good at showing how the percentage of deaths due to any particular cause varies from year to year. A good way of presenting this facet of the data is to draw a component bar chart in which the sections represent percentages rather than absolute physical quantities. One consequences of this is that all the bars must be of equal length. Such a bar chart is called a percentage component bar chart, and Fig. 3.6 is an example.

Clearly the percentage component bar chart cannot be used to show changes in absolute sizes; it is, however, very useful for showing changes in relative sizes.

3.9 Pie charts

A pie chart is a circle which is split up into segments like a pie cut into pieces from the centre outwards. Each segment represents one value (or possibly a range of values) taken by a variable. The *areas* of the segments are drawn in proportion to the physical quantities observed for each value. An example is given by Fig. 3.7. Example 3.4 shows how the correct areas are determined.

Example 3.4: Suppose that a variable can take three values, A, B and C. The observed frequences of A, B and C are 100, 200 and 300 respectively. Hence the areas representing these values should be in the ratio 100:200:300. It happens that the area of a segment is a constant multiple of the angle created

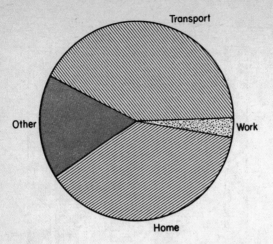

Fig. 3.7   A pie chart of accidental deaths in Great Britain, 1983. (Data from Table 3.1.)

by the segment at the centre of the circle. Hence the 360° of the circle must be split up into $100/600 \times 360 = 60°$, $200/600 \times 360 = 120°$ and $300/600 \times 360 = 180°$ for A, B and C respectively.

3.9.1 Multiple pie charts

As Fig. 3.7 shows, a pie chart is really only an alternative to a *single* bar of a component bar chart. The idea of a pie chart can be generalized to a multiple pie chart which is a set of circles, each one of which takes the place of a bar on a component bar chart. If this is done, then the areas of the individual circles should be in proportion to the overall physical quantities that they represent. The next example illustrates how this can be achieved.

Example 3.5:   The variable with values A, B, and C has been observed to have frequencies of 100, 200, and 300 respectively in 1985 and 400, 800 and 1200 respectively in 1986. Hence the overall total for 1985 is 600 and for 1986 is 2400 and the circles must thus have areas in the proportions 600:2400 or 1:4.

Now, the area of a circle is πr^2 where r is the radius and π is the constant 3.14159. . . . Let r_1 be the radius of the 1985 circle and r_2 the radius of the 1986 circle. Then it must be that:

$$4\pi r_1^2 = \pi r_2^2$$
$$\text{i.e.,} \quad 4r_1^2 = r_2^2$$
$$\text{i.e.,} \quad r_2 = \sqrt{4}\, r_1 = 2r_1$$

In general the squares of the radii need to be drawn in proportion to the overall physical quantities. For the example, this requires the radius of the 1986 circle to be twice that of the 1985 circle.

Although pie charts have a wide visual appeal, one disadvantage is that they are difficult to draw by hand, requiring a protractor and a compass. Furthermore, multiple pie charts take up a great deal of room when they involve many circles. Another objection is that it is rather more difficult to make comparisons of areas, which are two-dimensional, than of one-dimensional lengths. For these reasons, bar charts are almost always preferable to pie charts. Even so, pie charts are a useful device for enlivening a report which is already full of words, tables and bar charts, especially when a single pie chart will suffice and the facilities for producing it by computer are available.

3.10 Pictograms

Physical quantities are represented by rectangles on a bar chart and circles on a pie chart, but any shape could serve the same purpose. The general name for a diagram in which shapes represent quantities is a pictogram. If a pictogram other than the familiar bar chart or pie chart is used, it will usually employ a shape which is relevant to the subject depicted. Hence if the number of hospital beds available in a district were being portrayed the shape might be a stylized version of a bed.

There are two types of pictogram, the size-proportional and the size-constant. Due to their visual attraction they are particularly desirable when the display is used to educate or inform children. Their disadvantages are the difficulty of drawing them (although some graphics or 'painting' packages avoid this problem) and the lack of precision possible when determining physical quantities from even moderately complicated shapes.

1956   1982

Fig. 3.8 A size-proportional pictogram of number of nurses per available bed, NHS hospitals, 1956 and 1982. (From *Compendium of Health Statistics*, Office of Health Economics, 1984.)

3.10.1 Size-proportional pictograms

In a size-proportional pictogram the size of the shape is proportional to the physical quantity which the shape represents. An example is given by Fig. 3.8.

The particular problem with this type of statistical picture is that it is not always obvious whether lengths, areas or volumes are drawn in proportion to the physical quantities (Fig. 3.8 has the areas in correct proportion). As Huff (1954) (Chapter 6) demonstrates, this can give rise to some very misleading interpretations. A doubling in size in one dimension, for example, can look like an eightfold increase in three dimensions with a symmetrical figure, if the eye perceives the drawing as three-dimensional. This fact has not escaped the notice of advertisers and publicity agents! For this reason, the size-proportional pictogram is not recommended and if it is met it should be interpreted with caution. Although the bar chart and pie chart are clearly special cases of this type of diagram, they escape the general criticism since they are both abstract figures in which differences are represented in a generally accepted way.

1956

1982

Fig. 3.9   A size-constant pictogram of number of nurses per available bed, NHS hospitals, 1956 and 1982. (From *Compendium of Health Statistics*, Office of Health Economics, 1984.)

3.10.2 Size-constant pictograms

In a size-constant pictogram a shape of a certain size is declared to represent a specified number of units. The number of complete and part shapes that appear on the diagram then represents the physical quantity, as shown in Fig. 3.9. The number of nurses per available bed in 1982 was almost exactly three times the number in 1956. (This does not, however, imply that nurses' workloads have decreased by a factor of three over this period.)

3.11 Pictures of frequency distributions

In Section 3.3, a frequency distribution was defined as a set of values together with the observed frequencies of these values for a quantitative variable. Sections 3.8–3.10 described pictures which are appropriate when the variable is discrete (or qualitative). In this section, diagrams appropriate for continuous variables will be discussed.

Even though the pictures described here will not strictly be appropriate for discrete variables, it is sometimes useful to approximate a discrete variable by a continuous analogue, especially when the variable has been recorded as a grouped frequency distribution. This analogue is also used with those methods of Chapter 4 which assume that a continuous variable is being analysed.

The continuous analogue is derived from a discrete frequency distribution by using a *continuity correction* which simply joins adjacent classes at their midway point. Hence, the classes in Table 3.4 would be approximated thus:

> 1–2   becomes 0.5–2.5
> 3–5   becomes 2.5–5.5
> 6–15 becomes 5.5–15.5, etc.

There is some ambiguity about what to do with the lowest and highest classes, for example the class 0 in Table 3.4. The conventional way is to extend the pattern of the other classes, so that 0 becomes – 0.5 to 0.5.

Another problem with constructing diagrams from frequency distributions arises when the distribution has an open-ended extreme class. For example, if the only data on length of stay available were those in Table 3.4, there would be no way of accurately determining the highest length of stay and the extremity of the diagram would be unknown. The best that can be done under these circumstances is to take a guess at the extreme value using experience and common sense.

Since relative or percentage frequency distributions differ from ordinary frequency distributions only by a rescaling of the frequencies, any diagram would take the same shape regardless of the type of frequency distribution used.

3.11.1 Histograms

In appearance histograms often look very similar to bar charts. There are two differences. First, because histograms represent continuous data, adjacent rectangles are shown touching one another. Second, the *area* of the rectangle, rather than its length, is drawn in proportion to the frequency. When the class intervals of the frequency distribution are all equal, area is a constant multiple of the length and the rectangles in the histogram *can* be drawn exactly as for the bar chart. However, when the class intervals are not all equal then the widths of the rectangles will certainly not all be the same. In this case the length must be determined from the equation for the area of a rectangle, i.e.,

$$\text{area} = \text{length} \times \text{width}$$

that is, the length is the area divided by the width, where the area is determined from the observed frequencies and the scale of the diagram. Hence a class with an interval of 2 and a frequency of 1 would receive a rectangle half the length and twice the width of a class with an interval of 1 and a frequency of 1. The axis representing length should then be labelled as 'frequency per unit class interval' or 'frequency density' or some equivalent.

Example 3.6:   A histogram of the ages of patients taken from Table 3.5 appears in Fig. 3.10. Here the first class spans 30 years whilst the rest span only 10 years and hence the length of the first rectangle is

$$\frac{\text{observed frequency}}{30} = \frac{4}{30}$$

whereas for the other rectangles it is

$$\frac{\text{observed frequency}}{10}$$

For simplicity, the scale used in Fig. 3.10 multiplies these numbers by 10.

As already explained, when the data are discrete a continuity correction should be used, as illustrated by Fig. 3.11. This is drawn from Table 3.4

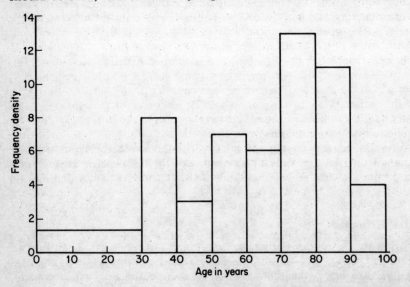

Fig. 3.10   A histogram of ages of a sample of inpatients. (Data from Table 3.5.)

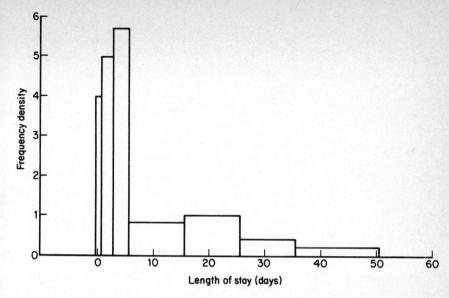

Fig. 3.11 A histogram of length of stay for a sample of inpatients. (Data from Table 3.4.)

which has an open-ended final class ('over 35 days'). This class has been guessed to go up to 50 days. Of course since the raw data are available in Appendix 1 this guess was not necessary, but if Table 3.4 were the *only* source of information some such approximation would be unavoidable.

One word of caution is appropriate when computer packages are used to produce histograms. For programming convenience or possibly due to misunderstanding, some packages will produce 'histograms' which are really bar charts, even when the variable is continuous. This is not a serious problem when the class intervals are all equal, but is not conventionally correct otherwise.

3.11.2 Stem and leaf diagrams

It has already been mentioned that diagrams suffer from inaccuracy, yet provide useful visual impact. One diagram which exhibits both accuracy and visual impact is the stem and leaf diagram. Essentially this is a histogram where the rectangles are built up to the correct length by numbers, rather like a size-constant pictogram where the shapes are numbers. Each data value is split into its stem, the first digit, and its leaf, the second digit. Thus 36 has stem 3 and leaf 6. The leaves are the numbers which make up the 'rectangles'. The stems label the classes.

An example is given in Fig. 3.12 where the first class, 0–4, has stem 0 and leaves 0,1,2,3,4 while the last, 65–69, has stem 6 and only one leaf, 5. Notice how in this example, each stem appears twice, once with leaves below 5, and once with those of 5 or greater, giving class intervals of length 5. This is simply to make the diagram more readable.

0	00001111112222333333344444444
"	556778
1	123
"	5899
2	2244444
"	
3	0001
"	
4	4
"	
5	4
"	
6	
"	5

Fig. 3.12 Stem and leaf diagram of length of stay for a sample of inpatients. (Data from Appendix 1.)

In outline, Fig. 3.12 is simply a histogram turned on its side; indeed it is simply Fig. 3.11 but with different class limits. The stem and leaf diagram shows more than just a histogram outline since it also records the actual data values themselves. Hence, for example, from Fig. 3.12 it is clear that in the 30–34 class all the observed lengths of stay are actually 31 or below. A histogram could not supply such precise information without becoming ridiculously large. Medians and modes (see Section 4.3) can easily be found from stem and leaf diagrams.

There are two limitations to stem and leaf diagrams. First, there is little flexibility in the choice of class intervals. Unlike those of histograms, they have to be kept equal and hence intervals of lengths that are whole number divisors of 10 (1,2,5, or 10) are the only possibilities for the hospital data. Second, to avoid confusing the display, all numbers should be represented by just two digits. This sometimes requires numbers to be rounded, for example 45661 would be represented as 4.6 thousand. This limits the accuracy.

Many modern statistical computer packages will produce stem and leaf diagrams from a single command. Sometimes enhancements are added to the display, such as declaring the scale for the stems or giving frequency counts alongside the diagram.

3.11.3 Frequency polygons and curves

Another alternative to the histogram is the frequency polygon in which relative frequencies are plotted against the mid-points of the corresponding class intervals and the points then joined by straight lines. Figure 3.13 gives an example.

Suppose that the class intervals used to create a frequency polygon were made very small. If a large number of observations were available, the outline of the polygon would appear less jagged. As the class intervals get smaller and smaller the outline becomes a smooth curve, called a frequency curve.

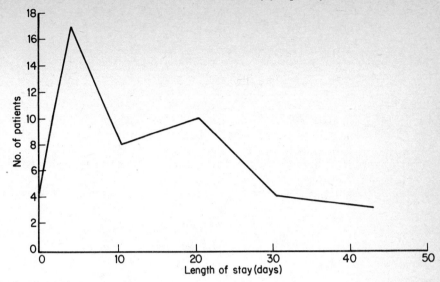

Fig. 3.13 A frequency polygon of length of stay for a sample of inpatients. (Data from Table 3.4.)

The most commonly occurring type of frequency curve has a bell-like shape known as the *normal* curve. Even when there are insufficient data to form a frequency curve, in some circumstances the normal curve may be a good approximation to the frequency polygon. Figure 3.14 shows the shape of the normal curve, which is symmetrical about its central peak and has tails which continually approach, but never touch, the horizontal axis.

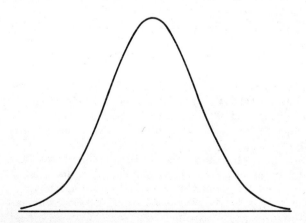

Fig. 3.14 The normal curve.

3.11.4 Ogives

A graph drawn from a cumulative frequency distribution is called an ogive. In this diagram, cumulative frequencies are plotted against *upper* class limits, joined either with straight lines or smoothed into a curve as in Fig. 3.15. In this diagram the upper class limits have been corrected for continuity.

Notice that the curve in Fig. 3.15 must peak at 1, since cumulative relative frequencies are used, and the peak must coincide with the highest observed value. As before, a guess of 50 days has been used for this value since Table 3.4 (the source) has an open-ended final class. The guess only affects the top right hand part of the curve and so is often not a particularly important issue when constructing the ogive.

Ogives from data following normal curves take a form somewhat like the letter 'S'.

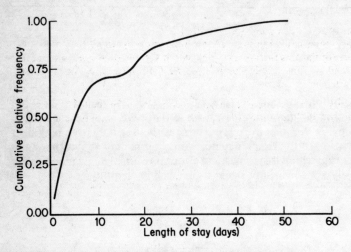

Fig. 3.15 Ogive for length of stay. (Data from Table 3.4.)

3.12 Maps

In essence any map can be regarded as a statistical picture, since maps represent quantitative information such as distance, height, etc. Clearly this is not an appropriate place to go into detail about cartography, but it is worth including a few words about two devices used in maps which can produce very effective visual displays of data.

The first device is to use shading to indicate the numerical values in the data appropriate to different geographical areas. For example, increasing density of shading may indicate increasing population density. The second device is to draw pictograms (of either type) at the geographic locations. For example, bars with sizes proportional to physical values could be placed with their bases touching the appropriate towns, perhaps to indicate the number of hospital beds in the towns. Maps used in this way can be a very powerful way of conveying information where comparisons are being made between different locations

Exercises

3.1 From the GP data on systolic blood pressure in Appendix 2 create:

(a) a frequency distribution;
(b) a cumulative percentage frequency distribution;
(c) a histogram;
(d) a stem and leaf diagram;
(e) an ogive.

3.2 Draw a stem and leaf diagram and a histogram for diastolic blood pressure from the GP data in Appendix 2.

3.3 Using the GP data again:

(a) draw a pie chart of body mass index (BMI);
(b) create a cross-classification table of smoking status against BMI;
(c) repeat part (b) but this time record the median systolic blood pressure in the table instead of the simple counts (see Section 4.3 for a definition of 'median').

Do these patients tend to be obese? Does smoking appear to be related to obesity? Does average systolic blood pressure appear to vary with smoking status and obesity?

3.4 Consider the following data:

Reported cases of industrial diseases, Great Britain

Cause	1979	1980	Year 1981	1982	1983
Poisoning	25	19	10	10	4
Ulceration	40	44	19	18	9
Compressed air illness	0	1	0	0	9
Gassing	195	157	93	69	40
Total	260	221	122	97	62

(Data from *Annual Abstract of Statistics*, Central Statistical Office, 1986.)

From these data draw:

(a) graphs for all 4 causes superimposed on the same set of axes;
(b) a layer graph;
(c) a compound bar chart;
(d) a component bar chart;
(e) a percentage component bar chart.

How good is each particular diagram at illustrating the important features of these data?

To compare the causes for their relative contributions it might be thought inappropriate to consider single years. Add the number of reported cases for each cause over the five years and draw

(f) a piechart of the five-year totals.

3.5 In a study of babies born in Berkshire, Neal and Tate (1972) give the following frequency distribution for birth weight of single births:

Birth weight/grams			Number
500	to less than	900	4
900	"	1300	9
1300	"	1700	20
1700	"	2100	37
2100	"	2500	161
2500	"	2900	561
2900	"	3300	1145
3300	"	3700	1187
3700	"	4100	622
4100	"	4500	163
4500	"	4900	33
4900	"	5300	3
Total			3945

(a) Draw a frequency polygon from these data. You should find that it resembles a normal curve.
(b) Draw an ogive from these data. This should have the characteristic 'S' shape of data that resemble a normal curve.

4
Summary statistics

4.1 Introduction

The tables and diagrams considered in Chapter 3 describe data in a compact and meaningful way. While such representations are essential for exploring the structure and uncovering underlying trends in the data, they are still too cumbersome for many purposes.

In order to talk about a set of data or to compare different sets of data simply and concisely it would be useful to have a single number, or possibly just a few numbers, which could be used to represent the entire set of data. Such numbers are known as *summary statistics*.

Summary statistics are used constantly when talking about health issues, or indeed any facet of everyday life. For example we might talk about the average number of operations performed each week by a particular gynaecologist. This average will have been calculated from records of the gynaecologist's workload over many weeks, and is thus a summary of the entire set of weekly observations.

This chapter will mainly be concerned with summary statistics for quantitative variables, that is to say measures of average and dispersion and their extensions. When the variable we are dealing with is qualitative the only technique covered in this chapter that will be relevant is the ratio.

4.2 Ratios

In a general sense a ratio is the result of the division of any two numbers. It is, however, usual to distinguish two particular kinds of division which are given the special names 'proportion' and 'rate'. A *proportion* is a ratio that compares a part with the whole and a *rate* is the result of the division of two numbers with different units of measurement (i.e., numerator per unit denominator). A further special kind of ratio is a *percentage*, but this is merely a proportion multiplied by 100 and so differs from a proportion only in term of scale of magnitude.

Ratios are useful summary statistics for data on a categorical variable, where the data take the form of a list of the categories together with their observed counts, or frequencies. The hospital ward data of Appendix 1 show an example of categorical data, the variable sex which has two outcomes, male and female. Altogether there are 56 observations, 5 of which are male and 51 female. The ward is nominally a female ward, so the incidence of male admissions is of interest. The data on sex can be summarized by stating:

   (i) the proportion of males in the ward, which is $5/56 = 0.089$, or
  (ii) the percentage of males in the ward, which is $0.089 \times 100 = 8.9\%$, or
 (iii) the ratio of men to women in the ward, which is $5/51 = 0.098$; otherwise expressed as 5:51 or 1:10.2.

In each case the values in the original data have been summarized by a single number which is a measure of relative importance.

When the data to be analysed consist of a sequence of observations on the same variable over time, often called a *time series*, ratios may be used to construct a set of *index numbers*, which provide a very useful measure of change in a variable over time.

The first step in constructing index numbers is to decide upon the base, or reference, time period. The index number for this base period is defined to be 100. For every other time period, the index number measures the relative change from the value of 100 in the base period. This is achieved by defining the index number as the value in the current period divided by the value in the base period, all multiplied by 100. That is:

$$i_c = \frac{x_c}{x_b} \times 100$$

where x_c is the value in the current period, x_b the value in the base period, and i_c the index number in the current period. Note that for the base period

$$i_b = \frac{x_b}{x_b} \times 100 = 100$$

Since the base period has an index of 100, all other index numbers have a simple interpretation in terms of percentage changes. For example, an index number of 105 means that the value of the variable has grown by 5% since the base period. Similarly an index number of 95 means a 5% decline.

Example 4.1: Table 4.1 gives some selected information on health personnel and inpatients treated in the United Kingdom.

    It is natural to take the earliest year, 1976, as the base period. Then, as an

Table 4.1   Health personnel and inpatients, UK (thousands). (Data from *Social Trends* (1985), Central Statistical Office.)

		Year			
		1976	1979	1980	1981
Personnel	Medical and dental	41.7	45.3	46.5	47.4
	Nursing and midwifery	429.6	449.2	466.1	492.8
Inpatient discharges and deaths	NHS	6430	6618	6954	7081
	Private	95	92	100	98

example of the calculation of an index number (1976 = 100), consider medical and dental personnel in 1979. The index number is 45.3/41.7 × 100 = 109 (to the nearest whole number).

The full list of index numbers is:

	1976	1979	1980	1981
Medical and dental	100	109	112	114
Nursing and midwifery	100	105	108	115
Discharges and deaths (NHS)	100	103	108	110
Discharges and deaths (private)	100	97	105	103

A useful way to present these results is on a graph such as Fig. 4.1. This shows that during 1976–1981 the medical/dental and nursing/midwifery professions grew overall by roughly the same amount (about 15%), although the medical/dental group initially expanded more quickly. The number of NHS discharges and deaths did not expand as much (only 10%), whilst private discharges and deaths barely increased at all (3%). Indeed the number of private inpatients actually decreased during two of the three intervals.

Notice that the type of index number considered here measures changes in a *single* variable only. It is also possible to define index numbers for several variables together, as will be shown in Section 4.4.2.

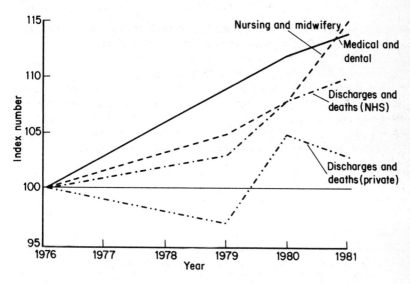

Fig. 4.1 Health index numbers, UK.

4.3 Averages

An average is a point of concentration around which the observed values of a variable lie. This definition implies that, in some way, an average is the centre of a set of data, and hence the other name for an average is a *measure of central tendency*. Averages can only be defined for quantitative variables (it would be difficult to imagine a useful definition of 'centre' for a categorical variable, such as 'sex', for which ratios were calculated in the previous section). For a quantitative variable such as age, 'average' is a familiar concept: most people would have some understanding of the term 'average age'.

Although many people talk about averages in everyday life, relatively few are likely to be confident about the exact definition of the term 'average'. There are, in fact, three types of average in common use: the mean, the median and the mode. Unfortunately they are not usually identical in value, which has led to the regrettable practice of considering all three and then reporting only that average which looks the best. (Various examples of such misuses of average are given in Huff (1954), Chapter 2.) To avoid misconceptions it is always best to state which average is being quoted, although in this book the common practice of using 'average' to denote 'mean', unless otherwise stated, will be followed. In Section 4.3.4, situations where a particular type of average is appropriate are considered.

First the precise meaning of each of the averages needs to be given:

The mean (strictly called the *arithmetic mean*) is the sum of the data divided by the number of values in the data.
The median is the positional mid-point of the data; half the values are less than the median and half are greater.
The mode is the most commonly occurring value in the data.

The calculation of each of these averages is straightforward when the data are available in their original raw form, that is when they consist of a list of numbers which have not been reduced by grouping or ordering in any systematic way. However, data are often available only in reduced form, typically a grouped frequency table such as Table 3.4. In such cases slightly more involved methods have to be used to compute averages which are approximations to the 'real' averages which could have been calculated if the raw data had been available.

4.3.1 Calculating the mean

From raw data

First count the total number of observations, then add up the observations and divide this by the count. It is useful to have a shorthand notation for this procedure. The observations are denoted by $x_1, x_2, \ldots, x_n$ where n is the total number of observations, and the mean of these observations is denoted by the symbol $\bar{x}$. Mathematicians use the symbol Σ, which is capital sigma, the Greek letter S, to denote 'the sum of'. Hence the procedure for calculating the mean can be represented as:

$$\bar{x} = \frac{\Sigma x_i}{n}$$

where i is a dummy variable which ranges over the values 1 to n.

Example 4.2:   Given the observations 5.2, 3.7, 2.4, 0.3, the mean is:

$$\frac{5.2 + 3.7 + 2.4 + 0.3}{4} = \frac{11.6}{4} = 2.9$$

Example 4.3:   For the hospital ward data the mean length of stay was calculated as:

$$\bar{x} = \frac{\Sigma x_i}{n} = \frac{660}{56} = 11.8 \text{ days}$$

Many calculators have a special button for obtaining $\bar{x}$. Those that do also usually have buttons for other statistical functions such as the standard deviation (see Section 4.6). The great advantage of this is that all the statistical functions can be obtained from a single entry of the data. To use all these buttons the calculator usually has to be put into 'statistics mode'.

All statistical computer packages have a command which allows automatic calculation of the mean from raw data.

From an ungrouped frequency table

Sometimes the data are available in the form of a set of numbers together with the frequency of occurrence of each number, such as Table 3.3. Let the numbers be denoted by $x_1, x_2, \ldots, x_m$ where m is the number of *distinct* outcomes in the data set, and let the corresponding frequencies be denoted by $f_1, f_2, \ldots, f_m$.

A little thought will show that the previous expression for $\bar{x}$ now becomes

$$\bar{x} = \frac{\Sigma f_i x_i}{\Sigma f_i} = \frac{\Sigma f_i x_i}{n}$$

where i ranges from 1 to m and where, as before, n is the total number of observations.

Example 4.4:   For the hospital ward data shown in Table 3.3, the mean number of operations performed per patient is:

$$\bar{x} = \frac{47 \times 0 + 8 \times 1 + 1 \times 3}{56} = \frac{11}{56} = 0.2$$

Some calculators with statistical functions will handle data from an ungrouped frequency table in this way, that is the xs and fs can be entered and the mean obtained automatically. Others require the data to be added as if it were raw data in a list, that is each x_i has to be entered f_i times. In fact this is not as bad as its sounds since the calculator will usually keep a number displayed leaving only the 'data entry' button to be pressed the required number of times.

When using a statistical computer package it is unusual not to be working with the raw data. If an ungrouped frequency table does form the input data then the mean will often have to be calculated in stages, that is $\Sigma f_i x_i$ will need to be calculated (usually trivial since such packages are very good at multiplying columns of figures) and then divided by Σf_i, the total of the 'f' column. Some packages will automatically calculate the mean by allowing each observation to be weighted by its frequency (see Section 4.4.2 for a discussion of weighted averages).

From a grouped frequency table

Sometimes data can only be obtained in the reduced form of a grouped frequency table as in Table 3.4. In such a case the exact value of $\bar{x}$ cannot be found. Instead an approximate mean is obtained by taking the mid-points of the class intervals to be the 'x' values, and then proceeding as for ungrouped frequency tables. This procedure would, in fact, give the exact value of $\bar{x}$ if the data were evenly spread within every class since then the mid-point would be the mean for that particular group of the table.

As described in Section 3.11, a continuity correction should be used when the data are discrete.

Example 4.5:   Suppose that the hospital ward data on length of stay were only available in the form of Table 3.4, which is reproduced below.

Length of stay (days)	Frequency
0	4
1–2	10
3–5	17
6–15	8
16–25	10
26–35	4
over 35	3
Total	56

To calculate the mean from this grouped frequency table the first step is to re-express the table with mid-points replacing class limits.

Mid-point of class	Frequency
0	4
1.5	10
4	17
10.5	8
20.5	10
30.5	4
43	3

As before, an upper limit of 50 days length of stay has been estimated. Then the mid-points make up the xs and the frequencies the fs in the formula

$$\bar{x} = \frac{\Sigma f_i x_i}{n}$$

for this re-expressed table. Now,

$$\Sigma f_i x_i = 4 \times 0 + 10 \times 1.5 + \ldots + 3 \times 43 = 623$$

and $n = 56$

Hence $\bar{x} = 623/56 = 11.1$ days.

Notice that this is close to the exact value of $\bar{x}$ calculated from the raw data in Example 4.3. Nevertheless, there is a difference and so it is worth stressing once again that if the original raw data *are* available, they should be used.

4.3.2 Calculating the median

From raw data

When calculating the median by hand from raw data the first step is to put the data values in rank order. The median will then be the middle value in this rank order list.

Example 4.6: Given the observations

$$7,1,9,3,4$$

the rank order list is

$$1,3,4,7,9$$

and hence the median is 4.

If the total number of observations is an odd number this is very easy; if it is even there is no true middle value and the median is then defined to be the arithmetic mean of the two most central values.

Example 4.7: Given the observations

$$7,1,9,3,4,3$$

the rank order list is

$$1,3,3,4,7,9$$

and hence the median is $\frac{1}{2}(3+4) = 3.5$.

In general if there are n observations the median will be:

for odd n: the $(\frac{n}{2})$th largest value
for even n: half the sum of the $(\frac{n}{2})$th and $(\frac{n}{2}+1)$th values.

Example 4.8: For the hospital ward data the rank order list for length of stay is:

$$0,0,0,0,1,1,1,1,1,1,2,2,2,2,3,3,3,3,3,3,3,4,4,4,4,4,4,4,4,5,5,6,7,7,8,11,12,$$
$$13,15,18,19,19,22,22,24,24,24,24,24,30,30,30,31,44,54,65$$

Since n = 56, which is even, the median is half the sum of the $\frac{n}{2}$ = 28th and $\frac{n}{2}+1$ = 29th values. Both these values are 4 and hence the median length of stay is 4 days.

Ordinary non-programmable calculators are no use for calculating medians since they cannot perform the necessary ordering of the data. Most statistical computer packages have commands which produce the median.

From an ungrouped frequency table

If the data are in the form of an ungrouped frequency table, part of the job of calculating the median has already been done! That is, the data set has already been ordered and it only remains to seek out the middle value or values and use the rules already given. So for the hospital ward data on the number of operations performed in Table 3.3, the median is once again half the 28th and 29th largest values. Since the first 47 values in the frequency table are all zero the 28th and 29th values, and hence the median, are all zero.

From a grouped frequency table

The exact value of the median cannot be calculated when data are only available in grouped form. Instead a method of approximation is used which estimates the n/2th largest value and takes this to be the median. Due to the approximate nature of the calculation no distinction is made between odd and even n.

The class which must contain the median is easily found by counting up the frequencies, as with an ungrouped frequency distribution. The exact position of the median within this class cannot be determined, so it is estimated by calculating where the median would lie if the data were evenly spread within the class. When the data are discrete a continuity correction should be used, as illustrated in Example 4.9.

Example 4.9:   Consider calculating the median from Table 3.4, reproduced in Example 4.5. Here n/2 = 28, and hence the 28th largest value is sought. As there are 14 values of 0 and 1–2 combined and yet 31 values of 0, 1–2 and 3–5 combined, the 28th largest value clearly lies in the class 3–5. With the continuity correction the first conclusion is that the median must lie between 2.5 and 5.5 days.

Overall there are 17 values in the class 3–5 and since 0 and 1–2 together contribute 14 values, the median is the 28 – 14 = 14th largest of these 17 values. Hence if it is assumed that the values are equally spread within this 'median class', the median must lie 14/17ths of the whole distance inside the class range. Taking into account the continuity correction this distance must be:

$$\frac{14}{17}(5.5 - 2.5) = 2.47$$

Hence the approximate value of the median is:

$$2.5 + 2.47 = 5.0 \text{ days (to one decimal place).}$$

Notice that this is close to the true value of 4 days derived in Example 4.8, but the error could have a substantial impact on further analyses.

Although a general formula exists for the calculation of an approximate median from a grouped frequency table, as shown below, this formula should *not* be memorized. It is much more important, and far safer, to understand the basic principles of the method, as described in Example 4.9.

The approximate median is given by

$$l_L + \frac{(\frac{n}{2} - C_L)}{(C_U - C_L)}(l_u - l_L)$$

where l_u and l_L are, respectively, the upper and lower class limits of the median class (possibly corrected for continuity), C_U is the cumulative frequency up to and including the median class, and C_L is the cumulative frequency up to the next class below the median class. As always, n is the total number of observations.

As Example 4.9 and the formula both suggest, it is often useful to first construct a cumulative frequency distribution when deriving the median from grouped data.

From ogives

An approximate evaluation of the median can easily be made from an ogive. By definition the median is at the mid-point of the cumulative frequency distribution so with an ogive drawn using a relative frequency scale, the median will be that value which corresponds to 0.5 (the middle value) on the graph.

The necessary steps in obtaining a median from an ogive are:

(i) draw a horizontal line from the value 0.5 on the vertical axis;
(ii) from the point where this line touches the curve draw a vertical line downwards;
(iii) find the point where this line touches the horizontal axis and take this to be the median.

It is recommended that a smoothed ogive should be used for these operations since this provides a continuity correction.

Example 4.10: From Fig. 3.15, the median length of stay for the hospital patients is estimated to be 5 days (the same answer as in Example 4.9).

4.3.3 Calculating the mode

From raw data the mode is found by first counting the frequency of each distinct value in the data set. The mode is then that value with the greatest frequency. Since an ungrouped frequency table already gives the frequencies it is a trivial matter to locate the mode from data presented in this manner. In Table 3.3, for example, it is obvious that the mode is 0 operations since 47 is the largest frequency.

Many text books go on to discuss how a mode can be estimated from a grouped frequency table (e.g., Thirkettle (1981), Chapter 1). However, the method suggested is likely to generate a considerable degree of error, and consequently it is better not to consider using the mode when data are available only in grouped form.

4.3.4 Which average?

As already stated, the mean, median and mode are liable to be different when calculated from the same set of data. For example, the hospital ward data have the following average lengths of stay:

Mean	11.8 days
Median	4 days
Mode	4 days

showing a substantial difference between the mean and the other two averages, although in this example the median and mode happen to be equal.

Given the potential for such differences, the advantages and disadvantages of the three types of average must be considered so that the most appropriate average may be used in any particular situation.

The *mean* is most often the preferred average because of its advantages:

 (i) it is what most people think of when they speak of 'an average';
 (ii) it is easy to calculate;
 (iii) it uses every value in the data, which seems intuitively correct;
 (iv) it has a useful mathematical representation which makes it suitable for further analyses.

It does, however, have certain disadvantages:

 (i) it does not usually correspond with an actual data value;
 (ii) it may be highly influenced by a few (or even only one) extreme values;
 (iii) when the only available data are grouped and have an open-ended extreme class then a poor estimate of the mid-point of the extreme class may lead to a poor approximation being obtained for the mean.

Of these disadvantages, (i) is illustrated by Examples 4.2–4.5. Example 4.5 also serves to illustrate (iii) since the guess of 50 days as the upper limit of stay was fairly arbitrary. If, for example, 45 days is used instead, the estimated mean drops from 11.1 to 11.0 days.

Disadvantage (ii) is the most serious drawback of the mean, and disadvantage (iii) is really just a consequence of this. Many sets of data contain extreme values or *outliers*. These may be perfectly valid observations, although not typical, which by their very magnitude will have a considerable influence upon the size of the mean. If the degree of influence of a data value is out of proportion to its real worth, then the mean presents a distorted picture of the average.

For example, a hospital ward may treat 40 people, all but one of whom have a completed stay of less than 3 days. Due to particular medical complications, the remaining patient may have stayed for over 30 days. Is it a fair reflection of the ward's efficiency to quote its mean length of stay when a single unusual case has inflated this average by a day or more? In fact, this example is a more extreme version of what has occurred in the hospital ward data in Example 4.3, since at least two of the values in that data set can be considered as outliers.

In general, the presence of outliers is most easily detected from a histogram, for example Fig. 3.11 which shows the outlying values at the right-hand end.

Now consider the advantages of the *median*:

 (i) it often corresponds to an exact data value;
 (ii) it is not affected by extreme values or open-ended extreme classes;
 (iii) it can be evaluated from only the first 50% of lifetime, or survival, data.

The first of these is an obvious consequence of the calculation method when there are an odd number of observations.

Advantage (ii) also becomes obvious when the methods of calculation are considered, for the median only depends upon the magnitude of the central value, or values, in the data set. The magnitudes of *all* other values are irrelevant. Thus the median of 1,2,3 is the same as the median of 1, 2, 3000. This is a useful property when, as in the hypothetical example of one unusually long length of stay in forty, extreme values would otherwise distort the average.

Advantage (iii) is applicable only in special cases where the data are lifetimes, or survival times, after some event. For example, suppose the problem was to find out how long, on average, people survive after open-heart surgery. Patients would then continue to be observed after their operation and their time to death recorded. To calculate the mean, every patient's time to death must be known since the mean uses every value in the data set. However, some patients may well live for many years and hence it will be a long time before the result is available. To calculate the median it is only necessary to wait until 50% of the patients have died. In such contexts the median is sometimes called a 'half-life'.

The disadvantages of the median are:

(i) it is not what most people think of when they speak of an 'average';
(ii) it only uses the central value, or values, in the data set;
(iii) it does not have a useful mathematical representation.

Finally consider the use of the *mode*. Its advantages are:

(i) when the data are ungrouped, it always corresponds to an exact data value;
(ii) it is not affected by extreme values or open-ended extreme classes;
(iii) it is simple to evaluate when the data are ungrouped.

However, the disadvantages of the mode are all those of the median plus:

(iv) it cannot necessarily be accurately calculated when the data are grouped;
(v) it is difficult to interpret when the data are not unimodal;
(vi) it is highly sensitive to the actual data collected.

Disadvantages (v) and (vi) require comment. Unimodal means 'single peaked', in other words data are unimodal if the histogram only has a single peak. Figure 3.10 for the ages of patients is clearly not unimodal. It is difficult to see what use the mode can be when a value far from the mode is almost as important in terms of frequency. In such cases it is more meaningful to specify local modes, for example with the length of stay data the overall mode of 4 days is a local mode for short stay patients and 24 days is a local mode for long stay patients (see Fig. 3.12). Of course it may also be interesting to calculate means and medians for short and long stay patients separately, using some recognized and well-defined criteria to distinguish long and short stay.

Disadvantage (vi) is a consequence of the fact that an observed set of data almost always represents a bigger, unobserved, set. In statistical jargon, a sample of data is used to represent a population. If the sample is small compared with the parent population then it is highly likely that, just by chance, a sample is selected which has a mode very different from the 'true' population mode. For example, with the length of stay data the value of 24 days almost reached the highest frequency. If, purely by chance, a few more patients with 24 days length of stay had happened to be sampled then 24 would have become the mode for that set of data. All other indications from the hospital data are that the mode is really somewhere between 0 and 4 days since this group together have relatively large frequencies (see Fig. 3.11).

In conclusion, the use of the mean is recommended unless the data have outliers which distort the mean as a measure of average, or unless the other disadvantages of the mean are crucial. In such cases the median should be used. Unless the circumstances specifically require it, the mode should not be used to represent the average. It is important that it is made clear which type of average is actually being reported.

4.4 Special averages

In addition to the mean, median and mode there are other averages which are used in certain special circumstances. In this section three such special averages will be considered, the *geometric mean*, the *weighted average* and the *moving average*.

4.4.1 The geometric mean

The geometric mean is used to find average amounts of increase (or decrease) when measured on a ratio scale. The geometric mean of the n values x_1, x_2, . . ., x_n is:

$$\sqrt[n]{x_1 x_2 \ldots x_n}$$

Example 4.11:   Using the data on nursing and midwifery personnel in Example 4.1, their numbers increased by a fraction:

$$\frac{466.1}{449.2} = 1.038 \text{ from 1979 to 1980 and}$$

$$\frac{492.8}{466.1} = 1.057 \text{ from 1980 to 1981}$$

The average fractional increase per year of nursing/midwifery personnel from 1979 to 1981 was thus:

$$\sqrt[2]{1.038 \times 1.057} = 1.047$$

(where $\sqrt[2]{}$ is the familiar square root, usually denoted just by $\sqrt{}$).

Notice that the geometric and arithmetic means are *not* the same, although in this example they happen to be very close.

4.4.2 The weighted average

When observations are available on a number of different variables it may be desirable to combine the variables to create an average for each unit of observation. For example, many authors have sought a single measure of the health of a community by combining such things as the death rate, the hospital inpatient attendance rate, the extent of overcrowding in homes, etc. (See Jarman (1983, 1984) and Fanshel and Bush (1970) for examples.)

Often when combining, or averaging, variables it is desired to give greater weight to certain variables simply because these are considered more important. Thus with v variables, $x_1, x_2, \ldots, x_v$, each variable is given a weight $w_1, w_2, \ldots, w_v$ and then the weighted average is

$$\frac{\Sigma w_i x_i}{\Sigma w_i}$$

The idea of a weighted average leads naturally to a definition of an *index number* for several variables combined. In Example 4.1 index numbers were generated for the number of medics/dentists and for the number of nurses/midwives separately. If, instead, indices for the total number of health care personnel were required, the variables 'number of medics/dentists' and 'number of nurses/midwives' need to be combined for each year. A simple way of combining them would be to take the arithmetic mean of the two figures. However, for some purposes it might be sensible to give a greater emphasis to one or other group, for instance doctors may get a weight of 3 and nurses a weight of 1.

The weighted average number of health care personnel in each year would then be:

$$\frac{3 \times (\text{number of doctors}) + 1 \times (\text{number of nurses})}{3 + 1}$$

To calculate index numbers from the yearly weighted averages the formulation of Section 4.2 is followed, that is each year's weighted average is divided by the 1976 weighted average and multiplied by 100.

Example 4.12: Using the weights of 3:1 as suggested, the weighted index for health care personnel in 1976 is:

$$\frac{3 \times 41.7 + 1 \times 429.6}{4} = 138.675$$

Using similar calculations, the weighted averages for the four years covered by the data are:

1976	1979	1980	1981
138.675	146.275	151.400	158.750

leading to the index numbers for health care personnel (1976 = 100) of:

1976	1979	1980	1981
100	105	109	114

The weakness of the method of weighted averages, and thus of the index numbers for several variables, is the arbitrary nature of the weights, as shown in this example. Are doctors really three times as important as nurses? Anyone's opinion of the relative importance of such personnel is likely to be highly subjective, and factual information is to be preferred. Probably the only factual information to hand is the salaries of the two groups, and it might be acceptable to use average salaries as the weights. However, even this is not entirely satisfactory since, for various reasons, wages may not reflect true worth. In many situations it is not easy to decide upon the appropriate weights.

Undoubtedly the most commonly used index number series is the *consumer price index*. Such an index is supposed to measure the relative price of a typical consumer's 'shopping basket' over time. To calculate the index a number of representative consumer items are selected and their prices, at a sample of shops, are monitored from month to month. The weighted average of prices for each year is calculated by weighting each price by the number of units of that item bought by the average consumer.

To illustrate this, suppose only two items were to be selected, although this is a vast over-simplification. Milk and bread are to constitute the 'basket' and a survey has shown that the average monthly consumptions per household are 35 pints of milk and 25 loaves of bread. In 1982, costs are 17p per pint of milk and 35p per loaf of bread and, in 1985, 21p and 40p respectively. The weighted average prices per month are:

$$\frac{35 \times 17 + 25 + 35}{35 + 25} = 24.50 \text{ for 1982 and}$$

$$\frac{35 \times 21 + 25 \times 40}{35 + 25} = 28.92 \text{ for 1985}$$

giving a price index of:

$$\frac{28.92}{24.50} \times 100 = 118 \text{ for 1985 (1982} = 100).$$

Similar indices could be constructed for such things as the cost of sterile supplies in a health district.

One further point to consider for the consumer price index is that consumer habits change over time, as new products become available and demand changes in line with changing prices. The survey of quantities consumed, and thus the weights, relate only to one time. If the quantity weights relate to the base period (1982 in the above example) the index is called a *Laspeyres* index, but if they relate to the current period (1985 in the above example) it is called a *Paasche* index. There has been much discussion as to which type of index is best. Sometimes a compromise is preferred; this is to take the geometric mean of the Laspeyres and Paasche indices to produce what is called a *Fisher* index. However, when a series of index numbers is to be produced successively over time, the Laspeyres (base weighted) index is the cheapest to produce since it requires no updating of the weights, and hence this is usually preferred.

Although the consumer price index is not directly a statistic of health, it is important that the concepts and methods used are understood because it is often used as a yardstick, or deflator, when comparing health spending over time. Care must be exercised here, since it is one thing to use the consumer price index to fix changes in pensions and divorce settlements to take account of inflation, but quite another to do so for health expenditure. It is quite likely that health expenditure items, such as medical equipment and drugs, increase in price at a far faster rate than do everyday consumer products such as bread and milk.

It is not appropriate to go into further detail on the construction or interpretation of index numbers since many books on economic statistics cover this thoroughly (e.g., Ilersic and Pluck, 1977). One general point always applies. It is essential that the indices should be relevant to the problem in hand, and their relevance can only be assessed if the method of construction is clearly stated.

4.4.3 Moving averages

The method of moving averages is a way of finding an underlying trend in a time series. Table 4.2 gives an example of a time series, which is illustrated by Fig. 4.2.

Data collected on a seasonal (three-monthly) basis often exhibit regular fluctuations with a cycle of length 4, as in the birth data. On the other hand, if data are recorded monthly they often exhibit a cycle of length 12, simply because patterns tend to repeat themselves from year to year. Frequently interest lies in determining the trend in the time series, that is the underlying tendency to growth or decline when seasonal fluctuations are removed. A moving average seeks to remove seasonality by averaging over each cycle, so that the effects of 'high' seasons and 'low' seasons within the cycle cancel out. The moving average then estimates the trend. By convention the average normally used is the arithmetic mean.

To illustrate the calculation of a moving average series, consider Table 4.2. The first four values in the series relate to Quarters I–IV, 1982. Averaging these gives:

$$\frac{176.7 + 180.2 + 185.9 + 176.5}{4} = 179.825$$

These figures span one cycle of the time series. The next cycle is Quarter II, 1982 to Quarter I, 1983 and the next moving average comes from averaging the values for these seasons. The process continues in this way. Each time, the starting point for the averaging process moves forward one time-point, which

Table 4.2 Live births in the United Kingdom (thousands). (Data from *Monthly Digest of Statistics* (July 1985), Central Statistical Office. ([1]indicates a provisional figure).)

Year	Quarter	Live births
1982	I	176.7
	II	180.2
	III	185.9
	IV	176.5
1983	I	175.3
	II	184.6
	III	187.4
	IV	174.1
1984	I	176.1
	II	180.7
	III	191.0
	IV	181.8
1985	I	189.0[1]

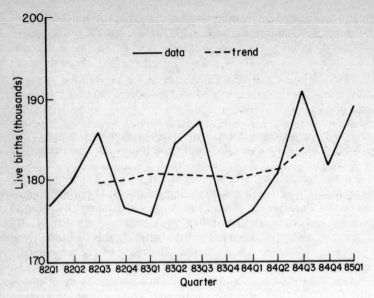

Fig. 4.2   Live births, UK.

explains the term 'moving average'. The result is called a four-point moving average or a moving average of order 4.

The number 4 was used here because the cycle length was four time-points, but if the data were recorded monthly, a 12-point moving average would almost certainly be more appropriate. It is possible, although unlikely, that the patterns in the data do not repeat on a year-to-year basis but instead, for example, repeat at 15-monthly intervals. It is always essential to first plot the data as in Fig. 4.2. The natural cycle, if it exists at all, should be obvious from the graph.

There is one further complication to deal with. In the example, each moving average is the average of values at four successive time points. It is natural to say that each moving average corresponds to the average time-point within the four used to construct it. This means, for example, that the first moving average corresponds to the mid-point between Quarters I and IV of 1982, which falls half way between Quarters II and III, and is not one of the original data points.

This will always be the case when averages are taken over an even number of data points, and is generally thought to be disadvantageous, since it is preferable to speak of 'the moving average for the third quarter' rather than 'the moving average for midway between the second and third quarters'. Hence it is conventional to average successive values in the moving average series to produce a centred moving average with time points corresponding to those in the original data set (see column 3 in Table 4.3). Of course the centring process is not necessary when the order of the moving average is odd. The moving average, or centred moving average, is called the *trend*.

Once the entire moving average series is obtained it can be useful to look at the effects of the seasons. Often it is assumed that, apart from random deviations, observations change by a constant proportion each time the same season comes around. This leads to the *multiplicative model* for time series:

$$y = T \times S \times R$$

where y is the observation, T the trend, S the *seasonality* and R the *residual*. The residuals account for random unexpected fluctuations, such as the effect of an epidemic of rubella for the birth data.

The seasonal components are estimated using an arithmetic mean once again, where a mean is calculated for each season from the set of values of y/T, the 'detrended series' for that season. The mean seasonality should be one, since in the multiplicative model this would correctly result in y = T × R once seasons have been averaged out. Hence the sum of the seasonal components (from quarterly data) should be four. In many cases this does not happen exactly, and consequently a correction factor is used to weight the components appropriately. Table 4.3 shows how this operates.

Table 4.3 A time series analysis of the birth data given in Table 4.2. Note: (2) is a four-point moving average of (1); (3) averages successive pairs of values from (2); (4) = (1)/(3); (5) average values from (4) for each season, using a correction factor as shown below; (6) = (4)/(5); and (7) = (1)/(5).

(0) Time period	(1) Data (y)	(2) Average of four quarters	(3) Centred moving average trend (T)	(4) Detrended series (y/T)	(5) Season-ality (S)	(6) Residual (y/ST)	(7) Deseason-alized series (y/S)
82 I	176.7				0.975		181.23
II	180.2	179.825			1.011		178.24
III	185.9	179.475	179.65	1.035	1.039	0.996	178.92
IV	176.5	180.575	180.03	0.980	0.975	1.005	181.03
83 I	175.3	180.950	180.76	0.970	0.975	0.995	179.79
II	184.6	180.350	180.65	1.022	1.011	1.011	182.59
III	187.4	180.550	180.45	1.039	1.039	1.000	180.37
IV	174.1	179.575	180.06	0.967	0.975	1.102	178.56
84 I	176.1	180.475	180.03	0.978	0.975	1.003	180.62
II	180.7	182.400	181.44	0.996	1.011	0.985	178.73
III	191.0	185.625	184.01	1.038	1.039	0.999	183.83
IV	181.8				0.975		186.46
85 I	189.0				0.975		193.85

Detrended (y/T) series split into seasons

Year	Quarter I	II	III	IV	Total
1982			1.035	0.980	
1983	0.970	1.022	1.039	0.967	
1984	0.978	0.996	1.038		
Average	0.974	1.009	1.037	0.974	3.994*
× 4/3.994	0.975	1.011	1.039	0.975	4.000

*Correction factor = 4/3.994

Once trend (T) and seasonality (S) have been found, forecasts can be made by extrapolating the trend line (perhaps by eye from a graph of the trend) and multiplying the result by the appropriate seasonal component. Before doing this, it is best to check the validity of the multiplicative model by examining the residuals (R). Section 8.3.2 discusses the analysis of residuals.

The deseasonalized series (y/S) represents the data adjusted for seasonality. Components of this series can thus be compared directly to identify time periods with unusually low or high values. Such deseasonalized figures are often quoted by the media to measure the current level of unemployment, trade and other economic variables. They are also useful with health data, for example to identify the times when there are births, notifications of diseases or GP consultations that are surplus to the normal patterns.

For more details about the analysis of time series see Yeomans (1968) or, at a more mathematical level, Chatfield (1980).

Example 4.13: A time series analysis of the birth data of Table 4.2 is given in Table 4.3.

The trend (T) shows a slight upwards movement in the number of live births during the period recorded. The seasonal components (S) are below 1 for the first and fourth quarters, indicating that births are relatively low at these times of the year; births are highest in the third quarter. The high deseasonalized value (y/S) for 1985 first quarter indicates that there were an abnormally high number of births in this quarter.

The reader should refer to Fig. 4.2 to determine whether these conclusions seem correct. The trend has been included on this graph.

4.5 Quantiles

As has been seen, averages are summary statistics used to represent the centre of a set of data. This is usually the single most interesting point of location within the data, but it is sometimes also very useful, for various reasons, to represent other points of location. This is achieved by generalizing the idea of the median.

The median divides the data up into two parts, the values below it and the values above it in terms of magnitude. Suppose that the lower half of the data were divided into two equal parts in the same way. The dividing point is then called the first *quartile* (denoted Q_1). If the upper half is then divided into two equal parts the dividing point is called the third quartile (Q_3). Hence the first quartile, the median and the third quartile divide the data into four equal parts in terms of magnitude.

This process can be continued by dividing up the quarters equally to split the data into eighths, and so on. There is, though, no particular reason for starting with the median (i.e., halves) at all; the same idea can be used to divide the data up into whatever fractions are desired. The most commonly used dividing points are:

  (i) the median: divides into halves;
 (ii) the quartiles (or *hinges*): divide into quarters;
(iii) the deciles: divide into tenths;
(iv) the percentiles: divide into hundredths.

The general name given to all such dividing points is *quantile* or *fractile*. As will be seen later, quantiles are useful when defining measures of dispersion but are often useful in their own right, especially when defining 'protection levels'.

Suppose, for example, that a new accident and emergency ward is being planned for a hospital. It is wasteful of resources to have empty beds and yet the ward must be big enough to cope with the majority of the demands made upon it. A sensible decision might be to elect to have enough beds to ensure that the ward is only short of beds 5% of the time (in such cases beds will presumably be borrowed from other wards). The required number of beds can then be estimated from past data on daily bed occupancy for accident admissions. From these data, the 95th percentile should be calculated, this being the value below which 95% of the bed occupancies lie, and thus above which only 5% lie.

4.5.1 Calculation of quantiles

From a data list

When calculating the median in Section 4.3.2 the simple process of dividing the data into halves became more complicated when there was an even number of data values. To specify rules for calculating any type of quantile, such complications need to be taken into account. Unfortunately there is no single agreed way of doing this, and various approaches have been devised. Some computer packages give the option of requesting a particular method of calculation, but most use only one approach. Normally the answers will be very similar, and so the choice of calculation method is unlikely to be crucial. The calculation method used here follows the formulation of Clarke and Cooke (1983).

Consider the pth q-tile, that is the pth largest of the set of summary statistics which divide the data into q equal parts. Suppose, as usual, that there are n values in the data set. As with the median, the pth q-tile is specified by considering two distinct cases:

(i) When $\frac{p}{q}n$ is a whole number (i.e., 1,2,3,4, etc.), the pth q-tile is defined to be the arithmetic mean of the $(\frac{p}{q}n)$th and $(\frac{p}{q}n + 1)$th observations in rank order.

(ii) When $\frac{p}{q}n$ is not a whole number, the pth q-tile is defined to be the rth observation in rank order when r is the value of $\frac{p}{q}n$ rounded up to the next biggest whole number.

Example 4.14: Consider the data set

$$12,4,3,23,28,15,13,2,8,23,7,18$$

In rank order this becomes

$$2,3,4,7,8,12,13,15,18,23,23,28$$

(i) The median has p = 1 and q = 2. Now n = 12 and hence $\frac{p}{q}n = 6$ (whole number).

The median is then the mean of the 6th and 7th largest values, i.e., $\frac{1}{2}(12 + 13) = 12.5$

(ii) The first quartile, Q_1, has $p = 1$ and $q = 4$. Hence $\frac{p}{q}n = 3$ (whole number) and $Q_1 = \frac{1}{2}(4 + 7) = 5.5$.

(iii) The seventh decile has $p = 7$ and $q = 10$. Hence $\frac{p}{q}n = 8.4$ (not a whole number). The seventh decile is then the 9th biggest value, i.e. 18.

Example 4.15:   For the hospital ward data on length of stay, the rank order list was given in Example 4.8. Using this list the quartiles can be found:

(i) Q_1 has $p = 1$ and $q = 4$ and, since $n = 56$ for these data, $\frac{p}{q}n = 14$. Therefore Q_1 is the mean of the 14th and 15th largest values, i.e. $\frac{1}{2}(2 + 3) = 2.5$.

(ii) The median has $p = 2$ and $q = 4$ and thus $\frac{p}{q}n = 28$. Hence, the median is the mean of the 28th and 29th observations, i.e. $\frac{1}{2}(4 + 4) = 4$, in agreement with Example 4.18.

(iii) Q_3 has $\frac{p}{q}n = 42$ and thus has value $\frac{1}{2}(19 + 22) = 20.5$

From an ungrouped frequency table

Treat as a data list and proceed as for raw data.

Example 4.16:   For the hospital ward data on the number of operations, the 95th percentile has $p = 95$, $q = 100$ and $n = 56$. Hence $\frac{p}{q}n = 53.2$. The 95th percentile is thus the 54th largest value which is easily seen from Table 3.3 to be 1.

From a grouped frequency table

As before, the method used for the median is generalized in an obvious way.

Example 4.17:   If the length of stay data were only available in the form of Table 3.4 then the quantiles would have to be estimated. For instance the first decile has $p = 1$ and $q = 10$ and thus $\frac{p}{q}n = 5.6$. This quantile must come somewhere in the class range 1–2 days since Table 3.4 shows that 14 people had stays below 3 days, but only 4 had stays below 1 day.

By analogy with Example 4.9 the first decile is approximated by

$$0.5 + \frac{(5.6 - 4)}{(14 - 4)}(2.5 - 0.5) = 1 \text{ day (to the nearest whole number)}$$

From ogives

Yet again, the method for medians is generalized.

Example 4.18:   To use Fig. 3.15 to estimate the first decile as above, a line should be drawn across from 0.10 on the vertical axis. A line is then dropped from the point where this horizontal line touches the curve. This vertical line crosses the horizontal axis at 1 day which is an estimate of the first decile.

4.6 Measures of dispersion

The average is a very useful summary of data because it specifies the value around which the entire data set is concentrated. It has the advantage of conciseness, but by itself it is often not sufficient to describe all of the important characteristics of the data.

Once the average has been dealt with, the next most important feature of the data is how the values spread themselves around this average. That is, are they tightly clustered or are they widely dispersed? Such information not only specifies how representative the average is of any single observation, but also specifies how similar the data values themselves are.

What is required is a measure of dispersion of the data about the average, sometimes also called a measure of *spread* or a measure of *variation*. In many cases the measure of dispersion is as important a summary statistic as the average. For example, consider a study of general practitioner to patient ratios in a health region. The average number of GPs per thousand head of population over all urban and rural areas in the region is 0.5, which is considered acceptable. However, when the figures are disaggregated it turns out that there are gross inequalities in that one town has 1.5 GPs per thousand head of population whereas one county borough has only 0.1. By its very nature, the average disguises the variation in the individual figures because in its calculation large and small figures tend to cancel each other out. If a measure of dispersion, perhaps over local authority areas, were quoted alongside the average, the summarization of the data would be much more meaningful.

Just as three different representations of average, each with different properties, were considered, three commonly used measures of dispersion will now be presented. Later the situations in which each of these is appropriate will be discussed. The three are:

The range – the difference between the largest and the smallest values in the data set.
The quartile deviation (or **semi-interquartile range**) – half the distance between the third and first quartiles.
The standard deviation – the square root of the average of the squared differences between the data values and the mean.

Of these three, the range is a very obvious measure of dispersion and the quartile deviation is essentially an improved version of this, since it is not influenced by outliers. The standard deviation requires some more thought, but of the three this is by far the most often used.

The philosophy behind the standard deviation is that dispersion can be considered as the differences between the data values and their average. An appropriate measure of dispersion might then be the mean of these differences, since this would specify how far, on average, each value is from the overall mean. This is not very useful since positive differences, that is the differences for values larger than the mean, exactly cancel out negative differences (for values smaller than the mean) and so the average of differences always turns out to be zero!

One way round the problem of cancelling of positive and negative values is to treat all negative differences as if they were positive. This leads to a measure of dispersion called the *mean deviation*. Another solution is to square all the differences, since all squares are positive and hence the problem of negatives cancelling positives can no longer be present.

The average of squared differences is called the *variance*. This is itself a measure of dispersion, but unfortunately the process of squaring the data also squares the units of measurement, and therefore the variance is measured in square units. To bring the units of measurement back to those of the original data the square root is taken and the result is called the standard deviation. It is important to remember that the variance is the square of the standard deviation. The standard deviation has superior qualities to the mean deviation, and for this reason the mean deviation is rarely used.

One final point about the standard deviation and the variance is that the averaging process requires division by n − 1 rather than n (where n, as usual, is the number of data values). This divisor is used because the n differences are subject to one constraint, that is the sum of the differences between the data values and the mean is zero. Hence if n − 1 of the differences are known then the nth difference can be derived mathematically without even observing it. In statistical jargon the differences are said to have n − 1 *degrees of freedom*.

Example 4.19:   Given the observations 3,8,5,8,6, the mean is

$$\frac{3 + 8 + 5 + 8 + 6}{5} = 6$$

The differences (observation minus mean) are −3,2, −1,2,0 with a sum and hence a mean of zero as predicted.

The variance is

$$\frac{(-3)^2 + 2^2 + (-1)^2 + 2^2 + 0^2}{4} = 4.5$$

The standard deviation is $\sqrt{4.5} = 2.12$

4.6.1 Calculating the standard deviation

From raw data

As is implied by the definition, the variance, which is denoted by s^2, is given by the formula,

$$s^2 = \frac{1}{n-1} \Sigma(x_i - \bar{x})^2$$

where $x_1, x_2, \ldots, x_n$ are the data values, as before, and $\bar{x}$ is the mean.

The standard deviation is the square root of this, and is thus denoted by s.

It is useful to have a special symbol to represent the sum of squared differences, or 'corrected sum of squares', and S_{xx} will be used here, i.e.,

$$S_{xx} = \Sigma(x_i - \bar{x})^2$$

whence

$$s^2 = \frac{1}{n-1} S_{xx}$$

To calculate S_{xx}, and hence s^2, by hand is very tedious unless another, less obvious but equivalent, formula is used. This formula is

$$S_{xx} = \Sigma x_i^2 - \frac{1}{n}(\Sigma x_i)^2$$

It can easily be shown, by algebraic manipulation, that the two formulations are equivalent.

Example 4.20: Given the observations 5.2, 3.7, 2.4, 0.3, Σx_i was found to be 11.6 and $\bar{x}$ to be 2.9 in Example 4.2.
Using the 'calculation formula',

$$\Sigma x_i^2 = (5.2)^2 + (3.7)^2 + (2.4)^2 + (0.3)^2 = 46.58$$

and thus

$$S_{xx} = 46.58 - \tfrac{1}{4}(11.6)^2 = 12.94$$

On the other hand, if the 'definition formula' is used, the differences, $x_i - \bar{x}$, need to be found, squared, and added up thus:

$$S_{xx} = (5.2 - 2.9)^2 + (3.7 - 2.9)^2 + (2.4 - 2.9)^2 + (0.3 - 2.9)^2$$
$$= 5.29 + 0.64 + 0.25 + 6.76 = 12.94$$

i.e., the answers for S_{xx} are the same by either method.
The variance, s^2, is given by

$$s^2 = \frac{1}{n-1} S_{xx} = \tfrac{1}{3} \times 12.94 = 4.31$$

and the standard deviation, s, is

$$s = \sqrt{4.31} = 2.1$$

In Example 4.20, the two formulae for S_{xx} involved the same amount of work so neither was really superior. However, the calculation formula is very much superior when n is even moderately large. When n is as large as 100 the calculation formula saves roughly 50% of the mathematical operations required!

Many calculators have a special button for obtaining s, or perhaps s^2, as defined above. However, there are two things to be careful of:

(i) some calculators use the symbol σ instead of s. This is unfortunate because σ actually means something different in standard statistical notation as will be seen in Chapter 7;

(ii) some textbooks define the variance to be $\tfrac{1}{n}S_{xx}$, and some calculators use this formula. Use of the denominator n – 1 is to be *preferred*, since in any practical situation this is always at least as good, and is often better, than using denominator n (see Section 7.2). Of course if the calculator only gives $\frac{1}{n}S_{xx}$ then s^2 can easily be derived by multiplying this by

$$\frac{n}{n-1}.$$

To use a calculator correctly and efficiently it is essential to read the instructions and practice calculations on a few small and simple sets of hypothetical data, such as that used in Example 4.20. In particular, ensure that the denominator n – 1 is used. Efficiency includes the ability to calculate the mean and whatever else the calculator offers along with the standard deviation, at a single entry of the data.

All statistical computer packages have commands for obtaining s and/or s^2.

Example 4.21:   For the hospital ward data the standard deviation of length of stay was calculated using a simple calculator, without statistical functions. The steps were:

$$S_{xx} = \Sigma x_i - \frac{1}{n}(\Sigma x_i)^2$$

$$= 18752 - \frac{1}{56}(660)^2$$

$$= 10973.429$$

and hence

$$s = \sqrt{\frac{1}{n-1}S_{xx}} = 14.1 \text{ days}$$

From an ungrouped frequency table

If the data values are $x_1, x_2, \ldots, x_m$ and the frequencies are $f_1, f_2, \ldots, f_m$ respectively, then the formula for S_{xx} reduces to

$$S_{xx} = \Sigma f_i(x_i - \bar{x})^2$$

Once again there is an equivalent formula which is easier to use for hand calculation:

$$S_{xx} = \Sigma f_i(x_i)^2 - \frac{1}{n}(\Sigma f_i x_i)^2$$

where $n = \Sigma f_i$, the total number of observations.
Then, just as before, the variance is:

$$s^2 = \frac{1}{n-1}S_{xx}$$

and the standard deviation is $\sqrt{s^2}$.

Example 4.22:   For the hospital ward data on number of operations in Table 3.3:

$$\Sigma f_i(x_i)^2 = 47 \times 0^2 + 8 \times 1^2 + 1 \times 3^2 = 17$$

and

$$\Sigma f_i x_i = 47 \times 0 + 8 \times 1 + 1 \times 3 = 11.$$

Hence $\qquad S_{xx} = 17 - \dfrac{11^2}{56} = 14.839$

whence $\qquad s^2 = \dfrac{1}{55} \times 14.839 = 0.2698$

and finally $\qquad s = \sqrt{0.2698} = 0.52$ operations.

Notice that this standard deviation is tiny compared with that calculated in Example 4.21. Reference back to the full data set in Appendix 1 should explain this difference; it is very obvious that the variable 'length of stay' has a far greater spread of values than the variable 'number of operations'.

Most calculators will not deal with the calculation of s from frequency data directly, although some will give $\Sigma f_i x_i$, which is a help. Some computer packages will produce s from frequency data, perhaps calling it a 'weighted variance'.

From a grouped frequency table

Just as for the mean, an approximate standard deviation is found by taking x_i to be the centre of the ith class interval for each i, and continuing as for ungrouped frequency tables.

Example 4.23: Suppose that the hospital ward data on length of stay was available only in the form of Table 3.4. In Example 4.5, the class midpoints were calculated. Using these the relevant results are:

$$\Sigma f_i(x_i)^2 = 0 \times 4^2 + 10 \times (1.5)^2 + 17 \times (4)^2 +$$
$$8 \times (10.5)^2 + 10 \times (20.5)^2 + 4 \times (30.5)^2$$
$$+ 3 \times (43)^2 = 14647$$
$$\Sigma f_i x_i = 623 \text{ (from Example 4.5)}.$$

Hence $\qquad S_{xx} = 14647 - \dfrac{(623)^2}{56} = 7716.125$

whence $\qquad s^2 = \dfrac{1}{55} \times 7716.125 = 140.293$

and finally $\qquad s = \sqrt{140.293} = 11.8$ days.

This should be compared with the true standard deviation found from the raw data in Example 4.21. Remember that the answer depends on the assumption made about the mid-point of the open-ended final class of Table 3.4.

4.6.2 Calculating the quartile deviation

By definition, the quartile deviation is calculated as:

$$\tfrac{1}{2}(Q_3 - Q_1)$$

In Section 4.5.1, methods for calculating quartiles have already been given, and so there are no further complications to detail here.

Example 4.24:   For the hospital ward data on length of stay the quartiles were found in Example 4.15. Using these results the quartile deviation is:

$$\tfrac{1}{2}(20.5 - 2.5) = 9 \text{ days}$$

4.6.3 Calculating the range

Since the range is simply the biggest value minus the smallest value in the data set it is easy to calculate. When the data are only available in the form of a grouped frequency table, an approximate range is given by the end point of the final class interval minus the start point of the first class interval. If either extreme class is open-ended then it is better not to consider the range at all.

Example 4.25:   In the hospital ward data on length of stay the longest stay is 65 days and the shortest is zero days. Hence the range is:

$$65 - 0 = 65 \text{ days}$$

4.6.4 Which measure of dispersion?

The three measures of dispersion discussed here will not be equal in every situation of practical importance, and it will be necessary to decide which measure to use.

The *standard deviation* is most often used because of its advantages:

(i) it uses every value in the data set, which is intuitively appealing;
(ii) it has a useful mathematical representation, which makes it suitable for further analyses;
(iii) it has a direct interpretation for the normal distribution (see Section 6.11).

However, it has the following disadvantages:

(i) it may be highly influenced by a few (or even only one) extreme values;
(ii) when the only available data have an open-ended extreme class then the approximation can be very poor.

Apart from advantage (iii), on which discussion is deferred until later, these properties are best understood by analogy with those of the mean.

The *quartile deviation* has the advantages that:

(i) it is not affected by extreme values or open-ended extreme classes;
(ii) it can be calculated from the first 75% of lifetime or survival data.

Its disadvantages are that:

(i) it only uses two (or, at most, four) data values in its derivation, and takes no account of the dispersion outside these values;
(ii) it does not have a useful mathematical representation.

These properties follow closely those for the median already discussed in Section 4.3.4.

The *range* has the advantages that:

(i) it is the most obvious measure of dispersion;
(ii) it is very quick and easy to calculate.

These make it a very attractive measure of dispersion for simple applications. However, it has a number of important disadvantages, being those of the quartile deviation plus:

(iii) it is very sensitive to extreme values;
(iv) it tends to increase as more data are collected;
 (v) it is highly sensitive to the actual data collected.

Disadvantages (iv) and (v) require further comment. To illustrate disadvantage (iv), consider collecting data on lengths of stay over a month, calculating the range, then collecting data for a further eleven months and again calculating the range, this time for the entire year. It is almost certain that the year's range is bigger than the month's range simply because, as one would intuitively expect, the longer one samples data the more likely one is to find a freak, high, value. One thing is certain, the year's range cannot be smaller than the month's range. However, both ranges are measuring the same thing, i.e., the dispersion of patients' lengths of stay.

Disadvantage (v) is closely connected with (iv). If the sample is small in comparison with the entire population then it is highly likely that, just by chance, a sample is selected with values falling within a much smaller range than that of the population overall. A different small sample would very likely have a much different range.

In conclusion, the use of the standard deviation is recommended except in cases where the data include outliers or where there are open-ended classes to deal with. In such cases the quartile deviation is to be preferred. The range is probably never the best measure to use, and yet its simplicity ensures that it is frequently calculated, if only as a rough check for the other measures of dispersion. It should always be treated with extreme caution.

4.7 Relative dispersion

Up to now, average and dispersion have been treated as separate entities, but sometimes it is useful to relate the two in the form of a ratio. By far the most commonly used ratio is the *coefficient of variation* (CV) which is defined as:

$$CV = \frac{\text{standard deviation}}{\text{arithmetic mean}} \times 100 = \frac{s}{\bar{x}} \times 100$$

Since the ratio is multiplied by 100, it expresses the standard deviation as a percentage of the mean, in the style of an index number.

Measures of dispersion relative to the average, such as the coefficient of variation, are most useful when comparing data sets or subsets within the data. For example, if one wished to compare salaries paid to consultants and house officers it would be meaningless to compare the dispersion in terms of, say, the standard deviation, simply because the entire range of consultants' salaries would lie well above the entire range of house officers' salaries, so even if consultants were all paid more or less the same, their salaries would almost certainly have a larger standard deviation because the scale of measurement is so much larger. On the other hand, the coefficient of variation allows for such differing scales of measurement by standardizing the dispersion in terms of the mean value.

Another situation where a measure of relative dispersion might be useful is when two sets of data are to be compared and yet the unit of measurement used in one set differs from that used in the other. Due to the definition, relative dispersion is a unitless quantity and so the units of measurement of the original data are irrelevant.

Example 4.26:   For the hospital ward data on length of stay the mean is 11.8 days and the standard deviation is 14.1 days and so the coefficient of variation is $\frac{14.1}{11.8} \times 100 = 120$ showing that the dispersion is relatively high for these data.

4.8 Skewness and kurtosis

Together the average and measure of dispersion tell the reader a lot about the data, since they specify not only the centre of the data but also how widely dispersed the data are around this centre. However, even as a pair, they cannot possibly specify all the features of the data. For instance the presence of outliers causes lack of symmetry, or skewness, to appear in pictures of the data, as seen, for example, in Fig. 3.11. Neither the average nor measure of dispersion measures skewness.

In all there are four features of quantitative data which are commonly considered: the average, dispersion, skewness and kurtosis. The last two are hardly ever mentioned in reports, although summary statistics which measure them will be given here; some statistical packages produce these directly. In most cases a rough idea of skewness and kurtosis is sufficient, and this can be evaluated from a histogram or, in the case of skewness, a boxplot, as will be described in Section 4.9.

4.8.1 Skewness

Figure 4.3 gives smoothed frequency polygons which show the three possible types of skewness. Curve (a) shows perfectly symmetrical data, curve (b)

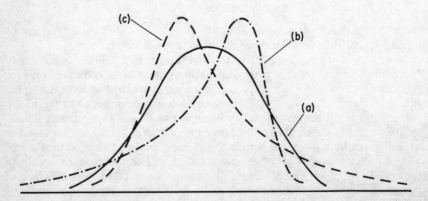

Fig. 4.3   Skewness.

shows left or negatively-skewed data and curve (c) shows right or positively-skewed data. In case (a) observations tend to fall with equal likelihood either side of the peak (the mode), but in (b) and (c) one side or other is favoured. Case (a) is the only case where all three averages, the mean, median and mode, are equal; as the data become more skewed they move further apart. In case (b) the median exceeds the mean whereas in case (c) it is the mean that is the bigger of the two.

The formula for calculating the coefficient of skewness is:

$$\frac{\Sigma(x_i - \bar{x})^3}{s^3}$$

which takes negative values if the data are negatively skewed, positive values if the data are positively skewed and is zero if they are symmetrical.

If the hospital ward data on length of stay are considered once again, it is clear from Fig. 3.11 that the data are positively skewed.

4.8.2 Kurtosis

Kurtosis is the degree of flatness of the peak, or equivalently the length of the tails, of a set of data. Generally it is used to compare the shape of data with roughly zero skewness against the shape of the normal curve.

Figure 4.4 shows smoothed frequency polygons exhibiting (a) the normal curve, (b) the peak flatter than the normal and (c) the peak sharper than the normal. Cases (a), (b) and (c) are sometimes called mesokurtic, platykurtic and leptokurtic respectively.

A summary statistic sometimes used to measure kurtosis is the coefficient of kurtosis defined as:

$$\frac{\Sigma(x_i - \bar{x})^4}{s^4} - 3$$

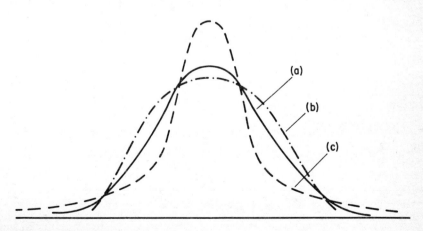

Fig. 4.4 Kurtosis.

This statistic is zero for mesokurtic (normal) curves, negative for platykurtic curves and positive for leptokurtic curves.

4.9 Boxplots

Boxplots, or 'box and whisker plots', are diagrammatic representations of the most useful quantiles. They provide a means of exploring the skewness in a set of data and of comparing the average and dispersion of different data sets or between subsets of the data.

A boxplot consists of a rectangle (the 'box') containing a plus sign, with arrows coming out of the left and right sides as shown in Fig. 4.5.

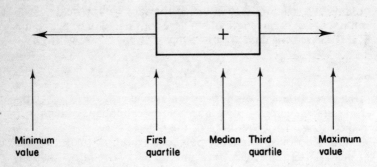

Fig. 4.5   A boxplot.

The most straightforward kind of boxplot has the ' + ' at the median, the left edge of the box at the first quartile and the right edge at the third quartile. The whiskers (arrows) terminate at the most extreme values in the data set.

If the data have no skewness, that is they are perfectly symmetrical, the ' + ' will lie exactly at the centre of the box and the whiskers will be of equal length. In Fig. 4.5 there is evidence of slight skewness to the left in the data because the left-hand end is longer than the right. A histogram would, of course, confirm this.

Example 4.27:   A boxplot for the hospital ward data on length of stay illustrates the severe skewness noted in Section 4.3.4:

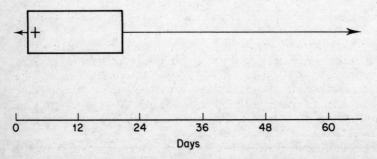

Fig. 4.6   Boxplot for hospital ward length of stay data. (Data from Appendix I.)

Boxplots are used to compare data sets by placing one boxplot directly above another (drawing each boxplot to the same scale). The relative positions of the ' + ' signs indicate the difference in average, and the relative widths of the boxes indicate the difference in dispersion (since the width is twice the quartile deviation). Difference in overall widths of the plots indicates difference in range. As with the stem and leaf diagram (see Section 3.11.2) the boxplot, unlike most other statistical diagrams, can easily be produced on an ordinary typewriter or word processor.

Example 4.28:   Figure 4.7 shows boxplots for two samples of data drawn to the same scale. Sample 1 exhibits greater spread and skewness but a smaller average.

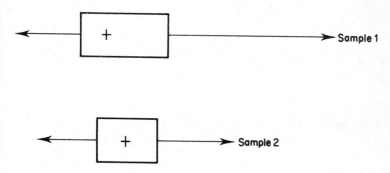

Fig. 4.7   Comparison of boxplots.

Most modern statistical computer packages produce boxplots. Sometimes the form of the boxplot differs from that used here: for example, to account for the problems inherent in using the range, the whiskers are sometimes truncated according to some rule (e.g., they might go to the 5th and 95th percentiles). The basic uses of the diagram remain the same.

4.9.1 Performance indicator boxplots

A somewhat different statistical picture, but one which has also received the name 'boxplot', is produced by the DHSS Performance Indicators Package (see Section 2.4.3). In this boxplot (Fig. 4.8) the box runs from the lowest to the highest values and the angled brackets cover the middle 80% of the data. An asterisk indicates one particular data point. Hence Fig. 4.8 could represent average lengths of stay for all district health authorities in England. The asterisk would represent the average length of stay for a chosen district; in the diagram this chosen district is amongst the 10% of districts with the highest lengths of stay. Such a presentation is extremely useful for seeing how a particular district compares with the rest. The DHSS package produces such boxplots for a wide range of indicators of health authority performance (see Section 4.11).

Elsewhere in this book, 'boxplot' will be taken to mean a diagram such as Fig. 4.5.

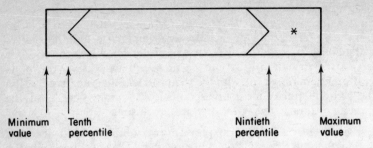

Fig. 4.8   A performance indicator boxplot.

4.10 Hospital bed statistics

One helpful set of summary statistics, used by health bodies in all parts of the world, are hospital bed statistics. These statistics rely entirely upon the methods of ratios and means described in this chapter.

4.10.1 Data and definitions

Hospital bed statistics use three sources of data:

(1) the number of patients occupying a hospital bed each night over an extensive period (typically a quarter or a year);
(2) the number of patients disposed (discharged, died or transferred) during the period;
(3) the number of beds available in the hospital.

A number of summary statistics are commonly calculated from the above data in order to measure the degree of utilization of hospital beds and the efficiency with which the beds and their occupiers are managed.

(1) *The number of bed-days used* This is the total amount of use that has been made of beds during the defined period, that is the sum of the occupied bed counts.
(2) *The average daily bed occupancy* This is the mean number of patients occupying a bed per day. It is calculated by dividing the number of bed-days used by the number of days in the period.
(3) *The average percentage bed occupancy* This is the percentage of available bed-days that were actually used during the period, that is one hundred times the number of bed-days used divided by the number of bed-days available. It may also be thought of as the mean percentage bed occupancy per day during the period.

In some countries the number of available bed-days is calculated by multiplying the 'bed establishment' (the official number of beds allocated to the hospital by the Ministry of Health) by the number of days in the period. Sometimes, however, the bed availability is *not* constant over the period but changes, due, for example, to temporary bed closures during redecorating. To account for such circumstances other countries (the United Kingdom is an example), use 'average daily bed availability' rather than 'bed establish-ment'. This can be derived by counting the total number of available beds (unoccupied as well as occupied) each night and subsequently calculating the mean of all the counts. This should give a more meaningful result.

In some cases a hospital may admit a patient when all the beds are already occupied, perhaps by using stretchers in corridors, squeezing two people (typically children) into one bed, or by laying mattresses between beds. In such cases, the percentage bed occupancy for that particular day will be greater than 100%, since 'beds' occupied will be greater than the true number of beds actually available. Percentage occupancies of over 100% are common in many developing countries.

(4) *The average length of stay* This is the mean time per patient (measured in days) that a hospital bed is occupied during the period. It may be calculated by dividing the number of bed-days used by the number of disposals during the period. Contrast this with the more direct approach used in Example 4.3 when individual patient records are available.

Given that the provision of health care usually involves the allocation of limited resources, this is essentially a measure of the efficiency of hospital care, since it specifies how quickly patients are dealt with. However, when comparing length of stay between hospitals, it should always be remembered that different illnesses will require different periods of hospitalization. For this reason, bed statistics are more useful if they are available separately for each specialty. Even then, comparisons between specialties in different hospitals should only be made when it can be demonstrated that the patients treated and the facilities offered are very similar.

(5) *The average turnover* This is the mean number of patients that have occupied any one bed during the period. It is calculated by dividing the number of disposals by the average number of available beds.

(6) *The average turnover interval* This is the mean length of time (measured in days) that a hospital bed is left empty between successive patients. It is calculated by subtracting the number of bed-days used from the number of bed-days available, and then dividing the result by the number of disposals during the period.

Generally this statistic is used to indicate the efficiency of hospital sche-duling procedures for non-emergency admissions. The optimum inpatient scheduling system is, at least in theory, one which results in there being no time delay in replacing departing patients, assuming that the demand for places in hospital is always greater than the supply. However, length of stay and turnover interval are not always independent, for example where a ward is only open for five days each week.

In cases where the daily number of occupied beds regularly exceeds the number of available beds (that is percentage occupancy is over 100%) this statistic is clearly not appropriate.

4.10.2 Formulae

Let n = total number of disposals during the period
d = number of days covered by the period
a = average number of available beds

A = number of bed-days available during the period
B = number of bed-days used during the period
b = average daily bed occupancy
P = average percentage bed occupancy
L = average length of stay
T = average turnover
I = average turnover interval

Then, from the definitions:

(1) b = B/d
(2) A = a × d
(3) P = B/A × 100
(4) L = B/n
(5) T = n/a
(6) I = $\dfrac{A - B}{n}$, provided that this is not negative.

Also, it may be shown that:

(7) L = $\dfrac{P}{100 - P}$ × I, provided that I is defined,

(8) L = $-I + \dfrac{d}{T}$, provided that I is defined,

(9) a = $\dfrac{L \times n \times 100}{P \times d}$

4.10.3 Estimating bed requirements

The last of these equations can be used to derive a formula for the actual number of acute hospital beds required for a specialty:

$$\text{Number of beds required} = \frac{\text{'correct' length of stay} \times \text{year's demand} \times 100}{\text{'correct' percentage occupancy} \times 365}$$

where correct percentage occupancy and correct length of stay can be estimated by professionals or from comparative statistics, and the year's demand can be measured by last year's disposals plus change in the elective admission list. However, this procedure has rarely been used in practice because of the subjective nature of these three assumptions.

More often, the bed requirement is calculated as:

Number of beds needed = population served × national bed norm

where the bed norm is, perhaps, the average number of beds per head of population in England and Wales overall. This is no less subjective but perhaps easier to comprehend!

A succinct discussion of this topic is given by Bailey (1962).

4.10.4 Presentation of bed statistics

The usual way of presenting bed statistics is in tables where the variables

Table 4.4 Oxford Regional Health Authority bed statistics, obstetrics and gynaecology. (Data from *Hospital Statistics*, Oxford RHA (1985).)

Specialty year	Average no of beds		Dischgs and deaths	Average % occupancy	Average length of stay	Average turnover	Average turnover interval
	Available	Occupied					
Obstetrics							
1980	648.2	483.5	33 932	74.6	5.2	52.3	1.8
1981	650.3	448.4	33 138	69.0	5.0	51.0	2.2
1982	646.4	432.3	32 396	66.9	4.9	50.1	2.4
1983	618.9	397.2	32 438	64.2	4.5	52.4	2.5
1984	598.4	399.2	33 676	66.7	4.3	56.3	2.2
Gynaecology							
1980	378.7	277.9	26 114	73.4	3.9	69.0	1.4
1981	376.2	271.0	25 575	72.0	3.9	68.0	1.5
1982	376.6	254.7	24 151	67.6	3.9	64.1	1.8
1983	359.1	251.3	25 030	70.0	3.7	69.7	1.6
1984	378.5	258.2	26 497	68.2	3.6	70.0	1.7

defined earlier index the columns, and hospitals, specialties or years index the rows. As an example, bed statistics for obstetrics and gynaecology from the Oxford Regional Health Authority (1985) summary statistics are given in Table 4.4.

A useful way of presenting these statistics visually is by use of Barber-Johnson (B-J) diagrams (see Barber and Johnson (1973)). These diagrams illustrate the relationships between the measures used to monitor the use of beds by exploiting equations 7 and 8 in Section 4.10.2.

The basic form of a B-J diagram is a plot of average length of stay (L) against average turnover interval (I). The plot will be either for various hospitals (or specialties) or for one hospital (or specialty) over time. This plot is in itself useful. When looking at one hospital over time, for example, horizontal movements show changing levels of efficiency in hospital scheduling, whereas vertical movements show changing levels of efficiency of hospital care, often interpreted as medical efficiency. This, of course, ignores the possibility of dependence between length of stay and turnover interval and takes no account of emergency admissions or other determining factors. The movements, and their naive interpretation, are summarized in Fig. 4.9.

The full version of the B-J diagram goes further. Lines are drawn on the graph at various values of average percentage bed occupancy (P) and average turnover (T) as in Fig. 4.10, which illustrates Table 4.4. Thus a single point on the diagram enables all four statistics (L, I, P and T) to be read off. This can be very useful when deciding how to attain policy goals. As an example, if it were decided that the percentage occupancy should go up by 5% the possible ways that this can be achieved by means of changing the values of length of stay and turnover interval can be seen. Such an analysis is sometimes preceded by deciding feasible regions within which the particular variables will be allowed to vary. It might be decided, for example, that length of stay should not, for medical reasons, be allowed to fall by more than 1 day.

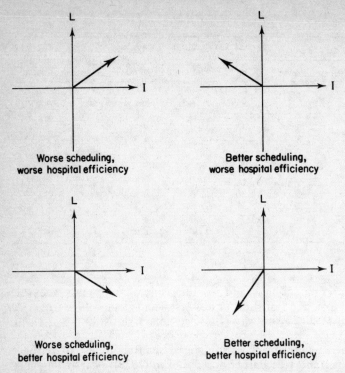

Worse scheduling,
worse hospital efficiency

Better scheduling,
worse hospital efficiency

Worse scheduling,
better hospital efficiency

Better scheduling,
better hospital efficiency

Fig. 4.9 Movements on the Barber-Johnson diagram. The intersection of the L and I lines gives the reference point, i.e., the hospital's plot at time 0.

4.10.5 Limitations of bed statistics

The limitations of these statistics can be thought of in two groups: the operational and the statistical. The former is not an appropriate area to go into here as it would include considering the quality and not only the quantity of hospital inpatient care, and also how inpatient care interacts with other health care provision, that is what effect do changing workloads in hospital wards have on other branches of the service?

Statistical limitations include the sole use of means. Bed workload data are frequently skewed, as indeed was the case with the length of stay data analysed repeatedly in this chapter. As has been noted, the mean is not the most appropriate average in such a situation: the median is superior because it is not affected by extreme values. Also, the dispersion of workload is not considered within the set of hospital bed statistics. One consequence is that a ward which always keeps patients in for 5 days is judged to be as efficient as one which has a wide spread of lengths of stay which happen to average out to 5. Whereas the second ward is most probably treating patients in the way individual cases suggest, the first is most probably scheduling patients in an automatic fashion regardless of their needs.

The great advantage of using only means to produce these hospital bed

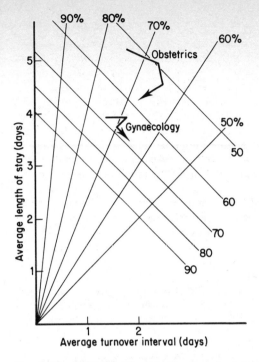

Fig. 4.10 A Barber-Johnson diagram of obstetrics and gynaecology, Oxford RHA, 1980–1984. The fanned lines give average percentage bed occupancy; parallel lines give average turnover during a year. (Data from Table 4.4. Note that sometimes the percentage occupancy and turnover values read from the graph do not correspond exactly with Table 4.4 due to rounding error.)

statistics is that the amount of data required is low, as is shown by the list in Section 4.10.1. However when individual patient records are available (say from a patient administration system) the database is very much richer. From such data, at the very least, histograms and boxplots should be produced to enhance the information communicated by the means.

One other important statistical limitation is that what is called 'average length of stay' is really the average number of nights spent in hospital. In fact patients stay for parts of days, but this is not reflected in the data. Notice that it cannot necessarily be said that 'number of nights' is 'length of stay' to the nearest whole number, although this is likely to be true for many patients. If the variable is used in this way then the use of decimal points in means is spurious accuracy. Similar arguments can be found for other bed statistics. Once again more data are required, in this case on admission and disposal times, if an improvement is sought.

4.11 Performance indicators

The hospital bed statistics described in Section 4.10 are examples of health service performance indicators. These are statistics covering a wide range of topics including hospital activity, manpower, estate and related areas such as ambulance and maternity services. They attempt to measure both the amount of service provided and the cost of providing it, and are used to compare performance at unit or district level with other similar services or with established norms.

Mention has already been made of the DHSS Performance Indicators Package (see Roberts (1986) for a brief description) which provides each district health authority each year with a large set of performance indicators for each DHA in the country. The package produces tables and pictures (such as Fig. 4.8) to make the indicators more understandable. This gives a useful way of assessing relative performance and of studying past trends but since the indicators are historic they cannot be used by managers for day-to-day planning.

The vast majority of the DHSS performance indicators are means, ratios, rates or simple counts, and suffer from the disadvantages already mentioned, particularly the lack of any related measures of dispersion or skewness within districts. They are also, of course, only as good as the data from which they are calculated. It is to be hoped that the introduction of modern information systems in districts will mean that the quality, accessibility and timeliness of the data from which they are obtained will be much improved.

Performance indicators should be interpreted in relation to the demands of the population served or to the type of service provided. For example:

(1) the number of admissions per year will often be more meaningful if broken down by age and sex groups to show which sectors of the population account for the greatest demand;
(2) the cost per case in two hospitals can only meaningfully be compared if the hospitals cover the same sets of specialties.

It may also be possible to take into account the characteristics of the area, such as the degree of industrialization or any particular health hazards, in assessing performance indicators relating to demand for services.

Too little emphasis has been placed on evaluating the outcome of the health care provided and while work remains to be done in developing useful indicators in this area, some of the currently available statistics, if properly used and interpreted in context, can go a long way to measuring the effectiveness of the service, as well as its efficiency.

Exercises

4.1 From the GP data on systolic blood pressure in Appendix 2 calculate:

   (a) the mean;
   (b) the median;
   (c) the mode;
   (d) the range;
   (e) the quartile deviation;
   (f) the standard deviation;
   (g) the tenth percentile;
   (h) the ninety-fifth percentile.

4.2 Use the frequency distributions created in Questions 3.1(a) and (b) to calculate approximate values for:

(a) the mean;
(b) the median;
(c) the standard deviation;
(d) the quartile deviation.

Compare these answers with the exact results from Question 4.1 above. Also:

(e) recalculate the median using the ogive created in Question 3.1 part (e).

4.3 (a) Draw boxplots for the systolic blood pressure and diastolic blood pressure variables of the GP data. Examine also the stem and leaf diagrams and histograms produced in the answers to the exercises of Chapter 3. Are the data on either of these variables skewed? Are there any obvious outliers?
(b) Calculate the mean minus the median for the two variables to see how far apart the two measures of average are in each case. Which measure of average (mean, median or mode) do you consider most appropriate for the data on these variables?

4.4 In the GP data find:

(a) the proportion of smokers;
(b) the ratio of smokers to non-smokers;
(c) the percentage of patients with a 'desirable' BMI.

4.5 Consider the GP data on systolic blood pressure separately for smokers and non-smokers. For each smoking group draw a

(a) stem and leaf diagram;
(b) boxplot (on the same scale).

Also calculate for each separate smoking group:

(c) the mean systolic blood pressure;
(d) the standard deviation of systolic blood pressure.

What does all this suggest about the difference in systolic blood pressure between smokers and non-smokers?

4.6 Gilchrist *et al.* (1987) describe an investigation of patients referred to a geriatric service in which the number of drugs taken by a selection of patients was found to be

Number of drugs	Number of patients
0	64
1	111
2	133
3	150
4	106
5	60
6–12	76

(a) Draw a histogram of these data (using a continuity correction).

Calculate:

(b) the mean;
(c) the median;
(d) the standard deviation;
(e) the quartile deviation

of the number of drugs taken per patient.

(f) Draw a boxplot for these data.

4.7 Using the data in Question 3.5 calculate:

(a) the mean birth weight;
(b) the standard deviation of birth weight.

4.8

Net ingredient cost of prescriptions dispensed by chemists and appliance contractors (£ million)

Year	England	Wales	Scotland	N. Ireland
1968	110.8	8.9	12.4	6.6
1970	133.5	10.1	15.2	7.6
1972	155.4	12.4	18.8	9.1
1974	205.0	16.3	24.6	11.7
1976	342.8	26.6	42.4	15.0
1978	517.6	39.4	62.9	22.4
1980	716.0	54.3	86.4	31.3
1982	977.4	72.2	112.0	42.7

(Data from *Compendium of Health Statistics*, Office of Health Economics, 1984.)

(a) For each country calculate index numbers (1968 = 100) of net ingredient cost. Produce a graph with the four index series on the same set of axes and consider what it shows about the relative increases in cost. How easy would it be to see these changes from a graph of the original data?
(b) Suppose that, instead of total costs, the table had given average costs per head of population. What weights would you then use to combine the averages from the four countries into an average for the United Kingdom overall?

4.9

The number of deaths in the United Kingdom (thousands)

Year	Quarter	Deaths
1981	I	182.4
	II	156.8
	III	146.7
	IV	172.1
1982	I	190.5
	II	160.1
	III	144.5
	IV	167.6

Year	Quarter	Deaths
1983	I	190.5
	II	157.5
	III	147.8
	IV	163.3
1984	I	179.8
	II	161.6
	III	140.4
	IV	163.1

(Data from *Monthly Digest of Statistics*, Central Statistical Office, July 1985.)

(a) Draw a graph of this time series.
(b) Estimate the trend by a four-point centred moving average.
(c) Estimate the seasonality as a *multiple* of the trend.
(d) Plot the trend on the graph of part (a). Extend the trend to 1985 first quarter by hand and thus produce a forecast for the number of deaths in 1985 I by adjusting the trend for seasonality.

4.10 During 1986 a hospital ward with 20 beds had complete bed availability each day and had no occasion to borrow beds from other wards. The midnight bed counts over the year produced the following frequency distribution.

Number of occupied beds	12	13	14	15	16	17	18	19	20
Number of days		2	14	31	64	116	78	24	18	18

(a) Draw a histogram of these data. Is the histogram symmetric?
(b) Calculate the average (mean) daily bed occupancy.
(c) Calculate the standard deviation of daily bed occupancy.
(d) Calculate the average percentage bed occupancy.
(e) Calculate the average length of stay, given that 850 patients were discharged, transferred or died in 1986.
(f) Calculate the average turnover.
(g) Calculate the average turnover interval.

Consider how much your answers tell you about the management of this hospital ward. Compare your answers with those given for Oxford in Table 4.4 (as far as possible). You should find that this ward has a higher percentage occupancy. How has it achieved this?

5
Demography

5.1 Introduction

Demography is the study of the size and structure of human populations. It is crucial in both the planning and evaluation of health activities since the population is the entity actually served by these activities. When a new hospital is planned, for instance, the primary factors in determining the bed requirement are the size and age/sex structure of the population to be served.

The subject of demography includes methods of collecting, analysing and presenting population data. In many cases these methods are no different from those applied to other types of data, as described elsewhere in this book. Nevertheless there are specific statistical techniques and definitions peculiar to demography.

Table 5.1 gives an example of demographic data which will be used subsequently to illustrate some of these techniques and definitions.

5.2 Sources of data

The fundamental source of data on the size of national and regional populations is the census of population, typically carried out at ten-year intervals. At its simplest the census is a headcount of people, although generally a series of questions on personal activity is asked of each participant. These headcounts can be of two types: *de facto*, in which only people present in the country at the time of the census are counted, and *de jure*, in which only people who are usually resident in the country are included. At a sub-national level the headcounts can be disaggregated into either the *de facto* or *de jure* form, provided that a question about the usual area of residence is included in the census. For health service purposes the *de jure* population is most useful. However the classification 'usually resident' does not have a universal definition; for example, students are sometimes treated as usually resident at their term-time address and sometimes at their parents' address. Figures from sources that use different definitions of 'population' are likely to be somewhat different, so that comparisons between them should only be made with care. For example the PP1 series of population estimates published by the Office of Population Censuses and Surveys (OPCS) use different definitions before and after 1981.

Table 5.1 Demographic data for England and Wales, 1983. Note that births to mothers aged below 15 are attributed to the 15–24 age group, and births to mothers aged over 54 are attributed to the 45–54 age group. (Data from OPCS.)

Age*	Mid-year population estimates (thousands)		Deaths		Live births by age of mother
	Males	Females	Males	Females	
Below 1	319.5	303.6	3 654	2 727	
1–4	1 280.4	1 216.0	604	489	
5–14	3 391.7	3 210.7	905	601	
15–24	4 108.7	3 955.1	3 215	1 226	245 911
25–34	3 433.0	3 385.7	3 071	1 655	335 074
35–44	3 242.7	3 205.4	5 581	3 708	47 487
45–54	2 751.8	2 746.4	15 632	9 786	662
55–64	2 699.2	2 903.2	47 315	27 792	
65 and over	2 948.9	4 551.7	209 442	242 205	
Total	24 175.9	25 477.8	289 419	290 189	629 134

*Age refers to age last birthday

A large number of detailed publications have been published by OPCS from recent censuses in the UK, including a set of county reports. OPCS have also compiled *small area statistics*, that is summaries of census data for small groups of households, electoral wards and local government districts, which have been made available in computer-readable form. Amongst many other uses, these have been employed to classify areas by health and social status (see Thunhurst (1985), Jarman (1983, 1984) and Morgan and Chinn (1983)).

Populations increase due to births and immigration and decrease due to deaths and emigration. Births and deaths are recorded on registration certificates, as are other *vital events* such as marriages. Data on vital events are called *vital statistics*. In the UK these are very complete, although this is not the case in most developing countries. International migration is estimated from the International Passenger Survey, a sample survey of passengers using the principal air and sea routes to and from the UK.

OPCS use census data, together with migration and vital statistics, to create national mid-year population estimates, such as those given in Table 5.1, for years between censuses. These estimates, and the vital statistics that helped to create them, are given in the *Monthly Digest of Statistics*, the *Annual Abstract of Statistics* and *Key Data* published by the Central Statistical Office. *Social Trends* includes a chapter on population which reviews and provides commentary on national trends, and *Regional Trends* presents figures disaggregated by regions of the country. National demographic data in fine detail appears in the regular series of Monitors published by OPCS, such as the Registrar General's Weekly Return for England and Wales (Series WR) and the quarterly record of deaths by cause (Series DH2).

At the district health authority (DHA) level population estimates can often be obtained from both OPCS and local authority information units. The differences, if any, between the figures are generally due to the use of different methods to estimate migration (or different definitions of 'resident population'). The national population census has included a question on internal migration (e.g., 'where were you living 12 months ago?'), but this information could be totally unrepresentative some years after the census. Consequently local sources of data, such as changes in registrations with Family Practitioner Committees (OPCS Series MN) or changes in school registrations and the electoral register in excess of those accounted for by births and deaths, are sometimes used to estimate migration. OPCS record births and deaths for DHAs in their SD52 and SD25 series respectively, whilst population estimates appear in the PP1 series of Monitors. District and regional health authority information departments should be able to supply these and other local demographic data.

At an international level the single best source of demographic information is the United Nations' *Demographic Yearbook*.

Some of the demographic information given in the publications listed will be in the form of ratios and rates. In the following sections the most commonly used of these will be explained. A standard reference for demographic terminology is the book by Shryock *et al*. (1976).

5.3 Ratios

One very simple, yet effective, description of a particular aspect of a population is a ratio (see Section 4.2). Probably the most commonly used is the *sex ratio*, that is the ratio of males to females. From Table 5.1 the sex ratio in England and Wales at 1983 was 24 175.9 : 25 477.8 or 1 : 1.05.

Other demographic ratios which are commonly used include the urban : rural ratio, the economically active : not economically active ratio and the child : woman ratio. A ratio is sometimes also used to describe an age differential, such as over 65 : under 65, which was 1 : 5.6 for England and Wales at 1983 (from Table 5.1).

Where appropriate, a proportion or percentage can often provide an attractive presentation. The under/over 65 age differential could, for example, be expressed as '15% of the England and Wales 1983 population were above 65 years of age'.

5.4 Frequency distributions for age/sex profiles

Although a ratio can be used to compare two portions of an age/sex profile it is frequently more useful to consider an exhaustive set of age and sex groups simultaneously in the form of frequency distributions of age for each sex. The first three columns of Table 5.1 give a simple example of such a presentation. As with other frequency distributions (see Section 3.3), relative frequencies or percentages are often used when comparisons are to be made, for instance to compare the populations of different district health authorities. When choosing the class limits it is usual to use five, ten or fifteen year intervals, usually with the first few years of life more finely differentiated due to the vastly different mortality experiences in this range. Unfortunately there is no absolute standard set of class intervals, and this may cause problems when data from different sources are to be combined or compared.

Age/sex distributions can be illustrated using a special type of histogram known as a *population pyramid*. In this display, the vertical axis represents age bands and the rectangles denoting population size (or relative population size) are drawn horizontally, with males and females shown separately on either side of the axis. Figure 5.1 shows two population pyramids for the UK population at the census dates of 1901 (when the population was 38 million) and 1981 (55 million) superimposed for the purposes of comparison. It is very clear from this how the UK population has aged, increasing the demand for geriatric care from the health and social services in recent years.

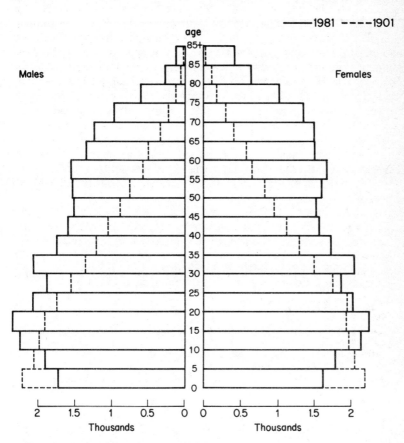

Fig. 5.1   UK population in 1901 and 1981.

5.5 Rates

Whereas ratios and age/sex distributions describe the static structure of a population, rates describe the dynamic aspects. As with ratios there are any number of rates that can be calculated to measure various characteristics of a population. Many of these have been defined as standards, particularly for the purposes of inter-regional and international comparisons; the most important ones will be described here.

5.5.1 Birth and fertility rates

The *crude birth rate* is the number of live births in a given year divided by the mid-year population estimate. Often this figure is multiplied by a thousand to avoid small numbers. From Table 5.1 the crude birth rate in England and Wales at 1983 was:

$$\frac{629\ 134}{24\ 175.9\ +\ 25\ 477.8} = 12.7 \text{ live births per thousand population}$$

(notice that in Table 5.1 the population is already expressed in thousands).

Although the crude birth rate gives an overall picture of the level of replacement of a population, it is not a sensible measure of fertility because its denominator includes males as well as old and young women who are not likely to give birth. Hence the *general fertility rate* is defined as the number of live births in a given year divided by the mid-year estimate of the number of females aged 15–44. Again, this is usually multiplied by a thousand. In England and Wales the general fertility rate was 59.7 per thousand in 1983 (from Table 5.1). In some countries a high percentage of mothers are below 15 or above 45 and consequently the ages used in the general fertility rate are sometimes extended, the most popular being 15–49 years old. In England and Wales, only 0.1% of all births in 1983 were to mothers outside the 15–44 year age range (and only 0.04% outside the 15–49 year range).

Since the ability of a population to reproduce is critically dependent upon the ages of women in the population, *age-specific fertility rates* are often quoted, where, for instance, the age-specific fertility rate for age 15–24 is:

$$\frac{\text{number of live births to mothers aged 15–24}}{\text{mid-year estimate of number of women aged 15–24}} \times 1000$$

For the 1983 England and Wales population, Table 5.1 shows this to be $\frac{245\ 911}{3955.1} = 62.2$ per thousand. The complete set of (ten-year) age-specific fertility rates (per thousand) for this population is:

age	15–24	25–34	35–44	45–54
rate	62.2	99.0	14.8	0.2

(but see the note to Table 5.1). This shows the considerable variation by age.

The major drawback with age-specific fertility rates is that they result in a set of numbers, rather than a single value, to represent fertility. To overcome this, rates may be combined to produce the *total fertility rate*. When the age class intervals used are all of width y years, the total fertility rate is the sum of the age-specific rates multiplied by y. Hence for the 1983 England and Wales population, the total fertility rate is $(62.2 + 99.0 + 14.8 + 0.2) \times 10 = 1762$ per thousand. It is usual to express this per woman rather than per thousand women, that is, 1.762 per woman.

The total fertility rate is better than either the crude birth rate or general fertility rate for comparing fertility between different populations, since it removes the effect of the age of the population. The crude birth rate is, for instance, bound to be lower in an area which contains a high proportion of elderly women, but this does not mean that people living in this area are inherently less fertile. The total fertility rate is an example of a *standardized rate* since it represents the result of exposing a standard population with the same number of women in each age group to a particular set of age-specific fertility rates.

A slight variation on the total fertility rate is the *gross reproduction rate*, which is exactly the same except that only female births are counted. This estimates the average number of daughters born to each woman in the population on the assumption that current fertility rates continue throughout their lifetimes, and they do not die before passing beyond childbearing age. This is supposed to represent the level of reproductivity of a population, but since some women *will* die before passing beyond childbearing age it is likely to be an overestimate. The *net reproduction rate*, described in Section 5.7.4, is a more meaningful measure of reproductivity.

Other birth rates sometimes used are those that relate the number of births to social or economic factors, such as nuptiality, religion, social class, race or employment. Nuptiality is particularly important since the vast majority of babies are born to married women. A change in the average age at marriage can thus have a considerable effect on the number of births.

5.5.2 Mortality rates

The *crude death rate* is the number of deaths in a given year divided by the mid-year population estimate. As with the crude birth rate, this figure is usually multiplied by a thousand. From Table 5.1 the crude death rate in England and Wales at 1983 was:

$$\frac{289\ 419\ +\ 290\ 189}{24\ 175.9\ +\ 25\ 477.8} = 11.7 \text{ per thousand}$$

(recall that the population is already expressed in thousands).

Although the crude death rate represents the average risk of mortality in a population it is not particularly good as a measure of health of a population since the chance of death varies so much with the age of the individual. The classic illustration of this is the comparison of an industrialized town in the north of England with a southern seaside resort. The crude death rate is likely to be higher in the resort simply because so many old people have retired to the seaside, and not because the resort is actually a less healthy place.

To account for age differentials, age-specific rates can be used. Since sex is also an important factor in determining the chance of death it is more usual to quote *age/sex-specific death rates*. For example, from Table 5.1 the death rate for 15–24 year old females is 1226/3955.1 = 0.3 per thousand (well below the overall average of 11.7). Table 5.2 shows the entire set of age/sex-specific death rates calculated from Table 5.1.

This population shows the usual mortality patterns of lower female rates at all ages and, for each sex, high mortality at early ages falling to a trough and then gradually rising again from the middle teens. The only exception to this smooth pattern is very high rates of mortality around the age of twenty due to accidental death, which occur in most developed countries. Of course Table

Table 5.2 Age/sex-specific death rates, England and Wales, 1983 (per thousand population).

Age	Male	Female
Below 1	11.4	9.0
1–4	0.5	0.4
5–14	0.3	0.2
15–24	0.8	0.3
25–34	0.9	0.5
35–44	1.7	1.2
45–54	5.7	3.6
55–64	17.5	9.6
65 and over	71.0	53.2

5.2 does not have fine enough detail (i.e., small enough age classes) for this to be seen.

To assess the relative importance of different causes of death, overall death rates are sometimes quoted for specific causes, as recorded on the death certificate. Cause-specific rates are usually defined as the number of deaths in a year due to the cause divided by the mid-year estimate of the total population. In 1983, 35 572 deaths in England and Wales were recorded as due to ICD code 162 (see Section 2.3.1), which represents malignant neoplasm of trachea, bronchus and lung. The overall death rate due to this cancer is thus:

$$\frac{35\ 572}{24\ 175.9\ +\ 25\ 477.8} = 0.7 \text{ per thousand (from Table 5.1)}$$

Since deaths by cause are influenced by age and sex it is probably more useful to calculate age/sex-specific rates for cause of death. A relatively older population should, for instance, exhibit a higher overall death rate due to lung cancer, because this normally affects older people (obviously tumours take time to grow).

Although age/sex-specific rates are more meaningful than crude rates, they are cumbersome when different populations, such as different occupational groups, different racial groups, different countries or the same country at different times, are to be compared. As with the total fertility rate, age/sex-specific death rates are usually combined into a single standardized measure of mortality. There are two commonly used methods of standardizing: direct and indirect. Each relates the mortality experience of a study population to a standard population. There is no standard population that is universally recognized; it would sensibly be chosen as a population similar in geography and time to those for which a comparison is required. So to compare district health authorities in England and Wales, a suitable standard population would be the population of the whole of England and Wales in the same year.

The *direct standardized death rate* is the number of deaths that would be expected in the standard population *if* the age/sex-specific death rates in the study population prevailed, divided by the size of the standard population. As usual this would normally be multiplied by 1000.

The *indirect standardized death rate* works the other way around. It considers the effect of applying the age/sex-specific death rates in the standard population to the study population, to obtain an expected number of deaths. The actual number of deaths in the study population is then divided by this expected number to produce the *standardized mortality ratio* (SMR) for the study population. If this ratio is above one the study population has higher mortality, after adjusting for age and sex, than the standard population. The indirect standardized death rate is defined to be the SMR times the crude death rate for the standard population. It is thus the basic measure of standard mortality weighted by a factor which represents the relative mortality of the population being studied.

The SMRs are themselves useful statistics and as a consequence the indirect rate is more popular. Another advantage of the indirect rate is that it does not require knowledge of age/sex-specific mortality in the study population. In some cases, particularly developing countries, age/sex-specific death rates are not accurately known.

It is useful to have formulae for the two standardized death rates. Let m_i be the death rate for age/sex group i in the study population, and $m_i^{(s)}$ the rate for the same age/sex group in the standard population. Let p_i and $p_i^{(s)}$ be the mid-year populations of the study and standard populations respectively in age/sex group i. Then:

$$\text{Direct standardized death rate} = \frac{\Sigma m_i p_i^{(s)}}{\Sigma p_i^{(s)}} \times 1000$$

$$\text{Standardized mortality ratio (SMR)} = \frac{\text{actual number of deaths in study population}}{\Sigma m_i^{(s)} p_i}$$

$$\text{Indirect standardized death rate} = \text{SMR} \times \text{crude death rate in standard population}$$

Notes:
(1) For purposes of presentation the SMR is often multiplied by 100, to make it an index number (standard population = 100).
(2) Standardized rates are sometimes quoted standardized for age only, although it is usually more meaningful to standardize for both age and sex.

Table 5.3 Estimated mid-year population (thousands) and deaths in 1983 for two health districts. (Data from OPCS.)

| | Milton Keynes | | | | East Berkshire | | | |
| | Population | | Deaths | | Population | | Deaths | |
Age	Male	Female	Male	Female	Male	Female	Male	Female
Below 1	1.3	1.2	17	12	2.6	2.4	28	12
1–4	5.5	5.2	1	1	9.8	9.9	4	5
5–14	12.2	11.4	2	4	24.7	24.1	7	2
15–24	10.8	11.8	12	1	31.9	28.5	26	8
25–34	12.0	12.6	13	6	27.5	25.9	29	11
35–44	10.2	9.5	11	3	24.8	24.3	36	27
45–54	6.3	5.8	41	25	20.5	20.4	100	51
55–64	5.1	5.0	85	60	19.0	19.3	248	163
65 and over	4.9	7.2	337	363	17.3	25.1	1165	1314
Total	68.3	69.7	519	475	178.1	179.9	1643	1593

Table 5.4 Age/sex-specific death rates for two health districts, 1983 (per thousand population).

| Age | Milton Keynes | | East Berkshire | |
	Male	Female	Male	Female
Below 1	13.1	10.0	10.8	5.0
1–4	0.2	0.2	0.4	0.5
5–14	0.2	0.4	0.3	0.1
15–24	1.1	0.1	0.8	0.3
25–34	1.1	0.5	1.1	0.4
35–44	1.1	0.3	1.5	1.1
45–54	6.5	4.3	4.9	2.5
55–64	16.7	12.0	13.1	8.4
65 and over	68.8	50.4	67.3	52.4

Example 5.1: Consider the data in Table 5.3. From this table, age-specific death rates for the two health districts are calculated by dividing deaths by population for each age/sex group in each district, resulting in Table 5.4.

Suppose that overall death rates are to be found for Milton Keynes. Then:

p_i are the population figures in the left half of Table 5.3
m_i are the rates in the left half of Table 5.4.

If the England and Wales population of 1983 is chosen as the standard population then:

$p_i^{(s)}$ are the population figures in Table 5.1
$m_i^{(s)}$ are the rates in Table 5.2.

Notice that all these figures are per thousand population.

Then,

(1) Crude death rate $= \dfrac{\text{Actual deaths}}{\Sigma p_i} = \dfrac{519 + 475}{68.3 + 69.7} = 7.2$

(using Table 5.3)

(2) Direct standardized death rate $= \dfrac{\Sigma m_i p_i^{(s)}}{\Sigma p_i^{(s)}}$

$= \dfrac{13.1 \times 319.5 + \ldots + 50.4 \times 4551.7}{24175.9 + 25477.8} = 11.4$

(using Tables 5.1 and 5.4)

(3) SMR $= \dfrac{\text{Actual deaths}}{\Sigma m_i^{(s)} p_i}$

$= \dfrac{519 + 475}{11.4 \times 1.3 + \ldots + 53.2 \times 7.2} = 0.976$

(using Tables 5.2 and 5.3)

(4) Indirect standardized death rate $= 0.976 \times 11.7 = 11.4$

All of these results come from simple operations such as cross-multiplying one column by another and summing up. Despite its simplicity the arithmetic is tedious and hence mistakes are likely to be made. A statistics or spreadsheet package is an ideal tool for such calculations.

Notes:

(1) In calculation 2 the rate does not have to be multiplied by 1000 since the divisor is already expressed in thousands.
(2) In calculation 4, 11.7 is the crude death rate for England and Wales (1983) calculated earlier (p. 105).
(3) When calculations are made using the rates from Tables 5.2 and 5.4 discrepancies may be found in the decimal place. This will be due to rounding errors – the results given above were calculated using several decimal places.

Similar results for East Berkshire are:

(1) Crude death rate = 9.0
(2) Direct standardized death rate = 10.9
(3) SMR = 0.929
(4) Indirect standardized death rate = 10.8

How can all this be interpreted? The SMRs are both below 1, so both districts have a lower mortality experience than the national average, when age and sex are taken into account. Within each district the two standardized rates are identical, or nearly so (to one decimal place). Consider the comparison of mortality in the two districts. Since the crude death rate is almost 2 per thousand higher in East Berkshire, it seems that a resident of Milton Keynes has a lower chance of death. This is misleading; the higher death rate in East Berkshire is due to the older population there. When the death rates are standardized for age and sex their order of magnitude changes; Milton Keynes' rate is now about 5 per hundred *higher*. Unlike the crude death rate, this is a fair comparison as it compares like with like.

The excess mortality amongst residents of Milton Keynes cannot be explained by the environment of the town since it is a new town predominantly inhabited by recent immigrants (this explains the relatively young age structure). More likely the excess mortality is due to the relatively poor environments previously experienced by the immigrants, which could (of course) have prompted their migration.

5.5.3 Young age mortality rates

The rate of death at young ages is of special interest because of the extra vulnerability of the very young. Young age mortality rates are, furthermore, fairly reliable indicators of health and social welfare.

Deaths at ages below 4 weeks old are generally attributable to the circumstances of the birth itself, whereas deaths at older ages are more likely due to such things as accidents or infections. For this reason separate mortality rates are defined for the two periods:

$$\frac{\text{neonatal}}{\text{mortality rate}} = \frac{\text{deaths aged below 4 weeks}}{\text{live births}} \times 1000$$

$$\text{post-neonatal mortality rate} = \frac{\text{deaths aged between 4 weeks and 1 year}}{\text{live births}} \times 1000$$

The infant mortality rate is the sum of the neonatal and post-neonatal rates, i.e.,

$$\text{infant mortality rate} = \frac{\text{deaths aged below 1 year}}{\text{live births}} \times 1000$$

As with fertility and birth rates and other mortality rates, the period over which births and deaths are recorded would usually be a calendar year. Notice that all these rates use live births as the denominator rather than a mid-year population estimate (used in other mortality rates). This is because, except perhaps in a census year, the births figure is more accurate (in fact it would be needed to obtain the mid-year population estimate).

From Table 5.1, the infant mortality rate for England and Wales in 1983 is $(3654 + 2727)/629\ 134 \times 1000 = 10.1$ per thousand. Table 5.1 does not record deaths below 4 weeks old and so the other two rates cannot be calculated from it. However, the *Annual Abstract of Statistics* (Central Statistical Office) records the neonatal mortality rate as 5.9 and the post-neonatal as 4.3 per thousand. Note that 5.9 + 4.3 is almost equal to 10.1; the difference is due to rounding error. Figure 5.2 shows the considerable improvements in these three young age mortality rates in the United Kingdom between 1974 and 1984.

The two other most commonly used young age mortality rates both concern stillbirths, that is foetal deaths occurring after 28 completed weeks of gestation. These are:

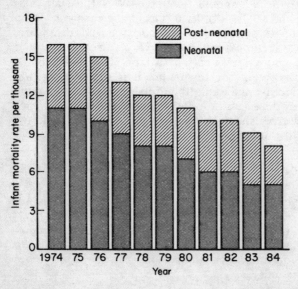

Fig. 5.2 Young age mortality rates, UK. (From *Annual Abstract of Statistics*, 1986, Central Statistical Office.)

$$\text{stillbirth rate} = \frac{\text{number of stillbirths}}{\text{number of live births and stillbirths}} \times 1000$$

$$\text{perinatal mortality rate} = \frac{\text{number of perinatal deaths}}{\text{number of live births and stillbirths}} \times 1000$$

where a perinatal death is a death below 1 week old or a stillbirth. *Social Trends* (Central Statistical Office) gives values of 5.8 and 10.5 per thousand for the stillbirth and perinatal mortality rates respectively for the United Kingdom in 1983.

5.5.4 Maternal mortality rates

The maternal mortality rate is a special kind of cause-specific mortality rate. Maternal deaths are those with codes in Chapter XI of the ICD (see Section 2.3.1). That is ninth revision codes 630–676: complications of pregnancy, childbirth and the puerperium. Then:

$$\text{maternal mortality rate} = \frac{\text{maternal deaths}}{\text{live births}} \times 1000$$

The denominator is likely to be close to the total number of confinements and gives a more meaningful base than the total mid-year population size used in other cause-specific rates. Clearly the maternal mortality rate is an indicator of antenatal and obstetric care.

In England and Wales there were 54 maternal deaths in 1983 giving a maternal mortality rate of $54/629\ 134 \times 1000 = 0.09$ per thousand (Table 5.1 records that there were 629 134 live births).

5.5.5 The rate of natural increase

Natural increase is population growth when migration is ignored. The rate of natural increase is defined as the difference between the crude birth rate and the crude death rate. From Sections 5.5.1 and 5.5.2 this was 12.7 – 11.7 = 1 per thousand for England and Wales in 1983. Often this is expressed as a percentage, e.g., $1/1000 \times 100 = 0.1\%$, which implies that the population is increasing at a rate of 0.1% per year (ignoring migration) based on 1983 figures.

Natural increase is of interest as an indicator of health but is not a sensible measure of population growth in a country where net migration is high. For example the population of the United Kingdom overall has recently experienced a net outwards migration of about one third of the size of the natural increase.

See Section 5.6.1 for a discussion of population growth taking account of migration.

5.5.6 Central and non-central rates

Many of the rates introduced so far use as their denominator the estimated population at the middle of the relevant time period. This denominator arises naturally when cross-sectional census data are used with vital statistics to produce the rate (see Section 2.5.1), which is the method usually employed. Such rates are called *central rates*.

Another possibility is to use the population at the start of the period, which is the natural choice to make when a cohort of people is followed through time (see Section 2.5.2). Rates with this denominator are called *non-central rates*, and, as will be seen in Section 5.7, these are the rates which appear in *life tables*. Consider the cohort of all people born on 1st January 1936. Suppose that 1000 of these were still alive on 1st January 1986 but 6 died during 1986. The non-central death rate for 50-year-olds would then be $6/1000 \times 1000 = 6$ per thousand.

Using the same example it is possible to estimate the central mortality rate for these people assuming that deaths are spread uniformly throughout the year (1986). If this is the case, then 3 of the 1000 that began the year will have died by mid-year, giving a mid-year population estimate of 997 and hence a central death rate of $6/997 \times 1000 = 6.02$ per thousand.

The symbol q_x is used for the non-central and m_x for the central age-specific death rate for people aged x. Let D_x be the number of deaths aged x during the year and P_x be the mid-year population aged x. By analogy with the example,

$$m_x = \frac{D_x}{P_x}$$

and

$$q_x = \frac{D_x}{P_x + \frac{1}{2}D_x}$$

Hence

$$q_x = \frac{m_x P_x}{P_x + \frac{1}{2}m_x P_x} = \frac{2m_x}{2 + m_x} \tag{1}$$

Notice that the multiplying factor of 1000 was omitted in the algebraic definitions of m_x and q_x. This is what is usually done in life tables where these formulae are employed.

Even when deaths do occur uniformly over a year, the formula (1) linking q_x to m_x will not be exact. This is because D_x is recorded as the number of people who died when aged x last birthday. Not all of these D_x were aged x at the start of the year; some will have started the year aged x − 1, had a birthday and then died during the year. On the other hand some of the people who were aged x at the start of the year and then died within the year died after reaching the age of x + 1. Generally these two errors cancel each other out to a considerable extent leaving (1) reasonably accurate.

5.6 Population projections

Demographic data are frequently used to make predictions about population size and structure in the future; indeed such predictions are essential to the rational planning of health services. Three methods for predicting populations will be outlined here. Predictions are, of necessity, always based on assumptions, the simplest assumption being that recent trends will continue in the future. In just about every case this assumption will not be absolutely correct, the largest errors being most likely to occur when projections are made far into the future. For this reason it is a good idea to also look at the effects of changes in the assumptions. 'How much will the predicted population structure alter if death rates increase by 5% over those in the recent past?' and 'What effect will this have on the predicted level of service required?' are questions that might sensibly be asked.

5.6.1 The overall growth rate method

The *annual overall growth rate* of a population is the change in population size in a year divided by the population size at the start of the year. Unlike the rate of natural increase this takes into account births, deaths *and* migration within the year. Simple population projections can be made assuming that this rate of growth continues in the future.

If the annual overall rate of growth is r and the population at the start of year one is P_1 then the population at the start of year two increases from P_1 by a factor of r, that is $P_2 = (1 + r)P_1$. Continuing in this way gives the equation for P_n, the projected population at the start of year n:

$$P_n = (1 + r)^n P_1$$

This is sometimes called the *compound interest formula*, since it can also be used, when P represents money, to quantify the accumulation of a monetary investment.

Of course the years for which projections are made do not have to coincide with calendar years; in fact it is more usual to make projections for the middle of years, in which case the start of each year in the calculation is the mid-point of a calendar year, as in Example 5.2.

Example 5.2:   The population of England and Wales was 49 601 thousand in mid 1982 and 49 654 thousand in mid 1983. If the overall growth rate experienced in the intermediate year continues, what will be the size of the population in mid 1990?

$$\text{1982/83 overall growth rate, r} = \frac{49\ 654 - 49\ 601}{49\ 601} = 0.00106853.$$

Now, 1990 is 7 years from 1983 and so:

$$\begin{aligned} P_{1990} &= (1 + r)^7 P_{1983} \\ &= (1.00106853)^7 \times 49\ 654 \\ &= 50\ 027 \text{ thousand} \end{aligned}$$

The same formula can be used to obtain an average annual overall growth rate when the population size is known at points distant in time.

Example 5.3:   The population of England and Wales at the 1961 and 1971 censuses was 52 709 and 53 788 thousand. What was the average annual percentage rate of growth in this period?

Let r be the average annual overall growth rate. Then:

$$P_{1971} = (1 + r)^{10} P_{1961}$$

and so
$$1 + r = (P_{1971}/P_{1961})^{1/10}$$
$$= (53\ 788/52\ 709)^{0.1} = 1.0020285$$

and so
$$r = 0.0020285.$$

The average annual percentage rate of growth was 0.2%

5.6.2 The curve fitting method

If the population is plotted against time on a graph a smooth curve can often be drawn passing very closely to all the points on the graph. This can then be extended, using personal judgement, to provide projections. In Section 8.3.1 it will be argued that a mathematical method of curve fitting is preferable to this subjective fitting by eye. The logistic function, or some modification of it, is frequently used as a mathematical model of population change (see Section 8.5.3).

The curve fitting method generally gives better results than the overall growth rate method when long-term projections are required.

5.6.3 The components method

A major drawback with each of the projection methods already described is that they do not account for the age and sex composition of the population. From Section 5.5 it is clear that age and sex are important determinants of fertility and mortality, so a method that takes these factors into account should be more accurate. Furthermore, projections of the age/sex distribution are more useful than projections of overall population size alone, particularly when paediatric, geriatric or gynaecological and obstetrical services are being planned.

The component method takes account of age and sex since it computes forecasts for each age/sex group separately, combining the forecasts for all the groups to produce an overall prediction of population size. The only complication is that people move into higher age groups as time goes by, so that someone aged 15–19 now will be aged 20–24 in five years time. This must be accounted for in the calculations.

A worked example of the component method is given by Pollard *et al.* (1981), Section 8.5.

5.7 Life tables

A life table is a convenient numerical representation of the mortality experience of a population. In its most basic form it shows the number of survivors, from an initial batch of births, at successive ages.

This information could be obtained by direct observation; for instance a set of people born in 1890 could have been followed throughout their lives until the last person died in 1986. The number of members of the cohort still alive at each age from 0 to 96 would then make up the basic life table. Such a life table is called a *cohort life table*.

Cohort life tables have two important drawbacks for human populations. First, they require considerable organization to keep track of all the individuals throughout their lifetimes. Second, they take an impractically long time to compile, 96 years in the hypothetical example used. The second drawback makes a cohort life table unsuitable to represent present-day expectations of mortality. The experience of people born in 1890 is unlikely to provide an accurate description of mortality in 1986 due, amongst other things, to the advances in medicine over the years.

Present-day mortality experience is represented by a *current life table*, which is the type presented in publications such as the *Annual Abstract of*

Statistics (Central Statistical Office). This uses current, or at least very recent, age-specific death rate data to construct a life table which shows the mortality experience now of people at different ages. The remainder of Section 5.7 will be concerned with current life tables.

5.7.1 Elementary functions

Suppose that the death rate specific to people aged x is m_x, as currently observed. Using Section 5.5.6, this can be converted into a non-central death rate, q_x, using the formula:

$$q_x = \frac{2m_x}{2 + m_x}$$

where q_x may be interpreted as the chance of dying during a year for people aged x at the start of the year.

Now, suppose that q_x was $\frac{1}{3}$. Then if, say, 300 people began the year at the age of x it would be expected that one third of them, that is 100, would die during the next year. In general, from a collection of l_x people aged exactly x, $q_x l_x$ can be expected to die during the next year, leaving $l_x - q_x l_x$ survivors at the end of that year. The expected number of deaths at age x is denoted by d_x, and the expected number of survivors to age x + 1 is denoted by l_{x+1}. Hence:

$$d_x = q_x l_x$$
$$l_{x+1} = l_x - q_x l_x = l_x - d_x$$

The chance of surviving for another year given survival to age x is denoted by p_x and:

$$p_x = 1 - q_x$$

(see Section 6.4.2).

Example 5.4: The following demographic data for ages 40–44 relate to a hypothetical health district:

Age	Population at 30th June 1986	Deaths in 1986
40	2900	8
41	2800	9
42	3000	10
43	3100	12
44	2900	12

The age-specific death rates per head of population, m_x, are simply deaths divided by population size. Converting these to non-central rates, q_x, results in:

Age	m_x	q_x
40	0.002759	0.002755
41	0.003214	0.003209
42	0.003333	0.003327
43	0.003871	0.003864
44	0.004138	0.004129

A partial life table between the ages of 40 and 44 can then be constructed using a 'starting value' of 100 000 people alive at age 40, e.g.

$$d_{40} = q_{40} \times l_{40} = 0.002755 \times 100\,000$$
$$= 275 \text{ (to the nearest whole number)}$$
$$p_{40} = 1 - q_{40} = 1 - 0.002755 = 0.997245$$
$$l_{41} = l_{40} - d_{40} = 100\,000 - 275 = 99\,725.$$

The 1986 life table for ages 40–44 is:

Age (x)	l_x	d_x	p_x	q_x
40	100 000	275	.997245	.002755
41	99 725	320	.996791	.003209
42	99 405	331	.996673	.003327
43	99 074	383	.996136	.003864
44	98 691	407	.995871	.004129

So, if there were 100 000 people now aged exactly 40, only 98 691 could expect to reach their 44th birthday: 275 would be expected to die before celebrating another birthday; 320 would be expected to die after their 41st but before their 42nd birthday, etc.

In Example 5.4, the life table has only been calculated for a small number of years of life merely to save space. Life tables usually start at age 0 (birth) and continue through to an age which few members of the population will exceed. A life table that includes every year of life is called a *complete life table*.

Notice that the starting value, the first number in the l (survivors) column, had to be given in the example. The starting value (l_0 in most life tables), is known as the *radix*, and can be chosen to be any convenient number. 100 000 is commonly used. By convention all l and d values are given to the nearest whole number.

5.7.2 The expectation of life

The elementary functions of a life table can be used to estimate the future life expectation at any given age, assuming that current mortality experience continues in the future.

Consider l_x people now aged x. In the next year it is expected that d_x of these will die, leaving l_{x+1} survivors. Assuming that, on average, each person who dies in the next year lives for half that year, then the expected number of person-years lived in the next year by the l_x people now aged x is

$$(1)l_{x+1} + (\tfrac{1}{2})d_x = l_{x+1} + \tfrac{1}{2}(l_x - l_{x+1}) = \tfrac{1}{2}(l_x + l_{x+1}) = L_x$$

Similarly in the second year from now it is expected that d_{x+1} will die and l_{x+2} will survive, of the people now aged x. Hence in this year, the l_x people now aged x are expected to live:

$$(1)l_{x+2} + (\tfrac{1}{2})d_{x+1} = \tfrac{1}{2}(l_{x+1} + l_{x+2}) = L_{x+1}$$

person-years. Continuing in this way, the expected total number of person-years lived until death by the l_x people now aged x is:

$$L_x + L_{x+1} + L_{x+2} + \ldots + = T_x$$

where the summation runs up until the last possible year of life, and L_i is defined by the equation $L_i = \frac{1}{2}(l_i + l_{i+1})$.

If l_x people can expect, between them, to live T_x years of life then, on average, each of the l_x people will live for T_x/l_x years. This is the expectation of life, denoted $\overset{\circ}{e}_x$, for someone now aged x. Of particular interest, especially as an indicator of the health of a community, is the expectation of life at birth, $\overset{\circ}{e}_0$.

When using the expectation of life it must be remembered that this is based on current mortality experience. If, say, a cure for cancer were found tomorrow, then these expectations could be seriously wrong.

Sometimes life tables include a column of L_x and T_x values as well as $\overset{\circ}{e}_x$. While the L and T functions are used primarily to obtain $\overset{\circ}{e}_x$, they have special interpretations of their own which may be of interest. For young ages the assumption that those dying live on average for half the year is inaccurate, and hence alternative formulae for L_x have been suggested for small values of x (see Shryock *et al.* (1976) Chapter 15).

5.7.3 Obtaining life tables

Life tables are published, generally separately for each sex, by many different countries. A summary of this information is given in the UN *Demographic Yearbook*. OPCS calculate complete life tables for England from census data; the English female life table calculated using 1981 census data is given as Table A.7 in Appendix 4. Abridged life tables for the United Kingdom overall, England and Wales, Scotland and Northern Ireland, showing life table functions for every 5th year of life are published in the *Annual Abstract of Statistics*. It should be stressed that published life tables sometimes use more accurate (but also more complex) methods of estimating some of the life table functions than those given in this chapter. Interpretation of the functions will still, however, be the same.

At a sub-national (e.g., district health authority) level, life tables may well have to be calculated from first principles, using population and death data, which are readily available, as in Example 5.4. A less accurate alternative is to treat the national life table as if it were also valid for the sub-national population. In developing countries, where even the basic death data are unreliable, a similar approximation is used when UN 'model' life tables are adopted.

Life tables do not have to be calculated on a geographical basis. For example, Woodward (1982) gives a life table for residents of mental handicap hospitals in England and Wales. In a wider context, though still very relevant to health management, Forbes (1971) shows how a life table suitable for man-power planning can be calculated. This life table measures length of stay in employment rather than life and has leaving employment rather than death as the terminating event (see also Bartholomew and Forbes (1979)).

The calculations needed to construct a life table from population and death data described in this chapter are all simple enough to be possible on the most basic calculators. Since the same operation has to be applied to every age value within a column this can be very tedious, especially for a complete life table. Spreadsheet and statistical packages are particularly efficient at such column operations and can, therefore, be an asset in this context.

5.7.4 Using life tables

Life tables have two major uses, to provide projections and to provide indicators. Examples 5.5–5.8 give some examples of projections using Table A.7.

Example 5.5:   What proportion of English females now aged exactly 20 can expect to live to the age of 50?

The number surviving to age 20 is l_{20} = 98 497
The number surviving to age 50 is l_{50} = 95 244

Hence the required proportion is $\dfrac{95\ 244}{98\ 497}$ = 0.97.

Example 5.6:   What is the expected age at death for an English female now aged exactly 20?

The expectation of life when aged 20 is $\overset{\circ}{e}_{20}$ = 58.124
Hence a woman now aged 20 would expect to be aged 20 + 58.124 = 78.1 years old at death.

Example 5.7:   What percentage of English females now aged exactly 20 can expect to die at the age of 80?

The number surviving to age 20 is l_{20} = 98 497
The number dying at age 80 is d_{80} = 3534

Hence the required percentage is $\dfrac{3534}{98\ 497}$ × 100 = 3.6%.

Example 5.8:   A health district currently has 989 women aged 30 alive within its boundaries. How many women aged 50-years-old can be expected in the health district in 20 years time? Assume no migration, and that women in this district will experience current national mortality patterns in the future.

Of l_x women now aged exactly x, l_{x+20} can expected to be alive, aged x + 20 exactly, in 20 years time. Hence a proportion l_{x+20}/l_x can be expected to be alive in 20 years time.

Women now aged 30 have an exact age somewhere in the range '30 but less than 31'. Assuming an even distribution of women in this age range, the average age of such women will be 30.5. Hence the proportion of women now aged 30 who can expected to be alive in 20 years time will be, approximately, $l_{50.5}/l_{30.5}$.

Neither $l_{30.5}$ nor $l_{50.5}$ is known. In general, if it can reasonably be assumed that ages are evenly spread within the year of age 'x but less than x + 1' $l_{x+\frac{1}{2}}$ can be approximated by $(l_x + l_{x+1})/2$. This gives the approximate result:

$$\frac{l_{50.5}}{l_{30.5}} = \frac{\left(\frac{l_{50} + l_{51}}{2}\right)}{\left(\frac{l_{30} + l_{31}}{2}\right)}$$

$$= \frac{\left(\frac{95\ 244 + 94\ 884}{2}\right)}{\left(\frac{98\ 105 + 98\ 054}{2}\right)}$$

$$= 0.9693$$

Hence the expected number of women in the district aged 50 years old in 20 years time is approximately $989 \times 0.9693 = 958.6$. (Notice that the assumption of an even distribution of women within a year of life is likely to be false in some cases, particularly at extreme ages.)

The indicators provided by life tables include $\overset{\circ}{e}_0$, which has already been mentioned as an indicator of health. A second example is the *net reproduction rate* which indicates whether a population is contracting (rate below one) or expanding (rate above one). The net reproduction rate is defined as the sum, over all possible ages, of the age-specific fertility rates for female births only times the chance of the daughter surviving to the age of her mother at the time of her birth. Example 5.9 shows how this chance can be estimated, for a particular year of age, from the life table. In practice it is more likely that 5 or 10-year age groups are used in the calculation of the net reproduction rate.

Example 5.9: What proportion of daughters born to English mothers aged 39-years-old can be expected to survive to the age of their mother at the time of their birth?

Of all births, a proportion l_x/l_0 can be expected to survive to exact age x. Mothers who give birth at the age of 39 can have exact age anywhere in the age group '39 but less than 40'. Assuming that their ages are evenly spread, a similar argument to that used in Example 5.8 gives an expected proportion of daughters who reach their mothers' delivery age of approximately:

$$\frac{l_{39.5}}{l_0} = \frac{\frac{1}{2}(l_{39} + l_{40})}{l_0}$$

Using Table A.7 this expected proportion is:

$$\frac{\frac{1}{2}(97\ 457 + 97\ 346)}{100\ 000} = 0.97$$

5.7.5 Abridged life tables

Instead of covering every single year of life, life tables are sometimes produced in abridged form where perhaps only every fifth or every tenth year of life is included. In some cases abridging is carried out simply to save space or to avoid unwanted complexity. Thus the abridged tables in the *Annual Abstract of Statistics* are merely excerpts from complete life tables. In other cases the life table is abridged because the data used to create it were only available in grouped form (e.g., five-year age bands).

Besides the l_x function, which is the same as for complete tables, abridged life tables may contain any of the following elementary functions:

$_nd_x$ = expected number of deaths aged between x and x + n – 1 last birthday;
$_np_x$ = chance of surviving to age x + n given survival to age x;
$_nq_x$ = chance of dying before reaching age x + n given survival to age x.

If the q function has to be estimated from central n-year age-specific death rates the approximate formula:

$$_nq_x = \frac{2n\,_nm_x}{2 + n\,_nm_x}$$

can be used, where $_nm_x$ is the central death rate for n-year age bands. For example, Table 5.2 gives $_nm_x$ per thousand, where n is usually 10. This formula assumes that deaths are evenly spread within each n-year period. This is not always true, particularly for deaths before the age of 5. See Shryock *et al.* (1976) for more precise methods.

Abridged life tables may also contain $\overset{\circ}{e}_x$, which has the same interpretation as for complete tables. The method of calculation is, however, slightly different. (Again, see Shryock *et al.* (1976) for details.)

5.8 Resource allocation

The demographic measures described in this chapter have many applications in health management and research. In this final section one particularly important application, which has received considerable attention, will be discussed: resource allocation. In particular the method of revenue allocation determined by the Resource Allocation Working Party (RAWP) in England will be described. In their report (DHSS, 1976) the allocation of capital funds is also covered, but this is omitted for the sake of brevity.

The re-organization of the NHS in 1974 created new health authorities (districts, regions and also, at that time, areas) with responsibilities for managing the health care of their resident population. RAWP adopted the philosophy that resources should be allocated proportionately to the size of the resident population, after adjusting this size to take account of local needs. This adjusted population will here be called the population of need. Then, given a national budget, the only figure needed to allocate resources to regions (the highest tier) is the population of need of that region. For each region the crude mid-year population estimate is changed into the population of need by a series of adjustments, but a constraint imposed on these adjustments is that the overall total population of need should equal the overall national mid-year population estimate.

Since different regions will make different demands on the separate parts of the health service, RAWP decided to look at seven areas of care separately: non-psychiatric inpatient, community, day and outpatient, ambulance, Family Practitioner Committee administration, mental illness inpatient and mental handicap inpatient. For each area of care the basic resident population is adjusted to take account of local needs, and then a final regional population of need is calculated as a weighted average of the seven adjusted populations. The weights used are the percentages of national revenue expenditure in the seven areas of care.

Non-psychiatric inpatient care will be considered as an example. This accounts for over 50% of NHS national revenue expenditure (RAWP's figures) and involves the most complex adjustment. This adjustment is performed in three basic stages, to take account of:

  (i) the age/sex composition of the resident population,
 (ii) local morbidity,
(iii) the cost of the underlying local need.

In addition, the final resource allocation should take account of cross-boundary flows and special factors such as teaching facilities, but see the RAWP report for such fine details.

Step (i) The population of a region is adjusted for age and sex to take account of the different health needs of the different age/sex groups. The national numbers of inpatient bed-days used by age/sex groups represent these needs. Thus, the expected number of bed-days used by age/sex group i in the region is the population size of group i in the region times the national number of inpatient bed-days used per person in group i. Summing over i gives the total expected number of bed-days used in the region, and this is taken as the adjusted population at the end of step (i).

Step (ii) The population is next adjusted for local morbidity, that is a factor supposed to represent sickness in the community due to environmental, occupational and other factors over and above the effects of age and sex. The RAWP report considers various measures of morbidity but, due partially to the lack of accurate data on sickness, RAWP finally decided to use the standardized mortality ratio (SMR) as a proxy measure for morbidity. In Section 5.5.2 the SMR was seen to be a measure of relative mortality adjusted for age and sex, so this would be ideal if mortality were the issue here. RAWP argue that patterns of mortality and morbidity are likely to be similar. Consequently the regional population is adjusted for morbidity by multiplying the result of step (i) by the SMR calculated using the national population as the standard.

Step (iii) The population is next adjusted for the cost of providing the particular mixture of inpatient care required, so that a region with a considerable need for an expensive service receives extra funding. To achieve this, RAWP suggested disaggregating inpatients by the 17 ICD chapters (see Section 2.3.1). Costs can then be collected, nationally, for treatment according to chapter. If expected bed-days used and SMRs are obtained for each chapter in steps (i) and (ii) then these can be multiplied by each other and then by costs per bed-day to provide an adjusted population for each chapter. Summing over chapters gives an overall adjusted population for the region.

Step (iv) A rescaling to make all the adjusted regional populations sum to the national population gives the populations of need and completes the adjustments.

Since costs by chapter were not available to them, RAWP omitted these from their final calculations, although the separation of bed-days and SMRs by ICD chapters was retained to achieve more sensitive adjustments for need. The RAWP report gives a summary formula for the population of need for non-psychiatric inpatient care, as well as a worked example for this and other areas of care.

One problem with using SMRs to represent morbidity is that some conditions rarely lead to death. In these cases, RAWP suggest omitting the SMR part of the adjustment, except in the case of ICD Chapter XI, conditions of pregnancy, childbirth and the puerperium. Here they suggest using a *standardized fertility ratio* (SFR) rather than an SMR. The SFR they suggest is as for the SMR except that only the female sex is involved and age-specific fertility rates are used, i.e.,

$$ \text{SFR} = \frac{\text{observed births in study population}}{\Sigma f_i^{(s)} p_i} $$

where $f_i^{(s)}$ is the age-specific fertility rate for age group i in the standard (national) population and p_i is the female population of age group i in the study population. When fertility rates were considered in Section 5.5.1 only live births were counted, since the growth of population was being measured. In this context the need for health care is being measured and hence it is appropriate to include stillbirths in the calculation.

The approach described here has formed the basis of resource allocation to the English regional health authorities, and then within regions to districts. Although it is not appropriate to quote them here, it is fair to record that various authors have criticized certain aspects of the RAWP approach, notably the use of SMRs to measure morbidity. (See, amongst others, Senn and Samson (1982), Forster (1977), Barr and Logan (1977), and Palmer *et al.* (1980).) Some of the English regional health authorities have modified the formula for use in allocation of resources to their districts, to include, for example, a deprivation weighting. At the time of writing, the entire RAWP methodology is being considered by a national RAWP formula review group.

Exercises

5.1 The tables at the top of the next page give selected demographic data for England and Wales in 1984 (OPCS data).

For England and Wales in 1984 calculate:

(a) the general fertility rate;
(b) the age-specific fertility rates (using 5-year intervals, and assuming all mothers aged below 20 are in the class interval 15–19 years and all aged 45 or over are in the class interval 45–49 years);
(c) the total fertility rate (using the 5-year interval figures from part (b));
(d) the crude birth rate (given that there were an estimated 24 244.2 thousand males in England and Wales in 1984).

Estimated mid-year female population (thousands)		Live births by age of mother	
Age	Population	Age	Births
0–4	1 527.0	Under 20	54 508
5–9	1 417.3	20–24	191 455
10–14	1 715.2	25–29	218 031
15–19	1 973.2	30–34	122 774
20–24	2 005.2	35–39	42 921
25–29	1 727.9	40–44	6 576
30–34	1 667.0	45 and over	553
35–39	1 817.2		
40–44	1 461.8	Total	636 818
45–49	1 384.6		
50–54	1 358.2		
55–59	1 396.8		
60 and over	6 068.0		
Total	25 519.4		

5.2 The table below gives demographic data for a Health District in 1983 (OPCS data).

Demographic data for Oxfordshire, 1983

Age	Mid-year population estimate (thousands)		Deaths	
	Males	Females	Males	Females
Below 1	3.4	3.2	27	26
1–4	13.8	12.6	1	7
5–14	34.7	32.0	9	8
15–24	57.3	47.5	36	8
25–34	38.7	37.1	24	11
35–44	34.2	33.5	51	35
45–54	27.0	26.3	135	97
55–64	24.6	25.5	324	197
65 and over	27.1	38.5	1704	1798
Total	260.8	256.2	2311	2187

For Oxfordshire Health District in 1983 calculate:

(a) the sex ratio;
(b) the crude death rate;
(c) the age/sex-specific death rates;
(d) the direct standardized death rate;
(e) the standardized mortality ratio (SMR);
(f) the indirect standardized death rate,

using the 1983 population of England and Wales (from Table 5.1) as the standard population.

Compare the various death rates with those for East Berkshire and Milton Keynes given in Section 5.5.2. Also:

    (g) plot the age-specific death rates for males in Oxfordshire (from part (c)), East Berkshire and Milton Keynes (from Table 5.4) superimposed on the same set of axes;

    (h) repeat part (g), but this time for females;

    (i) draw population pyramids for Oxfordshire, East Berkshire and Milton Keynes.

Use these diagrams to explain the differences between the crude, direct and indirect standardized death rates and the SMR in the three districts.

5.3 The population of Scotland at the 1951 and 1961 censuses was 5096.4 and 5179.3 thousand respectively. Calculate:

    (a) the average annual percentage rate of growth in the 1951–61 period;

    (b) the expected population size in 1971 if this rate of growth (from part (a)) were experienced each year between 1961 and 1971.

The population of Scotland at the 1971 census was 5229.0 thousand. Compare this with the figure obtained in part (b). Then calculate:

    (c) the true average annual percentage rate of growth for Scotland in the 1961–71 period,

and compare with the rate for England and Wales over the same period given in Section 5.6.1.

5.4 Using the following hypothetical data for a country, construct a male life table containing the l, d, p and q functions for the ages 30–34, given that l_{30} = 90 000 (radix, l_0 = 100 000).

Age	Mid-year population	No. of deaths in the year
30	211 402	284
31	211 014	303
32	198 444	275
33	201 717	296
34	200 240	259

5.5 Draw graphs of the following functions from the English female life table (Table A.7):

    (a) l_x;

    (b) d_x;

    (c) q_x

where in each case the function is to be plotted against x, the year of life (it is sufficient to plot only every 5th value except for the very early and late years of life). Comment on the patterns shown by your graphs.

5.6 Use the English female life table to find:

    (a) the expected proportion of women now aged exactly 80 who will live to be at least 100;

    (b) the expected proportion of women now aged exactly 60 who will die when aged 65 last birthday.

5.7 (a) A regional health authority had 2205 women aged 55 last birthday amongst its population in 1986. Assuming no migration, an even spread of women within the year of life, and using the English female life table as an approximation to the true life table for this population, how many women aged 65 last birthday can be expected in 1996?

(b) The same region found that 65-year-old women accounted for 5800 of the bed-days used in 1986, when there were 1961 women of that age living in the region. Assuming the same bed use per woman, use your answer to part (a) to find the expected number of bed-days used by 65-year-old women in 1996.

6
Probability and probability distributions

6.1 Definitions

There are very few things that can be predicted for certain in advance, from the simple problem of predicting the outcome of tossing a coin to the complex problem of predicting a patient's reaction to surgery. In such cases it is common to speak of the *chance*, *odds* or *probability* of a certain thing occurring.

A process whose result cannot be predicted for certain in advance is called a *random process*: for instance, when a die is rolled the result is not known beforehand. When a random process is observed to see what result it actually does produce this is called a *probability experiment* or *trial*, and the observed result is itself called the *outcome*. The set of all outcomes that could possibly occur is called the *sample space*, so when rolling a die the sample space is the set of possible outcomes 1, 2, 3, 4, 5 and 6.

Probabilities are defined for *events*, where an event can either be a solitary outcome, such as 'roll a 6', or a collection of outcomes, such as 'roll an even number', which requires one of the outcomes 2, 4, or 6 to occur.

Example 6.1: Consider the process of tossing a fair (that is, unweighted) coin. The side uppermost after the toss is either a head or a tail, but the outcome cannot be predicted for certain in advance, and so coin tossing must be a random process. Suppose that the interest lies in finding the probability of a head. The trial is a toss of the coin.

Here the sample space is the pair 'head and tail' and the event of interest is the single outcome 'head'.

Example 6.2: An urn contains 20 red and 30 blue balls, labelled R_1 to R_{20} and B_1 to B_{30} respectively, but otherwise identical. A ball is to be selected from the urn blindly. This is a random process, and the sample space is $R_1, R_2, \ldots,$ $R_{20}, B_1, B_2, \ldots, B_{30}$. The question, 'What is the probability of drawing a red ball?', is asked, and hence the event of interest is all the outcomes $R_1, R_2, \ldots,$ R_{20}, since the event 'red ball is selected' is any of the outcomes R_1 to R_{20}.

Example 6.3: A District Medical Officer owns a calculator which runs on batteries. The batteries have a limited, but not exactly predictable, lifetime and he is interested in finding the probability that they last for over 1000 hours of use. The sample space here is all possible lifetimes which, at least in theory, are all the times from 0 seconds onwards. The event for which a

Example 6.7:   A patient is given a course of treatment of 10 capsules. Eight of these are identical, but two are different, larger doses. The patient should take the capsule with the largest dose first, then the second largest dose and finally the eight identical capsules, on a daily basis.

Assuming that all the capsules are indistinguishable to the patient, what is the probability that he takes them in the correct order?

In this question order clearly counts, and so permutations are involved. The number of permutations of 10 capsules is simply 10!, and this gives n_T, the divisor for the probability. The numerator, n_E, where E is the event 'the capsules are successfully taken', is the number of ways in which the doses can be taken in the correct order. Now:

(i)   the number of ways in which the largest dose can successfully be taken = 1 (it should be the first),
(ii) the number of ways in which the second largest dose can successfully be taken = 1 (it should be the second),
(iii) the number of ways in which the eight identical doses can successfully be taken = 8! (any permutation of the eight is effectively the same).

Hence $$n_E = 1 \times 1 \times 8!$$

and $$P(E) = \frac{1 \times 1 \times 8!}{10!} = \frac{1}{10 \times 9} = 0.011.$$

6.4 Compound events

Up to now, probabilities have only been determined for single events, but in many situations probabilities that involve several events, compounded in some way, will be of interest. In genetic counselling, for example, it would be of interest to know the probability that a newborn baby is both male *and* has the gene for haemophilia.

6.4.1 'Or' probabilities and the addition law

Given two events, E_1 and E_2, it is often useful to determine the probability that E_1 occurs alone *or* E_2 occurs alone *or* they both occur, which will be called an 'or' probability. It turns out that:

$$P(E_1 \text{ or } E_2) = P(E_1) + P(E_2) - P(E_1 \text{ and } E_2)$$

where $P(E_1$ and $E_2)$, the probability of E_1 and E_2, is another form of compound probability called an 'and' probability (see Section 6.4.5). This equation is called the *addition law* of probability.

If areas are used to represent probabilities in a diagrammatic fashion the result is called a *Venn diagram*. This is a useful way of illustrating the addition law, as Fig. 6.1 shows. The left-hand circle represents $P(E_1)$ and the right, $P(E_2)$. Then $P(E_1$ or $E_2)$ is the area bounded by the two circles. This is clearly the sum of the areas of the individual circles less the intersection of the circles. Since the intersection (shaded) represents the part where both E_1 *and* E_2 occur; the addition law results.

When two events E_1 and E_2 *cannot* occur together, they are said to be

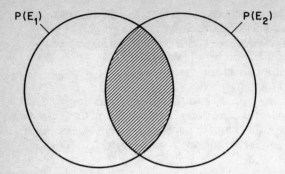

Fig. 6.1   A Venn diagram to illustrate the addition law.

mutually exclusive. Hence E_1 and E_2 mutually exclusive is equivalent to $P(E_1$ and $E_2) = 0$. Thus in the case of mutually exclusive events the addition law takes a simpler form:

$$P(E_1 \text{ or } E_2) = P(E_1) + P(E_2)$$

In a Venn diagram, mutually exclusive events would be represented by non-overlapping circles.

Example 6.8:   A World Health Organization (WHO) project team consists of four Americans, one Briton, two Germans and two Italians. If the team leader is chosen at random, what is the probability that he is from an English speaking country?

Let A be the event 'an American is chosen'

Let B be the event 'a Briton is chosen'

Then the event of interest is the compound event 'A or B'. Since there is only one team leader, A and B cannot occur together, so they are mutually exclusive and the simplified version of the addition law holds. That is

$$P(A \text{ or } B) = P(A) + P(B) = \tfrac{4}{9} + \tfrac{1}{9} = \tfrac{5}{9}$$

using the definition of probability for equally likely outcomes.

Since 5 of the 9 project team members are from English speaking countries, this result seems to agree with intuition.

6.4.2 Complementary events

If there are only two possible events, E_1 and E_2, and *either* E_1 occurs alone *or* E_2 occurs alone, then E_1 and E_2 are said to be complementary events. The events 'head' and 'tail' when a coin is tossed are complementary events, as are winning and not winning a lottery.

By convention, the complementary event to an event E is written as $\overline{E}$, which is read as 'not E'. Since it has been stated that either E or $\overline{E}$ occurs, these two events are mutually exclusive and so by the addition law:

$$P(E \text{ or } \overline{E}) = P(E) + P(\overline{E})$$

Now the probability of (E or $\overline{E}$) must be 1, since this is a sure event. Hence:

$$P(\overline{E}) = 1 - P(E)$$

Example 6.9: For the WHO team of Example 6.8, the probability of selecting a team leader from a country which is not English speaking is $1 - \frac{5}{9} = \frac{4}{9}$, since 'not English speaking' is clearly the complement to 'English speaking'.

6.4.3 Exhaustive events

If a number of events E_1, E_2, E_3, . . ., E_n are such that at least one of them must occur, then they are said to be *exhaustive*. Furthermore, if these events are mutually exclusive then, by applying the addition law repeatedly:

$$P(E_1 \text{ or } E_2 \text{ or } E_3 \text{ or } . . . \text{ or } E_n) = P(E_1) + P(E_2) + . . . + P(E_n) = 1$$

Example 6.10: Following on from the solution to Example 6.8, if
   G is the event 'a German is chosen',
and   I is the event 'an Italian is chosen',
then A, B, G and I are a set of mutually exclusive and exhaustive events and thus:

$$P(A) + P(B) + P(G) + P(I) = 1$$

6.4.4 Conditional probabilities

The occurrence of one event will often influence the probability of occurrence of a second event. For example, the probability of someone developing lung cancer is increased if that person smokes.

The probability of an event E_2 occurring given that E_1 has occurred is called the *conditional probability* of E_2 given E_1, written as $P(E_2|E_1)$. This is defined to be the probability that E_2 and E_1 *both* occur divided by the probability that E_1 occurs. That is:

$$P(E_2|E_1) = \frac{P(E_2 \text{ and } E_1)}{P(E_1)}$$

The logic behind this definition is most easily seen from a simple example.

Example 6.11: The urn of Example 6.2 actually contained some ivory and some plastic balls. 15 of the 20 red balls and 25 of the 30 blue balls were plastic, the rest were ivory. All this information may usefully be presented in the form of a two-way table such as Table 6.1.

A ball is selected at random and found to be made of plastic. What is the probability that the selected ball is red?

Since the selected ball is known to be plastic it must be one of the 40 balls in the first row of the table. Only 15 of these balls are red, so the probability

Table 6.1 Material and colour of balls in urn.

Material	Colour		Total
	Red	Blue	
Plastic	15	25	40
Ivory	5	5	10
Total	20	30	50

P(Red|Plastic) must be $= \frac{15}{40} = 0.375$ using the definition of probability for equally likely outcomes. What has happened is that the sample space has been restricted to only 40 balls, once it is known that the ball is plastic.

To see that the definition of conditional probability given above makes sense: P(Plastic) = 40/50 and P(Red and Plastic) = 15/50 since, respectively, 40 of the 50 balls are plastic and 15 of the 50 balls are both red and plastic. Hence:

$$P(\text{Red}|\text{Plastic}) = \frac{P(\text{Red and Plastic})}{P(\text{Plastic})} = \frac{15/50}{40/50} = \frac{15}{40}$$

as before. Clearly, the definition works because the 50 cancels from the top and bottom. In general, this type of cancelling always works out correctly. Notice that from Section 6.2.1, P(Red) = 0.4 but now P(Red|Plastic) = 0.375. Assuming that the material of a drawn ball is known before the colour is seen, the probability that it is red is reduced if it has been established that the ball is made of plastic.

6.4.5 'And' probabilities and the multiplication law

Probabilities of the type $P(E_1 \text{ and } E_2)$ have already been mentioned. In Section 6.4.1, such probabilities were represented by the intersection of circles in a Venn diagram and in Section 6.4.4 they formed part of the definition of conditional probability. In fact it is this definition which gives a formula for the calculation of 'and' probabilities. This formula is commonly known as the multiplication law:

$$P(E_2 \text{ and } E_1) = P(E_2|E_1)\, P(E_1).$$

When the events E_2 and E_1 are such that the probability of E_2 is the same whether or not E_2 has occurred, and vice versa, these events are said to be *independent*. In other words, E_1 and E_2 are independent whenever:

$$P(E_2|E_1) = P(E_2|\overline{E}_1) = P(E_2)$$

and $\qquad\qquad P(E_1|E_2) = P(E_1|\overline{E}_2) = P(E_1)$

In this special, simple case, the multiplication law reduces to:

$$P(E_2 \text{ and } E_1) = P(E_2)\, P(E_1)$$

which is the multiplication law for independent events.

In Example 6.11, the events 'red ball selected' and 'plastic ball selected' are definitely not independent since P(Red|Plastic) is *not* equal to P(Red), as was discussed.

Example 6.12:   A fair coin is tossed 3 times. What is the probability of getting a head followed by a head followed by a tail?

The result on any one toss has no influence on the result of subsequent tosses, that is, the probability of a head on the second toss must be the same whether or not a head appeared on the first toss. Hence, the events of interest here are independent.

Let H_i be the event 'a head occurs on the ith toss'. Then the probability required is $P(H_1 \text{ and } H_2 \text{ and } \overline{H}_3)$. By the multiplication law for independent events:

$$P(H_1 \text{ and } H_2 \text{ and } \overline{H}_3) = P(H_1) \times P(H_2) \times P(\overline{H}_3)$$
$$= \tfrac{1}{2} \times \tfrac{1}{2} \times \tfrac{1}{2} = \tfrac{1}{8}$$

Example 6.13:   Two student doctors, Smith and Jones, are asked to give their diagnoses for a set of symptoms. The two work completely independently of each other. Dr Smith has a 60% chance of getting the right diagnosis but Dr Jones only has a 30% chance. What is the probability that the correct diagnosis is made by one or the other of the doctors?
Let S be the event 'Dr Smith gets the diagnosis correct'
Let J be the event 'Dr Jones gets the diagnosis correct'
Then the probability required is P(S or J). By the addition law:

$$P(S \text{ or } J) = P(S) + P(J) - P(S \text{ and } J)$$

By the multiplication law for independent events:

$$P(S \text{ and } J) = P(S) \times P(J)$$

and hence $P(S \text{ or } J) = P(S) + P(J) - P(S) \times P(J)$
$$= \frac{60}{100} + \frac{30}{100} - \frac{60}{100} \times \frac{30}{100}$$
$$= \frac{72}{100} = 0.72$$

6.5 Probability distributions

In Sections 6.2–6.4 probabilities of particular events were evaluated. Often it is interesting to specify the probabilities of *all* possible events belonging to a particular sample space. When the events are all of a quantitative nature, the set of events together with their probabilities is called a *probability distribution*.

Example 6.14:   As in Example 6.12, a fair coin is tossed 3 times. What is the probability distribution for the number of heads?
The possible outcomes (where H = head and T = tail) are:

$$\text{HHH, HHT, HTH, HTT, THH, THT, TTH, TTT}$$

and each outcome is equally likely, since head and tail are equally likely. Thus, since there are eight outcomes in all, any individual outcome defines an event which has probability $\tfrac{1}{8}$. Now 3 of the events include exactly 2 heads (HHT, HTH and THH), 3 include exactly 1 head (HTT, THT, and TTH) whilst HHH is the only event with 3 heads and TTT the only event with no heads. All events are mutually exclusive, since it is impossible to get, say, HHH *and* HHT at the same time.
The addition law then gives the results:

$$P(3 \text{ heads}) = \tfrac{1}{8}$$
$$P(2 \text{ heads}) = \tfrac{1}{8} + \tfrac{1}{8} + \tfrac{1}{8} = \tfrac{3}{8}$$
$$P(1 \text{ head}) = \tfrac{1}{8} + \tfrac{1}{8} + \tfrac{1}{8} = \tfrac{3}{8}$$
$$P(0 \text{ head}) = \tfrac{1}{8}$$

Table 6.2   Probability distribution for the number of heads in 3 tosses of a fair coin.

x	P(X = x)
0	$\frac{1}{8}$
1	$\frac{3}{8}$
2	$\frac{3}{8}$
3	$\frac{1}{8}$
Total	1

It is useful to have a shorthand way of writing all the probabilities. The conventional way is to define a general event X which, for Example 6.14, is the number of heads in 3 tosses of a fair coin. Then the lower case letter x denotes any particular value that X takes. The probability distribution can either be written in the form of a mathematical equation (examples follow later) or in the form of a table. Table 6.2 shows the probability distribution for Example 6.14.

It will be seen that Table 6.2 has a similar appearance to a relative frequency distribution (see Section 3.3.2). The easiest way to understand the difference between a relative frequency and a probability distribution is to think of the definition of probability through limiting frequencies. If one were to actually pick up a coin, toss it three times and then repeat the process a further seven times, the frequencies of 0, 1, 2 and 3 heads would probably not be 1, 3, 3, 1 exactly, although they might well be fairly close. However if the process of three tosses were repeated hundreds of times the relative frequencies would almost certainly agree very closely with $\frac{1}{8}, \frac{3}{8}, \frac{3}{8}, \frac{1}{8}$. The probability distribution is the limiting form of the relative frequency distribution as the total number of observations gets infinitely large, just as a single probability was defined to be the limiting form of a proportion in Section 6.2.2.

The variable X which appears in a probability distribution is known as a *random variable*; that is, a variable which has a number of possible values, and the particular value taken cannot be stated for certain in advance. Just as with (relative) frequency distributions, random variables can either be discrete or continuous. Example 6.14 gave a simple example of a discrete random variable, and in Section 6.6 some general types of discrete random variables will be described. Continuous random variables are altogether more complicated, and will be deferred until Section 6.10.

6.6   Discrete random variables

As with any discrete variable, a discrete random variable can only take a limited set of values. Most often these values will either be 1, 2, 3, 4, etc. or 0, 1, 2, 3, 4, etc., as will be the case in all the examples that follow.

Discrete random variables can be classified into a number of types, each type arising under its own particular set of circumstances. In this section the most commonly occurring types will be derived, but there are many other types beyond those covered here. Each type has its own mathematical formula to represent the associated probability distribution.

One factor is common to all types of discrete random variable. Provided the outcomes are all mutually exclusive, Section 6.4.3 shows that:

$$\Sigma\, P(X = x) = 1$$

This equation states that 'the sum of the probabilities of all the individual events is one'. Note that the discrete random variable in Example 6.14 has its probabilities correctly summing to one.

6.6.1 The Bernouilli distribution

This is the simplest of all probability distributions, and arises when a single trial is carried out which can have one of only two possible outcomes. It does not matter what these outcomes are labelled, but it is convenient to call them 'success' and 'failure'. Let the probability of a success be π (the Greek letter pi).

The Bernouilli random variable, X, is defined to be the number of successes observed. Since there is only one trial the number of successes can only be either 0 or 1, and hence the probability distribution for X is:

$$P(X = 0) = 1 - \pi$$
$$P(X = 1) = \pi$$

The only unknown in this specification is π. Such a quantity is called a *parameter* of the distribution. The parameter, π, will take different values in different practical applications of the Bernouilli random variable, although because it is a probability, it must always have a value between 0 and 1.

Besides the simple example which follows, examples of the Bernouilli would include the sex of a newborn baby (e.g., success = female) and the response of a patient to an anaesthetic (e.g., success = no nausea).

Example 6.15: Toss a fair coin and regard the outcome 'head' as a success. Then P(success) = π = $\frac{1}{2}$, and if X is the number of heads observed then:

$$P(X = 0) = \frac{1}{2}$$
$$P(X = 1) = \frac{1}{2}$$

which specifies the probability distribution of the number of heads in the single trial.

6.6.2 The binomial distribution

Suppose, as before, that a trial can have two possible outcomes, which are success and failure. Further suppose that a specified number of such trials is to be carried out, and call this number n. Assume that the probability of a success on each trial is the same, and let this probability be π. Assume also that all trials are conducted independently.

The binomial random variable, X, is defined to be the number of successes observed in the n trials. X can clearly take any value between 0 and n inclusive. To derive P(X = x) for any x between 0 and n, the formula for combinations (see Section 6.3.1) will be employed.

Consider first the probability of getting an outcome with x successes in a *particular* order. For instance, the outcome $S_1S_2S_3 \ldots S_x F_1F_2F_3 \ldots F_{n-x}$ where each S represents a success and each F represents a failure. Since P(S) = π, and F is the complement to S, it must be that P(F) = $1 - \pi$ and thus by the multiplication law for independent events, the probability of this ordered outcome is $\pi^x(1 - \pi)^{n-x}$.

In general however, it does not matter about the order of the successes and failures. X simply records the number of successes *regardless* of where they come in the sequence. Any other ordered outcome with x successes will occur with the same probability as before, so all that is needed is to multiply the above probability by the number of ways of selecting x places from amongst the n in which to place the successes. There are $\binom{n}{x}$, the number of ways of choosing x from n, such ways. Hence:

$$P(X = x) = \binom{n}{x} \pi^x (1 - \pi)^{n-x}$$
$$\text{for } x = 0, 1, 2, 3, \ldots, n$$

In the case of the binomial there are two parameters, n and π.

Besides the simple example that follows, examples of the binomial would include the number of females amongst 100 newborn babies, and the number of patients out of 50 who did not experience nausea after having an anaesthetic.

Example 6.16:   Toss a fair coin three times and take the outcome 'head' to be a success. Then P(success) = $\pi = \frac{1}{2}$ and n = 3. Also π is constant from trial to trial and each trial is independent of all the others. Let X be the number of heads in the three tosses, then:

$$P(X = x) = \binom{3}{x} \left(\frac{1}{2}\right)^x \left(\frac{1}{2}\right)^{3-x} \quad \text{for } x = 0, 1, 2, 3$$

To actually work out P(X = x) for any value of x, this value must simply be substituted into the formula, i.e.:

$$P(X = 0) = \binom{3}{0} \left(\frac{1}{2}\right)^0 \left(\frac{1}{2}\right)^3 = \frac{3!}{3!0!} (1) \left(\frac{1}{8}\right) = \frac{1}{8}$$

$$P(X = 1) = \binom{3}{1} \left(\frac{1}{2}\right)^1 \left(\frac{1}{2}\right)^2 = \frac{3!}{2!1!} \left(\frac{1}{2}\right) \left(\frac{1}{4}\right) = \frac{3}{8}$$

$$P(X = 2) = \binom{3}{2} \left(\frac{1}{2}\right)^2 \left(\frac{1}{2}\right)^1 = \frac{3!}{1!2!} \left(\frac{1}{4}\right) \left(\frac{1}{2}\right) = \frac{3}{8}$$

$$P(X = 3) = \binom{3}{3} \left(\frac{1}{2}\right)^3 \left(\frac{1}{2}\right)^0 = \frac{3!}{0!3!} \left(\frac{1}{8}\right) (1) = \frac{1}{8}$$

Notice that these results agree exactly with Example 6.14 where these probabilities were derived from first principles.

Notes:
(1) Suppose $X_1, X_2, \ldots, X_n$ are n independent Bernouilli random variables all with the same parameter π. Then if $Y = \sum_{i=1}^{n} X_i$ the random variable Y has a binomial distribution.
(2) The Bernouilli is really only a special case of the binomial in which n = 1.

6.6.3 The geometric distribution

Suppose, once again, that a trial can have one of only two outcomes, success or failure. Assume that the probability of a success in each trial is the same, π, and each trial is independent of every other trial when repeated trials are carried out.

The geometric random variable, X, is defined to be the number of trials up to and including the first success. For the first success to be at the xth trial there must be a run of $x - 1$ consecutive failures followed by a success. That is, a sequence of the form $F_1F_2 \ldots F_{x-1}S$ must result. Since all trials are independent, the multiplication law states that:

$$P(X = x) = (1 - \pi)^{x-1} \pi$$
$$\text{for } x = 1, 2, 3, 4, \ldots$$

As with the Bernouilli distribution, the geometric has the single parameter π.

Examples include counting the number of newborn babies until the first female is born.

Example 6.17: Toss a fair coin and take the outcome 'head' to be a success. $P(\text{success}) = \pi = \frac{1}{2}$, is constant from trial to trial and also all trials are mutually independent, just as in Example 6.16. Let X be the number of trials until the first head, then:

$$P(X = x) = (\tfrac{1}{2})^{x-1}(\tfrac{1}{2}) = (\tfrac{1}{2})^x \quad \text{for } x = 1, 2, 3, \ldots$$

Again, particular values of x can be substituted into the formula to give particular probabilities, e.g.:

$$P(X = 1) = (\tfrac{1}{2})^0 (\tfrac{1}{2}) = \tfrac{1}{2}$$
$$P(X = 2) = (\tfrac{1}{2})^1 (\tfrac{1}{2}) = \tfrac{1}{4}$$

etc. That is, the chance of getting a head in the first trial is $\frac{1}{2}$, as would be expected, but the probability of not getting a head until the second trial is $\frac{1}{4}$, etc.

6.6.4 The uniform distribution

Suppose that a trial can have many possible outcomes, and let these outcomes be the numbers $1, 2, 3, \ldots, k$ so that there are k outcomes in all. Suppose that all of these outcomes are equally likely. Let X be the number of the outcome that is obtained when a single trial is carried out. X then has a uniform distribution.

By the definition of probability for equally likely outcomes it must be that:

$$P(X = x) = \tfrac{1}{k}$$
$$\text{for } x = 1, 2, 3, \ldots, k$$

The value k is the parameter of the uniform distribution.

Example 6.18: Roll a fair die. A die has six sides, each of which is equally likely to finish face upwards. Let X be the number of spots on the face showing, then:

$$P(X = x) = \tfrac{1}{6} \quad \text{for } x = 1, 2, 3, 4, 5, 6$$

This is a uniform random variable with k = 6.

Note: When success and failure are equally likely, a Bernouilli becomes a uniform random variable with k = 2, as in the case of tossing a fair coin.

6.6.5 The Poisson distribution

Consider a random variable, X, which is the total number of events of a certain type that occur within a given time interval, or within a given region of space (e.g., the number of employees getting married in a year, the number of bacteria in a culture). If the events occur randomly and independently of each other, and if the events occur uniformly insofar as the probability of a single event in a small interval (of time or space) is proportional to the size of that interval, then the random variable may be shown to have the *Poisson distribution*, for which:

$$P(X = x) = \frac{e^{-\lambda}\lambda^x}{x!}$$
$$\text{for } x = 0, 1, 2, \ldots$$

where e is the mathematical constant 2.71828 (to 5 decimal places) which is available on many calculators, and λ (the Greek letter lambda) is the parameter representing the mean number of occurrences of the event per unit of time or space.

Although the Poisson is a very important random variable, its distribution is quite difficult to derive, unlike the other random variables discussed so far. Hence the derivation is not included here, although it can be found in Chapter 12 of Folks (1981), and elsewhere.

Example 6.19:   Inpatients arrive at a hospital's Accident and Emergency Department at a mean rate of 0.1 per hour. Assuming that the conditions of the Poisson distribution (independent, uniform random occurrence) hold, what is the probability distribution of the number of arrivals tomorrow if the ward is open for the full 24 hours?

The mean number of arrivals in a 24 hour period is $\lambda = 24 \times 0.1 = 2.4$ (since 0.1 arrive per hour on average). Hence, if X is the number of arrivals in a 24 hour period:

$$P(X = x) = \frac{e^{-2.4}\,2.4^x}{x!} \quad \text{for } x = 0, 1, 2, \ldots$$

As in other cases, particular probabilities can be derived from the formula. For example, to calculate the probability that 4 patients arrive tomorrow, x is given the value 4 in the above formula, i.e.,

$$P(X = 4) = \frac{e^{-2.4}\,2.4^4}{4!} = 0.125$$

6.7 The cumulative distribution function

In many situations the probability that is of interest is not the probability of a particular outcome, $P(X = x)$, but instead that X exceeds a certain value, or is not more than a certain value.

The probability that X is not more than the value x is written as:

$$F(x) = P(X \leqslant x)$$

and F(x) is called the *cumulative distribution function*. The set of values of x and F(x) is clearly analogous to a cumulative relative frequency distribution (see Section 3.3.3).

If interest lies in the probability that X exceeds the value x, the law of probability for complementary events (Section 6.4.2) gives the result:

$$P(X > x) = 1 - P(X \leqslant x) = 1 - F(x)$$

Example 6.20:   The Accident and Emergency Department of Example 6.19 always has 4 empty beds available at the start of the day. Assume all patients who are admitted stay for the remainder of the 24 hour period and are then either discharged or transferred to another ward. The probability that 4 patients are admitted tomorrow has already been calculated, but what is of more interest is the probability that more patients arrive than there are beds to receive them. In other words, if X is the number of arrivals in a 24 hour period (once again), the value $P(X > 4)$ is required.

$$P(X > 4) = P(X = 4) + P(X = 5) + P(X = 6) + \ldots$$

where the number of terms to be added has no end. Even if this string of additions were only followed until the terms got very small, it is generally easier to use the law for complementary events:

$$P(X > 4) = 1 - P(X \leqslant 4) = 1 - F(4)$$

where

$$
\begin{aligned}
F(4) &= P(X = 0) + P(X = 1) + P(X = 2) + P(X = 3) + P(X = 4) \\
&= \frac{e^{-2.4} 2.4^0}{0!} + \frac{e^{-2.4} 2.4^1}{1!} + \ldots + \frac{e^{-2.4} 2.4^4}{4!} \\
&= e^{-2.4} \left(1 + 2.4 + \frac{2.4^2}{2} + \frac{2.4^3}{6} + \frac{2.4^4}{24} \right) \\
&= 0.0907 (1 + 2.4 + 2.88 + 2.304 + 1.3824) \\
&= 0.904
\end{aligned}
$$

Hence the probability that more patients arrive than there are beds to receive them is $1 - 0.904 = 0.096$. Another way of looking at this is to say that on almost 10% of days there will be a shortfall of beds. Presumably on such occasions beds must be borrowed from other wards.

6.8 Probability distributions as models

In the last sections, the idea of a discrete probability distribution has been developed, and before continuing this development it is useful to pause and think about the practical relevance of this topic.

In Section 6.5 a (discrete) probability distribution was introduced as the 'long-run' version of a relative frequency distribution, and this is a useful concept to keep in mind. However, it may well be asked why probability distributions are needed when it is much simpler to observe data and create a frequency distribution, as in Chapter 3. The answer is that it is rarely possible to collect *all* the data on a subject, for example all the lengths of stay of every patient that has been or will be admitted to a given hospital ward. The problem then is that the data collected may be special in some particular way. For example, the length of stay data in Table 3.2 have many repeated values of 24 days. This may well be purely by chance. If another hundred sets of data were collected from the same ward it could be that none of them would have even a single observation of 24 days. This is an important concept which is a fundamental issue in the study of statistics; the data collected are, almost always, merely one of many possible sets of data that *could* have been collected. Usually the goal is to generalize from the particular case.

In the case of probability distributions this generalization is essentially a smoothing process, just as where the jagged outline of a histogram becomes a smooth curve when enough data have been collected (see Section 3.11.3). In reality this smooth curve cannot be 'observed' but it can be approximated, or 'modelled', from the observed data.

A second advantage of the probability distributions met in Section 6.6 is that they have mathematical formulae. This enables general solutions to problems to be worked out, which then can be utilized once data have been collected. Typically this involves substitution of the value of one or more parameters, as illustrated by Examples 6.15–19. Using such formulae is much easier than having to deal with the individual numbers in a relative frequency distribution.

Despite these advantages it is, nevertheless, as well to always remember that the probability distribution is merely a model of real life. There are occasions, such as the simple coin-tossing examples (6.15–17), when the model is likely to be extremely good. On other occasions there may be reason to doubt the model. For example, is the Poisson model of accident admissions of Example 6.19 really appropriate? In particular, do accident cases really arrive independently, as required by the conditions of the Poisson? Is it not likely that some accident victims will arrive together, for example after car crashes? Also, it is almost certain that some patients will leave the Accident and Emergency Department before the end of the 24 hour period, so some beds may become available again. If an assumption of a particular probability distribution is made, as will be done repeatedly in Chapter 7, then any conclusions will necessarily be subject to this assumption. In Section 9.7.1 a method for deciding whether a probability distribution adequately fits a set of observed data will be discussed.

6.9   Mean and variance of a discrete random variable

In Chapter 4 summary statistics were used as shorthand descriptions of observed data. Just as means and variances were defined for frequency distributions, they will now be defined for probability distributions, and they can be used in an analogous way.

The mean of a random variable, X, will be given the symbol μ (the Greek letter mu). Sometimes μ is also called the *expected value* or *expectation* denoted E(x). In Section 4.3.1 the mean, $\bar{x}$, of a frequency distribution was calculated as:

$$\bar{x} = \Sigma\, x_i \frac{f_i}{n}$$

where the f_i/n are the relative frequencies. For discrete random variables μ is similarly defined:

$$\mu = \Sigma\, x_i\, P(X = x_i)$$

where x_i are all the possible values of the random variable X. Notice that $P(X = x_i)$ has simply taken the place of f_i/n, reflecting the fact that a probability is the limiting form of a relative frequency.

Similarly, the variance of a random variable is represented by σ^2, where σ is the Greek letter sigma. Now, for frequency distributions:

$$s^2 = \frac{1}{n-1} \Sigma\, f_i (x_i - \bar{x})^2$$

from Section 4.6.1 which is virtually the same as:

$$\Sigma\, (x_i - \bar{x})^2 \frac{f_i}{n}$$

except that the 'n' should really be 'n – 1' (see Section 7.2 for an explanation of the use of n – 1). By analogy, the variance of a discrete random variable X is defined to be:

$$\sigma^2 = \Sigma\, (x_i - \mu)^2\, P(X = x_i)$$

Example 6.21: Consider the process of tossing a coin three times and counting the number of heads. The resulting probability distribution was given in Table 6.1 and from this can be calculated:

$$\mu = \Sigma\, x_i\, P(X = x_i)$$
$$= 0 \times P(X = 0) + 1 \times P(X = 1) + 2 \times P(X = 2) + 3 \times P(X = 3)$$
$$= 0 \times \tfrac{1}{8} + 1 \times \tfrac{3}{8} + 2 \times \tfrac{3}{8} + 3 \times \tfrac{1}{8}$$
$$= \tfrac{12}{8} = 1.5$$

This says that the average number of heads obtained (in the 'long-run') when a coin is tossed three times is 1.5. Sometimes it is said that 1.5 heads can be *expected* from three tosses of a coin. Notice that this expectation has turned out to be fractional, although it is physically impossible to observe half a head! This can also happen with $\bar{x}$, as noted in Section 4.3.4. Also:

$$\sigma^2 = \Sigma\, (x_i - \mu)^2\, P(X = x_i)$$
$$= (-1.5)^2\, P(X = 0) + (-0.5)^2\, P(X = 1)$$
$$+ (0.5)^2\, P(X = 2) + (1.5)^2\, P(X = 3)$$
$$= 0.75$$

6.9.1 Formulae

By exploiting the mathematical expressions for the various probability distributions of Section 6.6, together with the mathematical expressions for means and variances, formulae for these summary statistics can be derived. These derivations are beyond the scope of this book, and hence only the results are given here. The formulae depend upon the parameter, or parameters, of the probability distribution, and are given in Table 6.3.

Table 6.3 Mean and variance for some discrete distributions.

Distribution	Parameter(s)	Mean	Variance
Bernouilli	π	π	$\pi(1 - \pi)$
Binomial	n, π	$n\pi$	$n\pi(1 - \pi)$
Geometric	π	$1/\pi$	$(1 - \pi)/\pi^2$
Uniform	k	$(k + 1)/2$	$(k^2 - 1)/12$
Poisson	λ	λ	λ

Example 6.22:   What is the average score when a die is rolled?

Since the result of the roll of a die has a uniform distribution with $k = 6$ (see Example 6.18) the mean score is:

$$\frac{(k + 1)}{2} = \frac{7}{2} = 3.5$$

Since each of the numbers from 1 to 6 is equally likely to appear, this seems an intuitively reasonable result.

Example 6.23:   When a coin is tossed three times how many heads can be expected?

Since the result of three coin tosses has a binomial distribution with $n = 3$ and $\pi = \frac{1}{2}$ (see Example 6.16), the expected number of heads is:

$$n\pi = 3 \times \frac{1}{2} = 1.5$$

Notice that this result agrees with the derivation in Example 6.21.

6.10 Continuous random variables

In Table 3.5 the continuous variable age was presented in the form of a grouped frequency distribution. It is rarely, if ever, sensible to present a continuous variable as an ungrouped frequency distribution unless all values are rounded, which is effectively the same as grouping. This is because the values taken are likely to be too sparse for any discernible pattern to emerge.

Similarly, continuous probability distributions are presented in a grouped form, this form conventionally being the cumulative distribution function. In fact, a continuous probability distribution cannot be represented in an ungrouped form because of the fact that $P(X = x) = 0$ *always*, regardless of the value of x. When first met, this seems a strange result, but one simple way of seeing that it is reasonable is to think of every value x occurring with the same chance. By the definition of probability for equally likely outcomes:

$$P(X = x) = \frac{1}{k}$$

where k is the number of outcomes for X. However, when X is continuous, k is infinity – there is no limit to the number of values for x. Thus:

$$P(X = x) = \frac{1}{\infty} = 0$$

where ∞ is the mathematical sign for infinity.

There is a particularly simple interpretation of the cumulative distribution function, $F(x) = P(X \leqslant x)$ which will now be useful. In Section 3.11.3 the outline of a histogram was represented as a frequency curve in the limit as more data were collected and class intervals shrank. The frequency curve is, in fact, a graph of the continuous random variable. Then just as the area of any rectangle in a histogram is in proportion to the associated frequency, the area between any two x values in a frequency curve is proportional to the associated probability. In fact, the overall area under a frequency curve must be one, since otherwise it could not be a true representation of probability. Hence, the area under the frequency curve *is* the associated probability.

The frequency curve can itself be represented by a mathematical equation. A more usual name for this is the *frequency density function,* denoted f(x). The more mathematically experienced reader will realize that f(x) is the derivative of F(x). Figure 6.2 illustrates the relationship between f(x) and F(x) for an arbitrary probability distribution which takes non-zero values between 0 and 15. In order to be able to use continuous random variables later in this book, it is sufficient for the reader to understand this diagram.

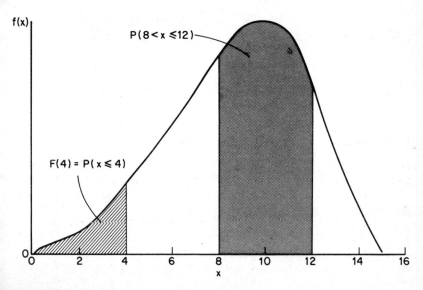

Fig. 6.2 Probability density function of a random variable X.

As with discrete random variables, there are very many types of continuous random variables. Three of the most important ones will be described, including the normal, which is by far the most important of all.

Since $P(X = x) = 0$ for all x, the mean and variance of a continuous random variable cannot be defined as they were for discrete random variables. The precise definition involves the use of mathematical calculus which is beyond the scope of this book. However, the mean and variance have exactly the same interpretation as for the discrete case, and the values will be given for the three random variables discussed. (Details of their derivation are given in more advanced texts such as Clarke and Cooke (1983), Chapter 13.)

6.10.1 The continuous uniform distribution

This is the continuous analogue of the uniform distribution. Now the range of values that the random variable X can take is continuous, but still X is equally likely to come anywhere within the range allowed for it.

Suppose that the range allowed for X is from a to b. Then it turns out that the probability density function for X is:

$$f(x) = \frac{1}{b-a} \text{ for } a \leqslant x \leqslant b,$$

$$= 0 \text{ for other values of x}$$

Also the cumulative distribution function is

$$F(x) = \frac{x-a}{b-a} \text{ for } a \leqslant x \leqslant b$$

$$= 0 \text{ for } x < a \text{ and } 1 \text{ for } x > b$$

The parameters of this distribution are a and b, as can be seen from the equations. The mean is $(b + a)/2$ and the variance $(b - a)^2/12$. The shape of

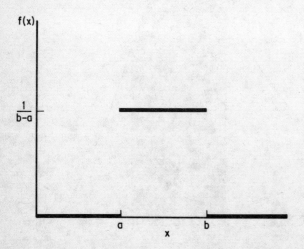

Fig. 6.3 The continuous uniform distribution.

the continuous uniform is shown by Fig. 6.3. Notice that this 'frequency curve' is actually three straight lines separated by gaps.

Example 6.24: A doctor measures patients' heights in centimetres, only recording heights to the nearest centimetre. What is the probability that a height is not overstated by more than 0.1 cm, if all errors in measurement are equally likely?

Let X be the error. The limits for X are −0.5 cm and 0.5 cm because of the practice of rounding to the nearest centimetre. The probability required is that X is less than or equal to 0.1, that is F(0.1)

Now
$$F(x) = \frac{x-a}{b-a} \text{ for } a \leqslant x \leqslant b$$

and so
$$F(0.1) = \frac{0.1 - (-0.5)}{0.5 - (-0.5)} = \frac{0.6}{1} = 0.6$$

6.10.2 The exponential distribution

This is a useful distribution for modelling lifetimes or times between events of the type which have a Poisson distribution. The exponential has parameter λ and probability density function:

$$f(x) = \lambda e^{-\lambda x} \quad \text{for } x \geqslant 0,$$
$$= 0 \quad \text{otherwise.}$$

The cumulative distribution function is:

$$F(x) = 1 - e^{-\lambda x} \quad \text{for } x \geqslant 0,$$
$$= 0 \quad \text{otherwise.}$$

The mean of the exponential is $1/\lambda$ and the variance $1/\lambda^2$. Figure 6.4 shows the shape of this distribution.

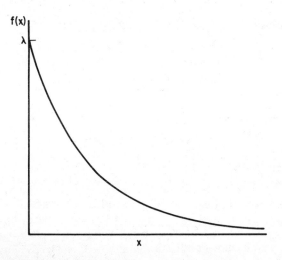

Fig. 6.4 The exponential distribution.

Example 6.25: In the Accident and Emergency ward of Example 6.19, what is the probability that no patients arrive during a period of 8 consecutive hours, assuming that times between patients arriving follow an exponential distribution?

From Example 6.19 the mean number of arrivals per hour is 0.1 and hence the mean time between arrivals must be $1/0.1 = 10$ hours. Thus $1/\lambda = 10$ which implies that $\lambda = 0.1$. If X is the time between arrivals, the probability required is:

$$
\begin{aligned}
P(X > 8) &= 1 - F(8) \\
&= 1 - (1 - e^{-8\lambda}) \\
&= e^{-0.8} = 0.45
\end{aligned}
$$

6.11 The normal distribution

The normal distribution is by far the most important of all probability distributions. There are two reasons why this is so. First, many physical processes follow, or at least approximately follow, the normal distribution. Second, a mathematical theorem (see Section 6.11.3) states that the mean of a sample of independent observations from *any* random variable will tend, as sample size increases, to have a normal distribution.

For the normal to be a reasonable model, the data must approximate to the normal curve illustrated by Fig. 3.14. That is, the data must:

(i) show a tendency to group around a central point (the mean);
(ii) take values either side of this point with approximately equal frequency;
(iii) have the frequency of values falling off as values get further from this point.

Sets of data from many applications have been found to have these properties, and the normal has been used successfully to model such things as heights, temperatures, hours of relief from pain, errors in measurement, etc.

6.11.1 Evaluating probabilities

Due to its importance, tables of probabilities have been produced for the normal distribution, such as Table A.1 in Appendix 4. Such tables simply remove the need to make specific calculations from the probability density or cumulative distribution functions. This is especially useful since these have quite complex forms; for instance the probability density function for the normal is:

$$
f(x) = \frac{1}{\sqrt{2\pi}\,\sigma}\, e^{-\frac{1}{2}\left(\frac{x-\mu}{\sigma}\right)^2} \quad \text{for } -\infty < x < \infty
$$

Here π is the mathematical constant 3.1415... usually associated with circles, and μ and σ are the parameters of the distribution. As would be expected, μ is the mean and σ the standard deviation (but π was used to mean something different earlier in this chapter). Notice that the normal can have x values of any size, over the complete continuous scale from minus to plus infinity.

Changing the value of μ and/or σ changes the position and/or shape of the

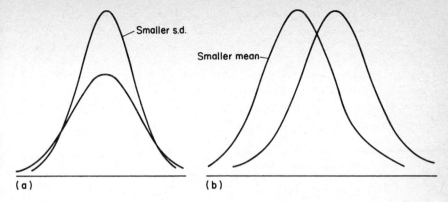

Fig. 6.5   The effect of changing the parameters of the normal distribution.

normal curve, as is illustrated by Fig. 6.5. Consequently, just as for any other probability distribution, the probabilities taken by a normal random variable change as these parameters change. This would seem to require a different table of probabilities for every different pair of values for μ and σ, which would be an enormous collection. Instead of this, a table is given for only one normal distribution, the *standard normal* with mean 0 and standard deviation 1. There is a simple method of transforming any other normal to this standard normal to enable the table to be used.

This method of standardization is to subtract the mean and divide by the standard deviation. The standard normal random variable is given the symbol Z. Hence, given a normal random variable X with mean μ and standard deviation σ:

$$Z = \frac{X - \mu}{\sigma}$$

The following examples show how the table of the standard normal and the transformation are used.

Example 6.26:   From the standard normal Z find:

(i) the percentage of values below 1.96;
(ii) the probability that Z lies above 1.64;
(iii) the probability that Z lies between zero and 1.96;
(iv) the probability that Z lies between -1.96 and 1.96;
(v) the value, z, such that 45% of all the positive values taken by Z lie between it and zero.

Answers:

(i) $P(Z < 1.96) = 0.975$ from Table A.1. Hence 97.5% of values taken by the standard normal are below 1.96.   0.975 is simply the area to the left of 1.96 in a graph of the standard normal.
(ii) $P(Z > 1.64) = 1 - P(Z \leqslant 1.64) = 1 - 0.9495 = 0.05$ (to 2 decimal places). This is the area to the right of 1.64.

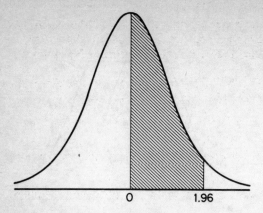

Fig. 6.6   Probability represented by the area under a standard normal curve.

(iii) $P(0 \leqslant Z \leqslant 1.96) = P(Z \leqslant 1.96) - P(Z < 0)$. This can be seen by thinking in terms of areas (see Fig. 6.2). The probability required is the area between two points, which can be found as the area to the left of the larger value minus the area to the left of the smaller (Fig. 6.6).

$$P(Z \leqslant 1.96) = 0.975$$
$$P(Z < 0) = 0.5$$

The latter is obvious because the normal is perfectly symmetric and so the mean and median must be equal. Hence

$$P(0 \leqslant Z \leqslant 1.96) = 0.975 - 0.5 = 0.475$$

(iv) $P(-1.96 \leqslant Z \leqslant 1.96) = P(Z \leqslant 1.96) - P(Z < -1.96)$
$$= 0.975 - 0.025 = 0.95$$

Alternatively, by symmetry, the probability that Z lies between -1.96 and 0 must be equal to the probability that Z lies between 0 and 1.96. Hence the probability required is twice that of part (iii), that is $2 \times 0.475 = 0.95$.

(v) The value z is required such that:

$$P(0 \leqslant Z \leqslant z) = 0.45$$

i.e. $$P(Z \leqslant z) - P(Z < 0) = 0.45$$

Hence $$P(Z \leqslant z) = 0.45 + 0.5 = 0.95$$

Then from Table A.1 (or from part (ii)), z takes the value 1.64.

Notice that in evaluating normal probabilities there is really no need to distinguish between $P(Z \leqslant z)$ and $P(Z < z)$ since the exact probability $P(Z = z)$ is zero in all cases.

Example 6.27:   A bloodbank is requested to provide an average of 144 pints of blood each day, with a standard deviation of 10 pints. The capacity of the bloodbank has been worked out assuming that not more than 160 pints per day will be requested. On what percentage of days will this 160 pint limit be exceeded?

Let X be the amount of blood requested in a day. Then X has mean 144 pints and standard deviation 10 pints. Experience has shown that a normal distribution is often a reasonable model for such quantities as X, so this will be assumed.

The probability required is $P(X > 160)$. To be able to use Table A.1 this probability must be transformed to a standardized form, i.e.:

$$P(X > 160) = P\left(\frac{X - \mu}{\sigma} > \frac{160 - \mu}{\sigma}\right)$$

(remember the same mathematical operation must be performed on both sides to retain the balance).

But now $\frac{X - \mu}{\sigma} = Z$, the standard normal; that is, the probability required is:

$$P\left(Z > \frac{160 - \mu}{\sigma}\right) = P\left(Z > \frac{160 - 144}{10}\right)$$
$$= P(Z > 1.6)$$
$$= 1 - 0.9452 \text{ (from Table A.1)}$$
$$= 0.05 \text{ (to 2 decimal places).}$$

Hence the figure of 160 pints will only be exceeded on 5% of days. 160 pints might thus be termed a 95% protection level.

6.11.2 Sampling distributions

In the last few sections, attention has been focused upon the distribution of values taken by a random variable, and it was seen that this is really an idealized version of the relative frequency distribution of a sample of observations, as met in Chapter 3. At this point these ideas can be taken a step further. When summary statistics were introduced in Chapter 4, these were summaries of a single set of data. If a second sample were taken it is unlikely that any of the summary statistics would be exactly the same as in the first sample, and in fact there are likely to be almost as many values taken by a summary statistic as there are different samples. These summary statistics themselves could be given a relative frequency distribution, and so they too have a probability distribution. This, like all others, has a mean and standard deviation, although it is more usual to call the standard deviation of a summary statistic the *standard error* (but the square of the standard error is still the variance).

For example, the lengths of stay in a hospital ward could be collected every month, and a mean calculated for each month's figures. After a year this would give 12 values of $\bar{x}$ all representing the same thing (assuming that there is no change in ward administration or pattern of use over time). This gives a frequency distribution of 12 values. Over many years this (relative) frequency distribution would tend toward the probability distribution. Even if time did have some important effect many samples could, for instance, be drawn even within a single year. If samples of size 30 were selected from amongst 400 inpatients using the ward in a given year there would be many millions of potential different samples. The exact number is $\binom{400}{30}$.

Of course, in reality only one sample would ever be taken; for instance if different samples were taken in different months they would simply be amalgamated into one big sample with a single $\bar{x}$. It does not matter that repeated sampling is not actually performed, it is sufficient to note that it *could* be. Hence the $\bar{x}$ (or otherwise) observed is really only one of a number of different $\bar{x}$s that could have been observed. As has been done elsewhere in this chapter, wherever a random variable is being discussed a capital letter is used, whereas a lower-case letter is used to denote a value taken by that random variable. Hence, in the observed sample, $\bar{X}$ takes the value $\bar{x}$.

The concept of summary statistics from samples having themselves a sampling distribution is fundamental to the understanding of statistical inference, which is the topic of the next chapter. It is also essential to the understanding of the *central limit theorem*.

6.11.3 The central limit theorem

This theorem states that, for a sample of independent observations, if the sample size, n, is big enough, $\bar{X}$ will follow a normal distribution very closely. The word 'central' is included in the title because this is the central, or crucial, result underpinning virtually all of standard statistical inference and, as already explained, it ensures the importance of the normal distribution.

In fact, whatever the distribution of X, if X has mean μ and variance σ^2, then $\bar{X}$ will, for large n, have a normal distribution with mean μ and variance σ^2/n. Notice that whilst the mean of the mean is still μ, the variance shrinks by a factor of $1/n$. Intuitively the latter result is reasonable since a mean is calculated by taking into account low and high figures which tend to cancel each other out. The mean should therefore have less spread than the original observations, and the bigger the sample size the more effective should be this process of cancellation.

In the statement of the central limit theorem, n has to be 'big enough'. It is not possible to say how big n has to be before a normal results, because it varies from case to case. Certainly if n is in the hundreds there is unlikely to be a problem whatever the distribution of X, but if X is itself normal, n even as small as 1 is sufficiently big. It really all depends on how close to the normal X itself is; a highly skewed X will probably require n to be more than 100 before $\bar{X}$ is approximately normal.

A point of warning is appropriate here. Some statistics textbooks will present the normal distribution as the panacea for all ills. Although it does give good results in a surprisingly large number of cases, there are some situations in which the normal is definitely not the appropriate distribution to use, such as in the examples of Sections 6.6 and 6.10 and where the sample size is too small for the central limit theorem to hold.

6.12 Distributions derived from the normal

As well as the normal distribution, there are three further distributions, all derived in some way from it, which are used in statistical inference and tabulated in Appendix 4. These distributions will merely be defined here; their use will be described in Chapters 7–9. To conserve space the tables in the appendix for each distribution are not complete. Intermediate values have to be estimated. More detailed tables are given by Fisher and Yates (1957).

6.12.1 *The chi-squared distribution*

If $Z_1, Z_2, \ldots, Z_n$ are n independent standard normal random variables, then:

$$Z_1^2 + Z_2^2 + \ldots + Z_n^2$$

follows a chi-squared distribution with n degrees of freedom.

The figure with Table A.3 shows the right-skewed shape of the chi-squared, written χ^2, distribution (χ is the Greek letter chi). The shape, and hence the associated probabilities, alter with degrees of freedom. As degrees of freedom increase so the spread of the χ^2 increases. In Table A.3, χ^2 probabilities are given; notice that they do depend upon the degrees of freedom.

6.12.2 The Student's t distribution

If Z is a standard normal random variable and χ_n^2 is a chi-squared random variable with n degrees of freedom, then, provided Z and χ_n^2 are independent:

$$\frac{Z}{\sqrt{\chi_n^2/n}}$$

follows a Student's t distribution, sometimes simply called a t distribution, with n degrees of freedom.

The figure with Table A.2 shows the shape of the t distribution. It is perfectly symmetrical and as the degrees of freedom increase, the spread of the central peak narrows. With degrees of freedom of 30 or more, the t is virtually indistinguishable from the normal. Table A.2 expresses probabilities for the t distribution. Notice how closely the probabilities agree with those in Table A.1 at the higher degrees of freedom.

6.12.3 The F distribution

If χ_n^2 is a chi-squared random variable with n degrees of freedom and χ_m^2 is independent of this and is a chi-squared random variable with m degrees of freedom, then:

$$\frac{\chi_n^2/n}{\chi_m^2/m}$$

follows an F distribution with (n, m) degrees of freedom.

Like the χ^2, the F distribution is right-skewed. Since there are two components to the degrees of freedom, the table of probabilities, Table A.4, needs to be indexed by both the first and second component. Notice that it is important to read these in the correct order.

Exercises

6.1   (a) Give the set of all possible outcomes (i.e., the sample space) when a card is drawn from a standard pack of playing cards.

    (b) Are these outcomes equally likely?

    (c) What is the probability of drawing an ace?

6.2 (a) Give the set of all possible outcomes of the score resulting from rolling two fair dice (that is the sum of the two upturned faces).

    (b) Are these outcomes equally likely?

    (c) What is the probability of scoring six?

    (d) What is the probability of rolling an even score?

6.3 A health authority librarian has the last 6 years' copies of 'Community Medicine' bound in 6 identical binders, one binder for each year. Unfortunately the year is omitted from the title embossed on the spine of each binder. When the binders are placed on the library shelves without checking, what is the probability that they line up in the correct order (in ascending years from left to right)?

6.4 A medical supplier provides thermometers in boxes of 40. A hospital storeman takes a sample of 4 at random from a box and tests them against a standard. He will accept the box if none of the sampled thermometers disagrees with the standard.

    (a) How many different samples of size 4 may be drawn from the 40 thermometers in the box?

    (b) If the box has 5 defective thermometers, how many different samples of size 4 that contain no defective thermometers may be drawn?

    (c) Use the results of (a) and (b) to determine the probability that the storeman accepts a box containing 5 defective thermometers.

    (d) Use similar arguments to determine the probability that the storeman rejects a box containing only one defective thermometer.

6.5 In a test a child with abnormally slow development is given a bucket containing 4 long and 8 short plastic rods of equal diameter. He is asked to pick out a long rod with his eyes shut. This he does correctly (although he is not told so) and the selected rod is taken away. He is then asked to select another long rod and, again, he is successful.

    (a) What is the probability that the two long rods have been correctly drawn purely by chance selection?

    (b) Suppose that the first long rod selected had been replaced in the bucket before the second selection was made. What would the answer to part (a) then become?

6.6 In each case below state, first, whether the events E_1 and E_2 are mutually exclusive and, second, whether they are independent.

    (a) E_1 = 'a six is scored'
        E_2 = 'an even number is scored'
        when a die is rolled.

    (b) E_1 = 'a six is scored'
        E_2 = 'an odd number is scored'
        when a die is rolled.

    (c) E_1 = 'a six is scored' when a die is rolled
        E_2 = 'a head results' when a coin is tossed.

    (d) E_1 = 'my brother has an accident in his car tonight'
        E_2 = 'my brother drinks alcohol tonight'.

    (e) E_1 = 'I have a nosebleed'
        E_2 = 'my GP is unavailable'.

6.7 An oil company plans to build a new refinery at a selected location. It is concerned about the situation of the proposed refinery in relation to nearby ambulance stations. The company feels that, in the event of a fire, it is essential that ambulances arrive within 10 minutes of an emergency call. Three ambulance stations are reasonably close by, and it has been calculated that they would have respective probabilities of 2/3, 3/5 and 1/4 of answering a call from the proposed refinery within 10 minutes. What is the probability that:

(a) all 3 ambulance stations answer a call within 10 minutes;
(b) none of the 3 stations answers a call within 10 minutes;
(c) at least one of the 3 stations answers a call within 10 minutes?

6.8 A certain disease has a prevalence of one in a thousand. A screening procedure is successful in detecting that a person has the disease in 99% of cases. When someone does not have the disease, however, the procedure wrongly identifies the disease one time in a thousand.

(a) What is the probability that someone who is identified by the screening procedure as having the disease does indeed have the disease?
(b) What is the probability that someone who is identified by the screening procedure as being free of the disease does, in fact, have the disease?
(c) Consider what information, in addition to (a) and (b), you would require to be able to decide whether the screening procedure should be used by a health authority.

6.9 An Accident and Emergency Department has enough beds to cope with 97% of daily workloads. Assume that workloads from day to day are independent.

(a) What is the probability that it is under-bedded on two days in a single week?
(b) How many days in a year can it expect to be under-bedded?

6.10 In a certain *in vitro* fertilization procedure there is a probability of 0.3 that a woman becomes pregnant after her eggs have been replaced. What is the probability that a woman has to have three batches of eggs replaced before she first becomes pregnant? What assumption has been used to derive this result? Is it likely to be true?

6.11 The number of telephone calls received at a switchboard is often modelled by a Poisson distribution. If a health authority receptionist receives an average of 120 calls per hour what is the probability that:

(a) at least one call is received in a minute;
(b) there is less than a 2 minute break between calls?

(Notice that the average number of calls *per minute* is 2.)

6.12 Patients with rheumatoid arthritis have a mean pulse rate of 81 per minute with a variance of 9. Assume that the pulse rate follows a normal distribution.

(a) What is the probability that a patient selected at random has a pulse rate of below 84 per minute?
(b) What proportion of patients have a pulse rate of between 72 and 90 per minute?
(c) When 100 patients are chosen to enter a clinical trial how many would be expected to have a pulse rate of above 85 per minute?

7
Inference

7.1 Samples and populations

In order to understand a physical process, such as the use of beds in a hospital, it is frequently beneficial to collect data on that process and analyse the data using the methods of Chapters 3 and 4. These types of analysis are sometimes collectively called *exploratory*, or *initial*, *data analysis*. Exploratory data analysis is often all that is required to explain the physical process satisfactorily. Sometimes, however, a more general analysis is called for – one which goes beyond the confines of the data themselves and makes general statements, or *inferences*, about the physical process in the universal case. In statistical jargon, a sample is used to make inferences about the population from which it has been drawn.

On some occasions an item of data is collected from every member of the population. This is when a census is carried out, for example the census of population. Censuses are relatively rare, not only because of economic considerations, but also because in many situations they are simply not possible. In evaluating the performance of a new drug to cure bronchitis, it would be impossible to measure the reaction of every person who has in the past, does in the present, and will in the future, suffer from bronchitis. Nevertheless a judgement of the efficacy of the drug in general is required, which can only be achieved by observing a sample of bronchitis sufferers, and then inferring a result for the entire population of bronchitis sufferers. In simple terms, then, inference implies the use of a sample to make an educated guess about the population from which it was drawn. In this chapter it will be assumed that the population is infinitely large, like the population of bronchitis sufferers, and that the sample is a simple random sample, that is every member of the population is equally likely to be sampled. Alternatives are considered in Section 10.9.

This chapter will deal with the two major techniques of statistical inference, estimation and hypothesis testing. These methods use the theory of probability to give a more formal analysis of sample data than is possible using Chapters 3 and 4. It is strongly recommended that their use should always be preceded by exploratory data analysis, since this gives a 'feel' for the data. Getting this feel is very important, both because it highlights important aspects of the data which may influence the inferential techniques used, and because it provides a check on the assumptions inherent in these techniques.

7.2 Introduction to estimation

Estimation is the process of using some summary statistic from the collected data to represent a feature of the parent population. Most often the feature to be estimated is the mean, a proportion or the variance, although the latter is usually only of indirect interest.

At the basic level, estimation is very simple. In Example 4.3 the mean length of stay for the sample of 56 patients from the hospital ward was 11.8 days, and this may be used as an *estimate* for the mean length of stay of patients staying in that ward. As was explained in Section 6.11.2, if a different sample of patients had been drawn the sample mean would almost certainly have been somewhat different from 11.8, and hence a different estimate for the population, or 'true' mean, would have resulted. This is an unfortunate but inevitable feature of any sampling process.

The crucial question when using an estimate is, 'how accurate is it?' Simply stating that the estimated mean length of stay is 11.8 days gives no clue about how well this estimates the true mean. It is usual to use estimates that are *unbiased*, such that the mean value of the estimate over many samples is the true value of the population parameter to be estimated. That is, on average the estimate is correct. The sample mean is an unbiased estimate of the population mean, as is the sample variance for the population variance and the sample proportion for the population proportion. Note that the sample variance is only unbiased if the denominator n – 1 is used (see Section 4.6).

'Unbiasedness', however, is a property concerned with repeated sampling. It says nothing about how accurate any one estimate calculated from a particular sample will be, and in reality only one sample and thus only one estimate will be available. The standard error, being the standard deviation of an estimate (itself a summary statistic), represents an average distance between an estimate and its mean, the true value (in the case of an unbiased estimate). So this at least represents the amount of error that an estimate can be expected to have. To many people the standard error is difficult to interpret and use so they prefer, instead, to specify limits of accuracy for the estimate. Since all estimation is done in a state of uncertainty, these limits of accuracy must be stated in the language of probability.

This leads to the distinction between two types of estimates. The simple, one number, estimate, such as the sample mean of 11.8 days in the example, is called a *point estimate*. The limits of accuracy of an estimate define an *interval estimate*, the most commonly used of which is the *confidence interval*.

7.3 Confidence intervals

The idea of a confidence interval is that two boundaries, the upper and lower limits of an interval, should be specified such that the interval they define has a predetermined probability of containing the true value of the unknown mean, proportion, variance or otherwise. The subject will be introduced using the particular example of finding a 95% confidence interval for an unknown population mean when the population variance is known. Other cases will follow.

In this section and in what follows, it is particularly important to distinguish between sample and population values. The notation that will be used agrees with previous chapters:

(i) $\bar{x}$ is the sample mean; μ is the population mean.
(ii) s is the sample standard deviation; σ is the population standard deviation (and the square of the standard deviation is the variance).
(iii) p is the sample proportion; π is the population proportion (or probability).

Also recall that capital letters denote random variables and small letters denote outcomes; for example where $\bar{x}$ is the value of $\bar{X}$ that is actually observed (only one of many possible realizations).

7.4 Confidence intervals for the mean

7.4.1 Confidence interval for μ when σ is known

Suppose that two numbers C_1 and C_2 can be specified such that

$$P(C_1 < \mu < C_2) = 0.95$$

That is, there is a 95% chance that the interval of numbers from C_1 to C_2 contains the value of μ. The interval (C_1, C_2) is then called the 95% confidence interval for μ.

Formulae for C_1 and C_2 can be derived using the central limit theorem (Section 6.11.3). This states that, assuming the sample size, n, is large enough, the sample mean, $\bar{X}$, is normally distributed with mean μ and variance σ^2/n. Hence, by Section 6.11.1, $\dfrac{\bar{X} - \mu}{\sigma/\sqrt{n}}$ has a standard normal distribution. From Table A.1 (see Example 6.26),

$$P(-1.96 < Z < 1.96) = 0.95$$

where Z as usual, represents the standard normal.

Hence, $$P\left(-1.96 < \frac{\bar{X} - \mu}{\sigma/\sqrt{n}} < 1.96\right) = 0.95$$

i.e., $$P(-1.96\,\sigma/\sqrt{n} < \bar{X} - \mu < 1.96\,\sigma/\sqrt{n}) = 0.95$$

i.e., $$P(\bar{X} - 1.96\,\sigma/\sqrt{n} < \mu < \bar{X} + 1.96\,\sigma/\sqrt{n}) = 0.95$$

Thus, the 95% confidence limits are:

$$C_1 = \bar{X} - \frac{1.96\,\sigma}{\sqrt{n}}$$

$$C_2 = \bar{X} + \frac{1.96\,\sigma}{\sqrt{n}}$$

That is, the 95% confidence interval (CI) is:

$$\left(\bar{X} - \frac{1.96\sigma}{\sqrt{n}}, \bar{X} + \frac{1.96\,\sigma}{\sqrt{n}}\right)$$

which is more simply written as

$$\overline{X} \pm \frac{1.96\,\sigma}{\sqrt{n}}$$

When a sample of data is collected and the particular outcome $\overline{x}$ calculated from it, the 95% CI derived from this sample is

$$\overline{x} \pm \frac{1.96\,\sigma}{\sqrt{n}}$$

From here on the expressions given for confidence intervals will be (like this one) those appropriate to sample data.

Example 7.1: A community based programme to control cardiovascular disease was carried out in a town. A multidisciplinary team consisting of health administrators, community physicians, general practitioners and surgeons devised a series of campaigns to promote healthy living amongst the town's population. These included 'fun runs', health food exhibitions, anti-smoking campaigns and the distribution of health education literature.

The team decided that one important measure of the outcome of the programme was the mean serum cholesterol concentration of the population. Ten years after the programme began, a sample of 50 people was selected in the town and their mean serum cholesterol concentration found to be 6.23 mmol/l. Assuming that the standard deviation of the concentration amongst the population was known to be 1.20 mmol/l, the 95% confidence interval for the mean serum cholesterol concentration in the town is:

$$\overline{x} \pm \frac{1.96\,\sigma}{\sqrt{n}}$$

$$= 6.23 \pm \frac{1.96 \times 1.2}{\sqrt{50}}$$

$$= 6.23 \pm 0.33 \text{ mmol/l}$$

Hence, there is 95% confidence that the interval 5.90 to 6.56 mmol/l contains the true mean concentration.

7.4.2 Interpretation

Before going on to other types of confidence interval it is useful to consider the interpretation of a CI, using Section 7.4.1. for illustration.

The proof in Section 7.4.1 showed that the interval $\overline{X} \pm 1.96\,\sigma/\sqrt{n}$ contains μ with a probability of 0.95. By reference to Chapter 6, this means that if a very large number of samples were taken, and the 95% CI for the mean calculated from each one, then 95% of these 95% CIs would actually contain the true value, μ. Notice that, because this definition is based on pro-bability, the interpretation of a CI has to be made in terms of repeated samples. Once again, it must be admitted that in reality only one sample will be used. By definition there is a 5%, or 1 in 20, chance that the particular CI obtained from this sample misses μ altogether. This is the price that must be paid for sampling.

It may be that the 1 in 20 chance is too big, and a smaller chance of being wrong is preferred. The 95% CI is by far the most widely used, but 90% and 99% are also commonly used, although any percentage is possible. Clearly 95% is more precise than 90% but less precise than 99%. When a different percentage is used all that is necessary is to change the value obtained from Table A.1 in the expression for the CI derived in Section 7.4.1. For example, to obtain a 99% CI, Table A.1 says that:

$$P(Z < 2.58) = 0.995 \text{ (to 3 decimal places).}$$

Thus, by the symmetry of the normal distribution:

$$P(Z < -2.58) = 0.005$$

and consequently $P(-2.58 < Z < 2.58) = 0.99$ (see also Example 6.26). Hence 2.58 in the 99% CI takes the place of 1.96 in the 95% CI, giving the 99% CI as:

$$\bar{x} \pm \frac{2.58 \, \sigma}{\sqrt{n}}$$

Example 7.2:   A 99% CI for the problem of Example 7.1 is:

$$6.23 \pm \frac{2.58 \times 1.2}{\sqrt{50}}$$

$$= 6.23 \pm 0.44 \text{ mmol/l}$$

Similarly a 90% CI is given by:

$$6.23 \pm \frac{1.64 \times 1.2}{\sqrt{50}}$$

$$= 6.23 \pm 0.28 \text{ mmol/l}$$

because $P(-1.64 < Z < 1.64) = 0.90$.

To completely understand confidence intervals it is important to consider the issues that affect their size. In Examples 7.1 and 7.2, it can be seen that the higher the percentage of confidence of being correct, the larger the confidence interval. In other words there is a greater confidence attached to a wider range of uncertainty.

The other factors that affect the length of the CI are the standard deviation, σ, and the sample size, n. Not surprisingly the CI is larger when σ is greater, since this means that there is inherently more variability, and thus more uncertainty, in the data themselves. There is nothing that can be done about this, since it is a feature of the phenomenon being observed. There is, subject to economic and other constraints, something that can be done about varying the sample size. In fact, the larger the sample size the smaller the CI. This is intuitively reasonable since as n gets bigger the sample represents more and more of the population, so that the uncertainty about the mean, or any other population parameter being estimated, obviously diminishes.

Notice that the CI for μ in Section 7.4.1 has the point estimate $\bar{x}$ at its centre. This seems reasonable because the underlying sampling distribution (i.e., distribution of sample means) is the normal, which is symmetric. There are cases, although not covered in this book, where the underlying distribution is skewed and consequently the CI is skewed in a similar way.

For these symmetric confidence intervals a general interpretation can also be given in terms of the variance of the estimate (or alternatively its square root, the standard error). If the estimate used is unbiased the confidence interval always has the form:

$$\text{estimate} \pm \text{ tabulated value} \times \sqrt{\text{variance of the estimate}}$$

i.e., estimate $\pm$ tabulated value $\times$ standard error

except where the true variance is unknown, in which case it is replaced by its own estimate (as in Section 7.4.3). In the particular case covered by Section 7.4.1:

> estimate $= \bar{x}$;
> tabulated value comes from the normal distribution;
> variance of $\bar{x} = \dfrac{\sigma^2}{n}\left(\text{i.e., standard error} = \dfrac{\sigma}{\sqrt{n}}\right)$

In other cases all these may change, as shown later. Formulae for variances are obtained using mathematical theory beyond the scope of this book.

7.4.3 Confidence interval for μ when σ is unknown

In Section 7.4.1, it was assumed that the population standard deviation, σ, was known. In reality this is hardly ever the case, and hence the method of Section 7.4.1 needs to be extended to deal with the situation in which the population standard deviation is unknown.

When σ is unknown, $\bar{x} \pm 1.96\,\sigma/\sqrt{n}$ cannot be calculated. Just as μ is estimated by $\bar{x}$, σ is estimated by s, so the obvious thing to do is to replace σ by s in the formula for the CI. Here a problem arises since $\dfrac{\bar{X} - \mu}{s/\sqrt{n}}$ is *not* normally distributed, and hence the derivation of the CI given in Section 7.4.1 breaks down. Instead $\dfrac{\bar{X} - \mu}{s/\sqrt{n}}$ has a Student's t distribution with $n-1$ degrees of freedom. In consequence the 95% CI for μ now becomes:

$$\bar{x} \pm t_{n-1} \times \frac{s}{\sqrt{n}}$$

where t_{n-1} is the value from Table A.2 such that 2.5% of the Student's t distribution on $n-1$ degrees of freedom lies to its right. Similarly for other percentages.

Example 7.3: In Example 7.1 it was assumed that the standard deviation of serum cholesterol concentration for the entire town was known to be 1.20 mmol/l. In reality this population standard deviation is not known. All that is known about the variability is that the standard deviation of the sample of 50 patients, referred to in Example 7.1, is 1.07 mmol/l. The 95% CI for the mean in this situation is,

$$\bar{x} \pm t_{n-1} \times \frac{s}{\sqrt{n}}$$

$$= 6.23 \pm t_{49} \times \frac{1.07}{\sqrt{50}}$$

Now, from Table A.2, $P(T_{60} < 2.000) = 0.975$ where T_{60} is the Student's t on 60 degrees of freedom. Here, as usual, a capital letter (T) denotes a random variable and a small letter (t) denotes an outcome of the random variable. Table A.2 gives values of $P(T < t)$ for different degrees of freedom. By the symmetry of Student's t, $P(-2.000 < T_{60} < 2.000) = 0.95$. Similarly, $P(-2.021 < T_{40} < 2.021) = 0.95$. Now, since 49 lies approximately half way between 40 and 60, it must be that:

$$P(-2.01 < T_{49} < 2.01) = 0.95$$

at least approximately. So t_{49} is approximately 2.01 (an exact value cannot be found from the table).

The 95% CI for μ is thus:

$$6.23 \pm 2.01 \times \frac{1.07}{\sqrt{50}}$$

$$= 6.23 \pm 2.01 \times 0.151$$

$$= 6.23 \pm 0.30 \text{ mmol/l}$$

That is, there is 95% confidence that the true mean concentration lies between 5.93 and 6.53 mmol/l. This is the true CI, not that calculated in Example 7.1, where σ was really only guessed.

In Section 6.12.2 it was noted that the percentage points of the t converge to the normal as n increases. So for large samples it will make little difference whether the t or normal tables are used. Indeed, in Example 7.3 the t value of 2.01 is not very different from the normal value of 1.96. The difference for small n can, however, be quite considerable.

7.5 Confidence intervals for the difference between two means when independent samples are taken

It commonly occurs that two populations or physical processes are to be compared and the comparison is made using means as summary statistics. Two hospitals are compared to see which has the higher average number of admissions per day, for instance, or two general practitioners are compared to see which has the shorter average consultation time.

The estimated difference between two means is simply the difference between the estimates of the means, that is $\mu_1 - \mu_2$ is estimated by $\bar{x}_1 - \bar{x}_2$, where '1' and '2' denote the two different populations. Confidence intervals can be given for $\mu_1 - \mu_2$, but as with the estimation of a single mean, the cases of known and unknown standard deviations must be considered separately. Here an extra complication is that when the standard deviations are unknown the cases of equal and unequal standard deviations have also to be considered separately.

Throughout Section 7.5 it will be assumed that two samples are taken, one from each of the populations, totally independently of each other. Section 7.15.1 considers a situation where this is not true. Also it will always be assumed that sample means have normal distributions, although Section 7.16 should be referred to for a discussion of this assumption.

7.5.1 Confidence interval for $\mu_1 - \mu_2$ when σ_1 and σ_2 are known

It may be shown that $\overline{X}_1 - \overline{X}_2$ has a normal distribution with mean $\mu_1 - \mu_2$ and variance $\sigma_1^2/n_1 + \sigma_2^2/n_2$. Hence the standardized normal random variable is:

$$Z = \frac{\overline{X}_1 - \overline{X}_2 - (\mu_1 - \mu_2)}{\sqrt{\sigma_1^2/n_1 + \sigma_2^2/n_2}}$$

When this is substituted into the equation

$$P(-1.96 < Z < 1.96) = 0.95$$

the 95% CI for $\mu_1 - \mu_2$ is found to be

$$\overline{x}_1 - \overline{x}_2 \pm 1.96 \sqrt{\sigma_1^2/n_1 + \sigma_2^2/n_2}$$

Example 7.4: In Examples 7.1–7.3 the cholesterol concentration after 10 years of the health promotion programme was considered. It is, however, impossible to judge whether the programme has been successful without knowledge of the concentration *before* the programme began (the 'baseline'). The real interest lies in the difference between average concentrations before and after the programme.

In fact before the programme was launched a sample survey was carried out in which 70 of the townsfolk were found to have a mean serum cholesterol concentration of 6.74 mmol/l.

Assuming that the standard deviations of concentration before and after are known for the whole town to be 1.12 and 1.20 mmol/l respectively, find a 90% confidence interval for the difference between before and after means.

Here,

	$n_1 = 70$	$n_2 = 50$
	$\overline{x}_1 = 6.74$	$\overline{x}_2 = 6.23$
	$\sigma_1 = 1.12$	$\sigma_2 = 1.20$

where the subscript '1' denotes 'before' and '2' denotes 'after'. The point estimate for $\mu_1 - \mu_2$ is:

$$\overline{x}_1 - \overline{x}_2 = 6.74 - 6.23 = 0.51 \text{ mmol/l}$$

The 90% confidence interval for this difference is:

$$\begin{aligned} \overline{x}_1 - \overline{x}_2 &\pm 1.64 \sqrt{\sigma_1^2/n_1 + \sigma_2^2/n_2} \\ &= 0.51 \pm 1.64 \sqrt{(1.12)^2/70 + (1.20)^2/50} \\ &= 0.51 \pm 1.64 \times 0.2161 \\ &= 0.51 \pm 0.35 \end{aligned}$$

Hence if the standard deviations are as stated, there is 90% confidence that the true decrease in serum cholesterol concentration during the health promotion programme lies somewhere between 0.16 and 0.86 mmol/l. In practice the standard deviations will not be known and Example 7.5 gives a much more realistic treatment of this problem.

7.5.2 Confidence interval for $\mu_1 - \mu_2$ when σ_1 and σ_2 are unknown but equal

When σ_1 and σ_2 are unknown, as they will be in most real life situations, two cases must be treated separately. The first of these is where it can be assumed that the two population standard deviations, although unknown, are nevertheless equal. In Section 7.11 a procedure for testing this assumption will be given.

In the case of equal σ_1 and σ_2 it is sensible to pool all the sample data to produce a common estimate of the two unknown standard deviations. It turns out that the most sensible pooled estimate to use for the common *variance* is:

$$s_p^2 = \frac{(n_1 - 1)s_1^2 + (n_2 - 1)s_2^2}{n_1 + n_2 - 2}$$

Then, as was the case in Section 7.4.3, this estimate is substituted for the population variances in the derivation of the confidence interval for $\mu_1 - \mu_2$. Again as in Section 7.4.3, the distribution followed is Student's t, but this time with degrees of freedom (d.f.) given by $n_1 + n_2 - 2$, the divisor for s_p^2.

That is:
$$\frac{\bar{X}_1 - \bar{X}_2 - (\mu_1 - \mu_2)}{\sqrt{s_p^2/n_1 + s_p^2/n_2}} = \frac{\bar{X}_1 - \bar{X}_2 - (\mu_1 - \mu_2)}{s_p\sqrt{1/n_1 + 1/n_2}}$$

has a Student's t distribution with $n_1 + n_2 - 2$ d.f.
The 95% confidence interval for $\mu_1 - \mu_2$ is thus:

$$\bar{x}_1 - \bar{x}_2 \pm t_{n_1 + n_2 - 2} \, s_p \sqrt{1/n_1 + 1/n_2}$$

where $t_{n_1 + n_2 - 2}$ has 2.5% of the Student's t with the appropriate d.f. to its right.

Example 7.5:   In Example 7.4 the two population standard deviations were assumed known, whereas in reality they were only guessed. The most appropriate CI to use in this situation is the one which takes into account the information available about variability from the before and after samples. Example 7.2 states that the 'after' sample standard deviation was 1.07 mmol/l; the 'before' value turned out to be 1.15 mmol/l. Calculate the 99% CI for $\mu_1 - \mu_2$.

Here,

$n_1 = 70$	$n_2 = 50$
$\bar{x}_1 = 6.74$	$\bar{x}_2 = 6.23$
$s_1 = 1.15$	$s_2 = 1.07$

using information and notation from Example 7.4.

There is evidence to suggest that the two population standard deviations are actually equal (see Example 7.13). The pooled estimate of this common variance is:

$$s_p^2 = \frac{(n_1 - 1)s_1^2 + (n_2 - 1)s_2^2}{n_1 + n_2 - 2}$$

$$= \frac{69 \times (1.15)^2 + 49 \times (1.07)^2}{70 + 50 - 2}$$

$$= 1.249$$

The 99% CI is:

$$\bar{x}_1 - \bar{x}_2 \pm t_{n_1 + n_2 - 2} \, s_p \sqrt{1/n_1 + 1/n_2}$$

$$= 6.74 - 6.23 \pm t_{118} \times \sqrt{1.249} \times \sqrt{1/70 + 1/50}$$

$$= 0.51 \pm t_{118} \times 1.117 \times 0.185$$

From Table A.2, $P(t_{120} < 2.617) = 0.995$. Hence, since 99% confidence is required, the value of t_{118} to be used in the formula above is approximately 2.62. The 99% CI thus becomes:

$$0.51 \pm 2.62 \times 1.117 \times 0.185$$
$$= 0.51 \pm 0.54 \, \text{mmol/l}$$

This says that the difference between means can be stated to lie between -0.03 and 1.05 mmol/l with 99% confidence. A minus sign means an *increase* in serum cholesterol concentration during the programme of health promotion.

7.5.3 Confidence interval for $\mu_1 - \mu_2$ when σ_1 and σ_2 are unknown and unequal

It is an unfortunate fact that in the situation of σ_1 and σ_2 unknown and unequal, there is no exact CI for $\mu_1 - \mu_2$ under the usual assumption of a normal distribution for $\bar{X}_1$ and $\bar{X}_2$.

The obvious thing to try in this situation is to substitute s_1 for σ_1 and s_2 for σ_2 in the formula given in Section 7.5.1. The problem is that

$$\frac{\bar{X}_1 - \bar{X}_2 - (\mu_1 - \mu_2)}{\sqrt{s_1^2/n_1 + s_2^2/n_2}}$$

does not have a known probability distribution, and hence the CI for $\mu_1 - \mu_2$ cannot be derived. Instead approximate procedures must be used.

If the sample sizes n_1 and n_2 are large then it is reasonable to assume that s_1 and σ_1 are not very different, and likewise for s_2 and σ_2. Thus the confidence interval can be estimated as in Section 7.5.1, for example

$$\bar{x}_1 - \bar{x}_2 \pm 1.96 \sqrt{s_1^2/n_1 + s_2^2/n_2}$$

is the approximate 95% CI.

If the sample sizes are small, the sample and population standard deviations are unlikely to be similar. In this case it has been found that Student's t, with a particularly complicated expression for degrees of freedom, provides a reasonable approximation to the correct distribution. Using this, the approximate CI is:

$$\bar{x}_1 - \bar{x}_2 \pm t_f \sqrt{s_1^2/n_1 + s_2^2/n_2}$$

where the d.f. are:

$$f = \frac{(s_1^2/n_1 + s_2^2/n_2)^2}{[(s_1^2/n_1)^2/(n_1 - 1)] + [(s_2^2/n_2)^2/(n_2 - 1)]}$$

for whatever percentage confidence is required.

7.6 Confidence intervals concerning proportions

Along with means, the other commonly estimated features of a population are proportions. Examples are the proportion of inpatients who live outside the health district, the proportion of the population hospitalized each year and the proportion of patients reacting favourably to a new treatment. In every case each member of the population can be thought of in some sense as either a 'success' or a 'failure'. For example a patient who has a favourable reaction to the new treatment is a success. The quantity of interest is the proportion of successes amongst the entire population. In some situations it can be useful to think of the population proportion as the probability (see Section 6.2 and Example 7.6).

7.6.1 Confidence interval for π

As with means and variances, the point estimate of the population proportion, π, is the observed sample proportion, p. The derivation of the confidence interval for π is quite complex, involving ideas which are not necessary to an understanding of how the CI should be used. However, this derivation does assume that the number of successes in a sample follows a binomial distribution, which means that successes must occur independently for the CI to be valid. The derivation also involves the central limit theorem to justify the use of the normal distribution, which means that n must be fairly big for the CI to be valid. If n is below 50 this CI could well be in serious error. In such cases the binomial distribution has to be used directly (see Armitage and Berry (1987), Section 4.7).

When n is large, P has a normal distribution with mean π and variance $\frac{\pi(1 - \pi)}{n}$. Hence

$$\frac{P - \pi}{\sqrt{\dfrac{\pi(1 - \pi)}{n}}}$$

has a standard normal distribution. Unfortunately this is not directly useful since it is impossible to go from a statement

$$P\left(- 1.96 < \frac{P - \pi}{\sqrt{\dfrac{\pi(1 - \pi)}{n}}} < 1.96\right) = 0.95$$

to a statement

$$P(C_1 < \pi < C_2) = 0.95$$

for any C_1 and C_2, as is desired. Since n is large it is, however, reasonable to assume that the sample and population variances are equal, and hence $\frac{\pi(1 - \pi)}{n}$ can be replaced by $\frac{p(1 - p)}{n}$ to give the approximate formula

$$P\left(- 1.96 < \frac{P - \pi}{\sqrt{\dfrac{p(1 - p)}{n}}} < 1.96\right) = 0.95$$

which is algebraically equal to

$$P\left(P - 1.96\sqrt{\frac{p(1 - p)}{n}} < \pi < P + 1.96\sqrt{\frac{p(1 - p)}{n}}\right) = 0.95$$

and hence the 95% CI for π is

$$p \pm 1.96\sqrt{\frac{p(1 - p)}{n}}$$

In Section 6.2.2 it was seen that as n gets larger the observed proportion of successes gets nearer to the probability of success in an infinitely large population. Now this can be interpreted as saying that the CI gets smaller as n increases, reflecting the decrease in uncertainty. Example 7.6 uses the data described in Section 6.2.2.

Example 7.6: Suppose that only the first 100 observations given in Example 6.4 were available. That is,. 100 births have been observed in a hospital, of which 48 were girls.

The point estimate of the probability of a girl birth is simply $p = \frac{48}{100} = 0.480$. The 95% CI is given by:

$$p \pm 1.96\sqrt{\frac{p(1 - p)}{n}}$$

$$= 0.480 \pm 1.96\sqrt{\frac{0.48 \times 0.52}{100}}$$

$$= 0.480 \pm 1.96 \times 0.0500$$

$$= 0.480 \pm 0.098$$

So, after only 100 observations, there is 95% confidence that the true probability of a girl birth is between 0.382 and 0.578. As always, increasing n decreases uncertainty. For instance, after the full 2000 observations of Example 6.4, the 95% CI becomes 0.462 ± 0.022 which suggests that the statement that the probability had, by this stage, converged to 0.46, made in Example 6.4, was possibility a little premature.

7.6.2 Confidence interval for $\pi_1 - \pi_2$

Just as with means there is often an interest in comparing different proportions, for example the proportion of cross-boundary flows in different years. Using similar arguments to those that have gone before, the point estimate for $\pi_1 - \pi_2$ is $p_1 - p_2$ and the 95% CI (for large n_1 and n_2) is

$$p_1 - p_2 \pm 1.96 \sqrt{\frac{p_1(1 - p_1)}{n_1} + \frac{p_2(1 - p_2)}{n_2}}$$

where the suffixes 1 and 2 denote the two entities being compared.

Example 7.7:   A second maternity hospital was monitored for a month. In that time, there were 120 births of which 66 were girls. Find a 99% CI for the difference between the probability of a girl birth in the two hospitals, using the full 2000 observations from Example 6.4 (which gave 924 girls).

Let the subscript '1' denote the hospital introduced in Example 6.4 and '2' denote the hospital introduced here.

Then,

$$n_1 = 2000 \qquad\qquad n_2 = 120$$

$$p_1 = \frac{924}{2000} = 0.462 \qquad p_2 = \frac{66}{120} = 0.55$$

The 99% CI for $\pi_1 - \pi_2$ is:

$$p_1 - p_2 \pm 2.58 \sqrt{\frac{p_1(1 - p_1)}{n_1} + \frac{p_2(1 - p_2)}{n_2}}$$

$$= 0.462 - 0.55 \pm 2.58 \sqrt{\frac{0.462 \times 0.538}{2000} + \frac{0.55 \times 0.45}{120}}$$

$$= -0.088 \pm 2.58 \times 0.0468$$

$$= -0.088 \pm 0.121$$

There is 99% confidence that the true difference is between -0.209 and 0.033 where a negative sign indicates a higher probability of a girl birth in hospital 2.

7.7   Introduction to hypothesis testing

Hypothesis testing involves making an assertion, or hypothesis, about a population and then collecting a set of data to see whether this gives evidence against the hypothesis. This should be contrasted with estimation where no prior claim about the population is made, but rather some feature of the population is estimated in a state of confessed ignorance. Examples of hypotheses are a claim from a drug manufacturer that a new tablet is more successful (than an existing one) in clearing headaches, a statement that one town has a lower incidence of morbidity than another, or an assertion that smoking is associated with lung cancer.

As with estimation, if data are collected from the entire population the hypothesis can be tested easily and precisely, but in many cases data are only available from a sample, for example a sample of headache sufferers who have taken both the old and the new treatments. Hypothesis testing is concerned with testing assertions using sample data: that is, making a decision about whether or not the information contained in the sample should cause a rejection of some prior hypothesis about the population. Consider, for example, the hypothesis that newborn babies are equally likely to be male or female, and suppose that the available data are those from Example 6.4, that is from 2000 births, 924 girls resulted. The question is then asked, 'Is the observed proportion of girl births, 0.462, so different from the hypothesized proportion of 0.5 that the difference is unlikely to have arisen merely by chance?'. The word 'unlikely' is important here: there can never be any guarantee that any observed difference is not purely due to the chance selection of a particular sample. Even if boys and girls *are* equally likely, it is not impossible to take a (very unfortunate) sample consisting almost entirely of one sex.

7.7.1 Null and alternative hypotheses

In hypothesis testing, a specific hypothesis, such as 'the probability of a girl birth is $\frac{1}{2}$', is formulated and the data are used to try to *reject* it. This postulated hypothesis is called a *null hypothesis*. It is important to realize that if the data do not lead to a rejection of the null hypothesis then all that can be concluded is that there is insufficient evidence to reject it.

It is not possible, however, to talk about rejecting a hypothesis unless it is stated what will then replace it. To complete the specification of the problem to be solved, the *alternative hypothesis* should be stated. The simplest alternative hypothesis is the general alternative which is simply the negation of the null hypothesis. For example, the null hypothesis that the probability of a girl birth is $\frac{1}{2}$ could be tested against the alternative that the probability is something other than $\frac{1}{2}$. This would be written:

$$H_0 : P(\text{girl birth}) = \tfrac{1}{2}$$
$$H_1 : P(\text{girl birth}) \neq \tfrac{1}{2}$$

where H_0 represents the null and H_1 the alternative hypothesis.

As well as this general alternative, two slightly more particular alternative hypotheses will be considered. The first of these is the 'less than' alternative, for example where the other side of the argument that each sex is equally likely at birth is that girls are less likely. This would be written

$$H_0 : P(\text{girl birth}) = \tfrac{1}{2}$$
$$H_1 : P(\text{girl birth}) < \tfrac{1}{2}$$

The second is the 'more than' alternative, for example where the alternative is that girls are more likely, written

$$H_0 : P(\text{girl birth}) = \tfrac{1}{2}$$
$$H_1 : P(\text{girl birth}) > \tfrac{1}{2}$$

Testing procedures, which will be slightly different for each of the pairs of hypotheses above, can be devised for different features of the population. However, just as with estimation, the features most commonly used are means, proportions and variances. The birth example specifies hypotheses about a proportion.

7.7.2 The critical region

Once the hypotheses have been formulated a decision must be made about the criterion which will be used to reject or fail to reject H_0. The criterion used is to calculate some appropriate measure from the sample data, called a *test statistic*, and reject H_0 whenever this takes a value which is unlikely if H_0 is true. For instance, in the births example, one simple test statistic is p, the sample proportion. H_0 is rejected whenever p is far from $\frac{1}{2}$. Clearly the nearer p gets to the hypothesized figure of $\frac{1}{2}$ the weaker is the evidence against H_0, but a value (or values) for p must be decided such that H_0 is rejected only when p is as extreme as this value. This value is called the *critical value* for the test, and the range of values of p which lead to a rejection of the null hypothesis is called the *critical region*.

A formal derivation of the critical region will be given later, but for the sake of illustration suppose that the hypotheses to be tested are

$$H_0 : \text{P(girl birth)} = \tfrac{1}{2}$$
$$H_1 : \text{P(girl birth)} < \tfrac{1}{2}$$

Now any value of p greater than $\frac{1}{2}$ clearly supports H_0 more strongly than H_1 so these values must lie outside the critical region. Even if the probability of a girl birth really is $\frac{1}{2}$ the sample proportion of girls is quite likely to be somewhat different, although this difference will tend to be smaller as sample size increases. Nevertheless, whatever the sample size, there will be a range of values that p could take just below $\frac{1}{2}$ where the evidence in support of H_0 is still quite strong, and does not suggest the rejection of H_0 because the deviation from $\frac{1}{2}$ seems likely to be due simply to chance. The largest value of p which still leads to rejection of H_0 is the critical value (see Fig. 7.1).

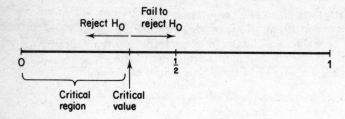

Fig. 7.1   One-sided ('less than') test for the births example.

On the other hand suppose the hypotheses were

$$H_0 : \text{P(girl birth)} = \tfrac{1}{2}$$
$$H_1 : \text{P(girl birth)} \neq \tfrac{1}{2}$$

Then a value of p far from $\frac{1}{2}$ on either side should lead to rejection of H_0. This time there should be two critical values, near to, but on either side of, $\frac{1}{2}$, as in Fig. 7.2.

In Figs 7.1 and 7.2 there are some values of p which cause rejection when H_1 is P(girl birth) $\neq \frac{1}{2}$, but not when it is P(girl birth) $< \frac{1}{2}$. In general, different alternatives can lead to different conclusions so it is important to specify the correct hypotheses at the outset. Alternatives of the 'less than' and 'more than' types are called *one-sided* or *one-tailed*, whereas '$\neq$' alternatives are called *two-sided*, or *two-tailed*; the reason for these names is obvious

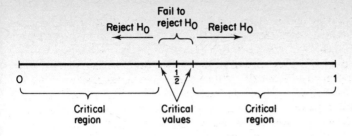

Fig. 7.2 Two-sided test for the births example.

from Figs 7.1 and 7.2. The corresponding tests with these alternatives are similarly called one- or two-sided (tailed) tests.

7.7.3 Significance and error

Although critical values were explained in Section 7.7.2, nothing was said about how to actually calculate them. This is achieved by finding the value of the test statistic such that it or any more extreme value would only be achieved with a very small probability if H_0 were true. The probability used is usually expressed as a percentage and called the *significance level*. The most common significance level is 5% although 10% and 1% are also often used.

Look at Fig. 7.1 again. Suppose the 5% significance level is to be used. Then the critical value will be such that when the probability of a girl birth really is $\frac{1}{2}$, there is a probability of $5/100 = 0.05$, or 1 in 20, that the sample proportion is less than or equal to the critical value. Then, when p is calculated from the sample, H_0 is rejected at the 5% level of significance if, and only if, p is at or below this critical value.

According to the definition just given, the significance level is the chance of being in error by rejecting the null hypothesis when it is really true. Uncertainty is inherent in any inferential procedure, and the significance level is one statement about the degree of uncertainty in the conclusion drawn as a result of the hypothesis test.

There is, however, another type of possible error in hypothesis testing, that is the error of failing to reject the null hypothesis when it is really false. This is called Type II error, whereas the error of wrongly rejecting the null hypothesis is called Type I error. The significance level is thus (one hundred times) the probability of a Type I error occurring.

Although the procedure for fixing critical values, and thus the test procedure, depends only on Type I error, it is essential to realize that, with all else fixed, the probability of Type I error decreases as the probability of Type II error increases (and vice versa), which means that the significance level cannot be lowered without incurring costs elsewhere. This can be formally proved, but can be seen from Fig. 7.1, where moving the critical value to the left leaves less room for wrongly rejecting H_0, but conversely more scope for failing to reject it when it is really false. The choice of which significance level to use depends on the relationship between the two probabilities of error since if the significance level chosen is too small, the chance of a Type II error could become unacceptably large; 5% significance has been found to be a reasonable compromise in most situations. It is *not*, however, sensible to use 5% in all situations, for example where Type I error represents an erroneous conclusion that a new drug produces no adverse effects. In this case the significance level chosen should be very small.

Because of the significance levels, hypothesis tests are sometimes called *significance tests*.

7.8 Hypothesis tests for the mean

The procedures followed in all hypothesis tests are very similar, the only difference being that different test statistics and probability distributions are needed in different situations. In this section hypothesis tests will be developed for means. As in Section 7.4 it will be assumed that sample means follow normal distributions.

7.8.1 Test of $H_0 : \mu = \mu_0$ when σ is known

As with confidence intervals the starting point for hypothesis tests will be the most trivial, and yet unusual, situation in which the population variance is known. Again, as with CIs, this case provides a basis for the other, more practically relevant, cases to follow.

The problem is to test the null hypothesis that μ takes some numerical value, denoted by μ_0. To find critical values for a given, say 5%, level of significance, a test statistic which has a known probability distribution when $\mu = \mu_0$ is required. By Section 6.11.3, when H_0 is true $\overline{X}$ has a normal distribution with mean μ_0 and variance σ^2/n. Hence

$$\frac{\overline{x} - \mu_0}{\sigma/\sqrt{n}}$$

comes from a standard normal distribution, and this will consequently be the test statistic.

This test statistic has the form

$$\frac{\text{estimate} - H_0 \text{ value}}{\sqrt{\text{variance of estimate}}}$$

or, more simply,

$$\frac{\text{estimate} - H_0 \text{ value}}{\text{standard error}}$$

Apart from the case considered in Section 7.11, all test statistics in this chapter have this form (unless the variance is unknown and has to be replaced by an estimate).

The critical value for 5% significance is such that the probability of the test statistic taking a value this extreme when H_0 is true is 0.05. To determine the critical value the alternative hypothesis must be specified.

Consider the alternative $H_1: \mu > \mu_0$. If H_1 is true the real mean of $\overline{X}$ will be greater than μ_0 and hence the test statistic will tend to take bigger values.

Hence the critical value should be chosen so that 5% of the standard normal distribution is at least as big as it. Table A.1 says that

$$P(Z < 1.64) = 0.95$$

and thus
$$P(Z \geqslant 1.64) = 0.05$$

which means that 1.64 is the 5% critical value for this one-sided test.

The test procedure is simply to calculate the test statistic from the sample and reject H_0 whenever the test statistic is greater than or equal to 1.64 (see Fig. 7.3).

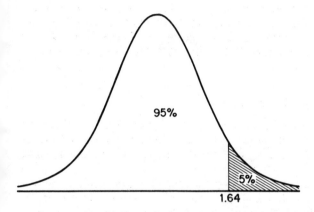

95%

5%

1.64

Fig. 7.3 One-sided ('greater than') 5% test. The shaded area indicates the critical region.

With the other one-sided alternative, H_1: $\mu < \mu_0$, a similar procedure is followed. By the symmetry of the normal distribution the critical value will now be -1.64 and H_0 is rejected whenever the test statistic turns out to be -1.64 or less.

Consider finally the alternative H_1: $\mu \neq \mu_0$. Now H_0: $\mu = \mu_0$ should be rejected whenever the test statistic takes a value far from zero on either side. Due to the symmetry of the normal distribution the critical region is taken to be the same size in each tail, that is for 5% significance, the critical region splits up into two halves, each of 2.5%. Since, by Table A.1,

$$P(Z < -1.96) = 0.025$$

then also
$$P(Z \geqslant 1.96) = 0.025$$

and the critical region must be as in Fig. 7.4.

H_0 is rejected whenever the test statistic is either equal to or less than -1.96 or equal to or greater than 1.96.

Example 7.8: Consider the problem and assumptions of Example 7.1. Suppose it was thought, by comparison with other studies, that the mean serum cholesterol concentration should be 5.9 mmol/l on completion of the health promotion programme. If the mean is above this the programme will be judged to have failed in this particular aspect. Test, at the 5% level, whether the programme has achieved its target.

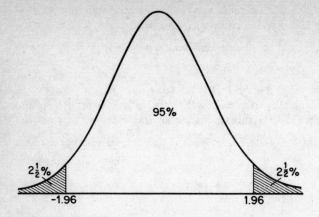

Fig. 7.4 Two-sided 5% test. The shaded area indicates the critical region.

The hypotheses to be tested are

$$H_0 : \mu = 5.9$$

$$H_1 : \mu > 5.9$$

From Example 7.1, $\sigma = 1.20$, n = 50 and $\bar{x} = 6.23$. Hence the test statistic is

$$\frac{\bar{x} - \mu_0}{\sigma/\sqrt{n}} = \frac{6.23 - 5.9}{1.2/\sqrt{50}} = 1.9445$$

Since 1.9445 is greater than 1.64, H_0 is rejected; there would be less than a 1 in 20 chance of getting as extreme a result as 1.9445 if H_0 were true. The conclusion is that the programme has not succeeded in achieving its target.

The 5% level of significance has been used throughout so far for simplicity. If 10% is used, Table A.1 gives critical values of 1.28 or -1.28 for the one-sided tests and ± 1.64 for the two-sided test. If 1% is used these become 2.33 and -2.33 for one-sided and ± 2.58 for two-sided tests. Other significance levels follow similarly.

7.8.2 Test of $H_0 : \mu = \mu_0$ when σ is unknown

As was seen in Section 7.4.3, in the more usual case of σ unknown, σ can be replaced by its sample estimate s, and then all goes as for the case of σ known except that the distribution required is now t on n – 1 degrees of freedom. The test statistic is thus

$$\frac{\bar{x} - \mu_0}{s/\sqrt{n}}$$

which is to be compared with critical values derived from t with n – 1 d.f.

Example 7.9:   Repeat Example 7.8 except now do *not* make the unrealistic assumption that σ is known. Use the sample information from Example 7.3, i.e., n = 50, $\bar{x} = 6.23$, s = 1.07 and $\mu_0 = 5.9$. The test statistic is

$$\frac{\bar{x} - \mu_0}{s/\sqrt{n}} = \frac{6.23 - 5.9}{1.07/\sqrt{50}} = 2.1808$$

From Table A.2, $P(T_{49} < 1.68)$ is approximately equal to 0.95. Since 2.1808 is greater than 1.68, H_0 is rejected at the 5% level.

7.9 Hypothesis tests for the equality of two means when independent samples are taken

Now consider tests for the equality of two population means, μ_1 and μ_2. As in Section 7.5 the sample means $\bar{X}_1$ and $\bar{X}_2$ will be assumed to be normally distributed and the two samples will be assumed to have been drawn independently (see Section 7.15.1 for another situation). Just as in Section 7.5 three cases need to be distinguished.

7.9.1 Test of $H_0 : \mu_1 = \mu_2$ when σ_1 and σ_2 are known

Once again the easiest way to begin to understand this group of tests is with the unlikely case of the population variances known. When $H_0 : \mu_1 = \mu_2$ is true, reference to Section 7.5.1 shows that $\bar{X}_1 - \bar{X}_2$ has a normal distribution with mean $\mu_1 - \mu_2 = 0$ and variance $\sigma_1^2/n_1 + \sigma_2^2/n_2$. Hence the test statistic

$$\frac{\bar{X}_1 - \bar{X}_2}{\sqrt{\sigma_1^2/n_1 + \sigma_2^2/n_2}}$$

should be compared with critical values derived from the standard normal distribution, exactly as specified in Section 7.8.1.

7.9.2 Test of $H_0 : \mu_1 = \mu_2$ when σ_1 and σ_2 are unknown but equal

In the case of σ_1 and σ_2 unknown but equal, Section 7.5.2 gives the method to be used. First the pooled estimate of variance, s_p^2, should be calculated. The test statistic is then

$$\frac{\bar{X}_1 - \bar{X}_2}{s_p\sqrt{1/n_1 + 1/n_2}}$$

which is an observation from a t distribution with $n_1 + n_2 - 2$ d.f. if H_0 is true.

Note that a test for the equality of the two variances is given in Section 7.11.

Example 7.10: Test the hypothesis that the health promotion programme has had no effect upon the mean serum cholesterol concentration in the town against the alternative that it has indeed had some effect. Use the information given in Example 7.5.

The relevant information from Example 7.5 is

$$n_1 = 70 \qquad n_2 = 50$$
$$\bar{x}_1 = 6.74 \qquad \bar{x}_2 = 6.23$$
$$s_1 = 1.15 \qquad s_2 = 1.07$$

and the hypotheses are

$$H_0 : \mu_1 = \mu_2 \text{ (no effect)}$$

$$H_1 : \mu_1 \neq \mu_2 \text{ (some effect)}$$

where 1 is 'before' and 2 'after' the programme.

From Example 7.5 the pooled estimate of variance is $s_p^2 = 1.249$. The test statistic is

$$\frac{\bar{x}_1 - \bar{x}_2}{s_p\sqrt{1/n_1 + 1/n_2}} = \frac{6.74 - 6.23}{\sqrt{1.249 \ (1/70 + 1/50)}} = 2.46$$

Since $P(T_{120} < 1.98) = 0.975$ (from Table A.2)

the critical values which split T_{118} into 2.5%, 95% and 2.5% must be approximately -1.98 and 1.98. Since 2.46 is greater than 1.98 the null hypothesis is rejected at the 5% level of significance. Notice, however, that if 1% significance is used, Table A.2 gives critical values of, again approximately, ± 2.617. Since the test statistic's result of 2.46 is between -2.617 and 2.617, H_0 is not rejected at the 1% level of significance.

This illustrates the point that the conclusion reached will depend upon the significance level adopted, which will be considered further in Section 7.12. 1% is quite an extreme level of significance, so the overall conclusion here might be that there is evidence against H_0 but this evidence is not exceptionally strong; on balance the programme probably has had an effect (in fact, a good one).

Another point that is worth making about Example 7.10 is that the use of the two-sided alternative, as specified in the question, might not truly be appropriate. It may be that interest lies only in a *decrease* of serum cholesterol concentration over time, rather than a change in either direction. The alternative $\mu_1 > \mu_2$ should then be used and the test *will* then be significant at the 1% level. This shows that it is essential not only to use the correct hypotheses, but also to state clearly what these hypotheses are in any report of findings.

7.9.3 Test of H_0: $\mu_1 = \mu_2$ when σ_1 and σ_2 are unknown and unequal

Once again this test follows from the theory of the corresponding section on confidence intervals, this time Section 7.5.3. Thus the test statistic is

$$\frac{\bar{x}_1 - \bar{x}_2}{\sqrt{s_1^2/n_1 + s_2^2/n_2}}$$

which *approximates* to an observation from

(i) a standard normal, when n_1 and n_2 are large;

(ii) t with f d.f. (where f is given in Section 7.5.3), when n_1 or n_2 is small.

Thus the test statistic needs to be compared with critical values from different distributions, depending on sample sizes.

Since this test procedure is only approximate it would not be sensible to place much reliance on any result which is only marginally significant.

7.10 Hypothesis tests concerning proportions

7.10.1 Test of $H_0: \pi = \pi_0$

In this section a formal procedure will be given for the type of situation illustrated by the births example in Section 7.7 where it was naively suggested that the sample proportion, p, would provide a suitable test statistic. When H_0 is true, the sample proportion has a normal distribution with mean π_0 and variance $\pi_0(1 - \pi_0)/n$ for large n (see Section 7.6.1). This normal distribution is not tabulated and so critical values cannot simply be read off. As usual the solution to the problem is to standardize and use the test statistic

$$\frac{p - \pi_0}{\sqrt{\dfrac{\pi_0(1 - \pi_0)}{n}}}$$

which is to be compared with critical values from Table A.1. As explained in Section 7.6.1, the justification for this procedure is based on a normal approximation to the binomial distribution which is good for reasonably large n.

Example 7.11: Use the sample data of 924 girls born from 2000 births to test at the 5% level of significance the hypothesis that the sexes are equally likely at birth against the general two-sided alternative that the sexes are not equally likely. The hypotheses are

$$H_0 : \pi = \tfrac{1}{2}$$
$$H_1 : \pi \neq \tfrac{1}{2}$$

where π is the probability of a girl birth. The test statistic is

$$\frac{p - \pi_0}{\sqrt{\dfrac{\pi_0(1 - \pi_0)}{n}}} = \frac{\dfrac{924}{2000} - \dfrac{1}{2}}{\sqrt{\dfrac{\dfrac{1}{2} \times \dfrac{1}{2}}{2000}}} = -3.399$$

From Table A.1 the critical values at 5% are ± 1.96 and since -3.399 is less than -1.96, H_0 is rejected at the 5% level.

7.10.2 Test of $\pi_1 = \pi_2$

If the null hypothesis that two population proportions are equal is true, then the difference between sample proportions $P_1 - P_2$ has a normal distribution with mean $\pi_1 - \pi_2 = 0$ and variance

$$\frac{\pi(1 - \pi)}{n_1} + \frac{\pi(1 - \pi)}{n_2} = \pi(1 - \pi)[1/n_1 + 1/n_2],$$

where π is the common proportion, (i.e., $\pi_1 = \pi_2 = \pi$). Since π is not specified explicitly in the null hypothesis, no further progress can be made towards developing a suitable test statistic unless it is approximated by some sample-based estimate. The best estimate, when H_0 is true, is the proportion of 'successes' in the two samples combined. The combined proportion will be denoted by p_c which is calculated as

$$p_c = \frac{s_1 + s_2}{n_1 + n_2} = \frac{n_1 p_1 + n_2 p_2}{n_1 + n_2}$$

where s_1 and s_2 are the number of successes in the two samples.

Then, for large n_1 and n_2, $P_1 - P_2$ is approximately normally distributed with mean 0 and variance $p_c(1 - p_c)(1/n_1 + 1/n_2)$. Standardizing gives the test statistic

$$\frac{p_1 - p_2}{\sqrt{p_c(1 - p_c)(1/n_1 + 1/n_2)}}$$

to be compared with critical values derived from the standard normal distribution.

Example 7.12:   In Example 7.7 data were given for two maternity hospitals. Test the hypothesis that the probability of a baby being born female is the same in the two hospitals. From Example 7.7,

$$n_1 = 2000 \qquad n_2 = 120$$
$$p_1 = 0.462 \qquad p_2 = 0.55$$

No specific alternative hypothesis was stated in the question, and indeed there seems no reason to have any prior supposition that either hospital produces more females. Hence the general two-sided alternative will be used; that is

$$H_0 : \pi_1 = \pi_2$$
$$H_1 : \pi_1 \neq \pi_2$$

where π_1 and π_2 are the probabilities of girl births in hospitals 1 and 2.

The combined estimate of the common proportion is

$$p_c = \frac{n_1 p_1 + n_2 p_2}{n_1 + n_2} = \frac{(2000 \times 0.462) + (120 \times 0.55)}{2000 + 120}$$

$$= 0.467$$

(Notice that this is simply the total number of girl births observed in the two hospitals divided by the total number of births.) The test statistic is

$$\frac{p_1 - p_2}{\sqrt{p_c(1 - p_c)(1/n_1 + 1/n_2)}} = \frac{0.462 - 0.55}{\sqrt{0.467 \times 0.533 \, (1/2000 + 1/120)}}$$

$$= -1.8767$$

Now the critical values for the two-sided test at 10% are ± 1.64 and at 5% are ± 1.96. Since -1.8767 is less than -1.64 but greater than -1.96, H_0 is rejected at the 10% level, but not at the 5% level. There is only weak evidence to reject the hypothesis of equal probabilities; the probability that the observed difference arose purely by chance is less than 1 in 10 but greater than 1 in 20.

7.11 A hypothesis test for the equality of two standard deviations

Although measures of spread play a vital role in all aspects of inference it is quite rare to want to create a confidence interval or carry out a test concerning such measures for their own sake. When considering problems concerned with two means when the standard deviations are unknown, in Sections 7.5 and 7.9, exact procedures are only available for the case of equal population standard deviations. For this reason different methods were given for the cases $\sigma_1 = \sigma_2$ and $\sigma_1 \neq \sigma_2$, and it is therefore useful to have a test for the equality of the two standard deviations, even if only as a preliminary to other tests. A test of the hypotheses

$$H_0 : \sigma_1 = \sigma_2 \qquad\qquad H_0 : \sigma_1^2 = \sigma_2^2$$
$$\text{or equivalently}$$
$$H_1 : \sigma_1 \neq \sigma_2 \qquad\qquad H_1 : \sigma_1^2 \neq \sigma_2^2$$

will now be developed.

It may be shown that the ratio of the two sample variances has an F distribution (see Section 6.12.3) when H_0 is true and thus makes a suitable test statistic. Now the F distribution is not symmetrical about zero, unlike the normal and t distributions used so far in this chapter. Hence the critical values will not be plus and minus the same value for the two-sided test. It happens that values in the left-hand tail (below the peak) are always below one, and hence as long as the test statistic is forced to be above one it *must* be bigger than the lower critical value. This is easily achieved by always dividing the larger variance by the smaller. Then only the larger (right tail) critical value has to be read from F tables; H_0 is rejected whenever the test statistic equals or exceeds it. Remember that when a two-sided test is to be done, for example, at the 10% level of significance, the larger critical value will have 95% of the probability below it (see Fig. 7.5).

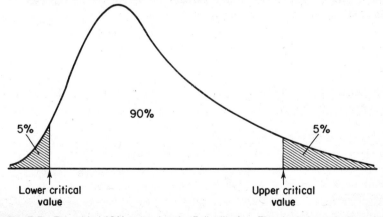

Fig. 7.5 Two-sided 10% test using the F distribution. The shaded area indicates the critical region.

The hypothesis test procedure is summarized by:

(i) if $s_1^2 > s_2^2$ calculate s_1^2/s_2^2 and compare with F on $(n_1 - 1, n_2 - 1)$ d.f.;
(ii) if $s_1^2 \leqslant s_2^2$ calculate s_2^2/s_1^2 and compare with F on $(n_2 - 1, n_1 - 1)$ d.f.

Notice how the degrees of freedom depend, as usual, on the sample sizes.

Example 7.13:   In Examples 7.5 and 7.10 the population standard deviations were assumed to be equal, given sample standard deviations $s_1 = 1.15$ from $n_1 = 70$ observations and $s_2 = 1.07$ from $n_2 = 50$ observations. Is there any evidence to doubt this assumption?
Since $s_1^2 > s_2^2$ the test statistic is

$$\frac{s_1^2}{s_2^2} = \frac{(1.15)^2}{(1.07)^2} = 1.1551$$

The d.f. for the F distribution are $(n_1 - 1, n_2 - 1) = (69, 49)$. Table A.4 says that 5% of the values taken by F on (60, 50) d.f. are greater than or equal to 1.58, and thus the 10% critical value for the two-sided test will be close to 1.58. Certainly 1.1551 is well below 1.58 and so H_0 is not rejected, even at the 10% level. Thus there is not even weak evidence to show that the population variances are unequal.

7.12 P values

In the last four sections results of hypothesis tests have been quoted at the 10%, 5% or 1% levels of significance. Sometimes a particular null hypothesis has been rejected at one level, but not at a more extreme level, as was the case in Examples 7.10 and 7.12. The more extreme (smaller) the significance level at rejection the more evidence there is against H_0.

As previously stated, 10%, 5% and 1% are really quite arbitrary measuring points, and it is perfectly acceptable to use any other significance level. In every hypothesis test there must be some significance level such that H_0 is rejected at that level, but not at any more extreme level. This significance level, expressed in the form of a probability, is called the *P value* for the test. The P value can be read from statistical tables as the probability of

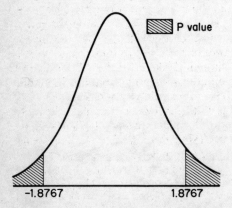

Fig. 7.6   P value (shaded area) for Example 7.12.

getting a value of the test statistic which is as extreme as that actually observed, when H_0 is true.

Example 7.14: In Example 7.12 the test statistic was observed to be -1.8767 and, when H_0 is true, the test statistic comes from the standard normal distribution. Since the test was two-sided the P value will be the size of the area indicated in Fig. 7.6.

Using rounded figures from Table A.1, $P(Z < 1.8767) = 0.97$ and hence $P(Z \geqslant 1.8767) = P(Z \leqslant -1.8767) = 0.03$. The P value is thus $0.03 + 0.03 = 0.06$.

Notice that $0.1 > P > 0.05$ and so the test would be rejected at the less extreme 10% but not at the more extreme 5% significance level, as found in Example 7.12.

The great advantage of quoting the P value is that precise information is conveyed. For example a statement that a test is significant at 5% only says that the evidence against H_0 is moderately strong. It could be that the test was also significant at a level as extreme as 0.01%, which is strong evidence indeed to refute H_0. The other advantage is that P values avoid the temptation to put absolute faith in the result of a 5% test of significance, as is done all too often. 5% is *not* a magic figure and, bearing in mind the meaning of significance as a type of error caused by sampling variation, it cannot make sense to totally disregard a hypothesis with a P value of 0.051, and yet take one with a P value of 0.049 as conclusively proved. At the end of the day, a decision as to the level of Type I error that is acceptable must be made whenever a hypothesis test is done.

The tables in Appendix 4 do not always allow the precise P value to be found. More complete tables (such as Fisher and Yates (1957)) do exist, and most statistical computer packages will give exact values.

7.13 Confidence intervals and hypothesis tests

There is a close relationship between these two major strands of statistical inference, but though the same probability distributions, and often the same values from statistical tables, feature in the two topics, the difference is in application. Estimation is concerned with finding a likely value for some unknown quantity, whereas hypothesis testing is concerned with testing an assertion. It is important to differentiate the two, and to understand where each should be applied. Gardner and Altman (1986) have argued that hypothesis tests have often been used in situations where confidence intervals would have been more appropriate, during medical investigations.

One explicit relationship between CIs and hypothesis tests arises when the test is two-sided. The null hypothesis will then be rejected at $(100 - X)\%$ significance whenever the hypothesized value falls outside the X% confidence interval. This is demonstrated by Examples 7.5 and 7.10.

In Example 7.5 the 99% CI for $\mu_1 - \mu_2$ was evaluated as 0.51 ± 0.54 which clearly includes 0. Hence the test of $H_0: \mu_1 = \mu_2$, equivalent to $H_0: \mu_1 - \mu_2 = 0$, in Example 7.10 failed to reject H_0 at the $100 - 99 = 1\%$ level of significance.

Due to the approximation made for the variance (see Section 7.6.1) there can be slight discrepancies in this relationship when proportions are involved. These are rarely important, for instance in Example 7.6 the 95% CI (with 2000 observations) for π was 0.462 ± 0.022 which does not include 0.5. Hence in Example 7.11 it should be expected that H_0: $\pi = 0.5$ would be rejected at 5%, as was indeed the case.

For quick reference a list of the most useful formulae associated with CIs and hypothesis tests is given in Appendix 3.

7.14 Determination of sample size

Before collecting a sample of data it is important to decide upon the sample size to be used. This should be chosen so as to obtain results to the required level of accuracy. Too small a sample can lead to results which are so imprecise as to be useless, whilst too large is wasteful of resources. Often the desirable level of accuracy turns out to require a sample size that is far too big for the available resources, and then lost accuracy has to be traded off for savings in money and time.

The accuracy required from a sample can be expressed in the language of confidence intervals or hypothesis tests, each of which will be considered here. In any given situation the reader must decide whether it is estimation or testing that is to be done, and then either Section 7.14.1 or 7.14.2 should be used accordingly. All the assumptions of normal distributions used previously in this chapter will be used once more.

7.14.1 From confidence intervals

Sample size can be determined by specifying in advance the amount of error that can be tolerated in an estimate to a stated degree of confidence. For example, when estimating the mean serum cholesterol concentration in a town it could be acceptable to be 95% confident that the estimate is no more than 0.5 mmol/l in error.

In general the sample size can be calculated from the expression for $\pm$ error in the various cases covered by Sections 7.4–7.6. For means these expressions include standard deviations and these need to be given numerical values before any progress can be made. This can be done by reference to literature on previous, similar studies, or by taking a small 'pilot' sample (see Section 10.6). Since this is all to be done before the main sample is selected the situations where sample standard deviations substitute for population standard deviations are not applicable. Here the population standard deviations are always taken as 'known', even though this is often merely an educated guess.

The expression for a CI for a mean is

$$\bar{x} \pm z\,\sigma/\sqrt{n}$$

from Section 7.4.1, where z is the value from the standard normal distribution at the prescribed level of confidence. If the acceptable error in an estimate is given the symbol ϵ (the Greek letter epsilon) then it must be that

$$z\,\sigma/\sqrt{n} = \epsilon$$

when n is chosen to be no larger than is necessary.

i.e.
$$\sqrt{n} = \frac{z\sigma}{\epsilon}$$

i.e.
$$n = \left(\frac{z\sigma}{\epsilon}\right)^2$$

Similarly, when estimating the difference between two means, Section 7.5.1 shows that

$$z\sqrt{\sigma_1^2/n_1 + \sigma_2^2/n_2} = \epsilon$$

When the simplifying assumption that the two sample sizes will be equal ($n_1 = n_2 = n$) is made, this reduces to

$$z\sqrt{(\sigma_1^2 + \sigma_2^2)/n} = \epsilon$$

and thus
$$n = \left(\frac{z}{\epsilon}\right)^2 (\sigma_1^2 + \sigma_2^2)$$

Example 7.15: When mean serum cholesterol concentration is to be estimated to an accuracy of 0.5 mmol/l with 95% confidence the formula for sample size becomes

$$n = \left(\frac{z\sigma}{\epsilon}\right)^2 = \left(\frac{1.96\sigma}{0.5}\right)^2$$

since 95% of standard normal values lie between ±1.96. To actually calculate n a number is needed for σ. Previous studies of cholesterol levels (Puska *et al.* (1983)) suggest that σ is unlikely to exceed 1.4 mmol/l. Hence the formula becomes

$$n = \left(\frac{1.96 \times 1.4}{0.5}\right)^2 = 30.12$$

That is n needs to be no bigger than 31.

Determining sample size when estimation involves proportions follows similar patterns. For estimating a single proportion Section 7.6.1 gives the CI

$$p \pm z\sqrt{\frac{p(1-p)}{n}}$$

which means that if

$$z\sqrt{\frac{p(1-p)}{n}} = \epsilon$$

n is no larger than is necessary (where z and ϵ are as before). Now p is unknown before the sample is drawn so it needs to be 'guessed' to arrive at the formula

$$n = \left(\frac{z}{\epsilon}\right)^2 p(1-p)$$

It may be shown that the maximum error occurs when p is 0.5 and so in the absence of good information to help guess at p, or perhaps simply to play safe, n may be taken as

$$n = \left(\frac{z}{\epsilon}\right)^2 (0.5 \times 0.5) = \frac{1}{4}\left(\frac{z}{\epsilon}\right)^2$$

Similarly, when estimating the difference between two proportions, Section 7.6.2 shows that

$$z\sqrt{p_1(1-p_1)/n_1 + p_2(1-p_2)/n_2} = \epsilon$$

When sample sizes are equal ($n_1 = n_2 = n$)

$$z\sqrt{(p_1(1-p_1) + p_2(1-p_2))/n} = \epsilon$$

and thus

$$n = \left(\frac{z}{\epsilon}\right)^2 (p_1(1-p_1) + p_2(1-p_2))$$

To play safe put $p_1 = p_2 = 0.5$, giving $n = \frac{1}{2}\left(\frac{z}{\epsilon}\right)^2$

Example 7.16: What sample size is needed to compare the proportion of girls born in two hospitals so as to be 99% confident that the error is no more than 0.02 (i.e., 2%) with the same sample size at each hospital?
From the formula, playing safe,

$$n = \frac{1}{2}\left(\frac{z}{\epsilon}\right)^2 = \frac{1}{2}\left(\frac{2.58}{0.02}\right)^2 = 8320.5$$

Hence to satisfy the requirements n must be at least 8321 at each hospital. It is possible that the investigators will not be prepared to take over 16 000 sample observations, and hence in this case the requirements might have to be relaxed. Where proportions are concerned it is not unusual to find a large sample size requirement, although in this example the worst possible guess at p_1 and p_2 was used and prior knowledge might enable a more reasonable guess to be made.

7.14.2 From hypothesis tests

An alternative approach arises when a hypothesis test is to be performed with the probabilities of both Type I and Type II error specified (see Section 7.7.3). P(Type I error) is simply the significance level expressed as a probability, as used in all hypothesis tests. Rather than specifying P(Type II error) it is more usual to use its complementary probability, called the *power*, i.e.,

$$\text{power} = 1 - P(\text{Type II error})$$

Power is thus the probability of rejecting H_0 when H_1 is true. This can be evaluated for any particular outcome of the alternative hypothesis.

Alternatively, and more relevant here, an equation for power can be derived and then solved to find a value for n.

Example 7.17: In Example 7.8 the hypotheses tested at 5% significance were

$$H_0: \mu = 5.9$$
$$H_1: \mu > 5.9$$

Assuming, as in Example 7.8, that $\sigma = 1.2$, what sample size is needed to give a power of 0.99 when $\mu = 6.49$? (Note that power must always be evaluated for a specific value of μ under the alternative hypothesis.)

H_0 will be rejected whenever the test statistic $\dfrac{\bar{X} - \mu_0}{\sigma/\sqrt{n}}$ is 1.64 or greater.

Hence the probability that H_0 will be rejected is

$$P\left(\frac{\bar{X} - 5.9}{1.2/\sqrt{n}} \geqslant 1.64\right)$$

If $H_1: \mu = 6.49$ is true, $\bar{X}$ has a normal distribution with mean 6.49 and variance σ^2/n. Hence, when H_1 is true, $\dfrac{\bar{X} - 6.49}{1.2/\sqrt{n}}$ has the standard normal distribution.

Now
$$P\left(\frac{\bar{X} - 5.9}{1.2/\sqrt{n}} \geqslant 1.64\right)$$

$$= P\left(\frac{\bar{X} - 5.9 + (5.9 - 6.49)}{1.2/\sqrt{n}} \geqslant 1.64 + \frac{5.9 - 6.49}{1.2/\sqrt{n}}\right)$$

$$= P\left(\frac{\bar{X} - 6.49}{1.2/\sqrt{n}} \geqslant 1.64 + \frac{5.9 - 6.49}{1.2/\sqrt{n}}\right)$$

When H_1 is true this becomes an expression for the power, and the left-hand side becomes a standard normal, i.e.,

$$\text{power} = P\left(Z \geqslant 1.64 + \frac{5.9 - 6.49}{1.2/\sqrt{n}}\right)$$

Now the requirement is that, for $H_1: \mu = 6.49$, power should be 0.99 hence

$$P\left(Z \geqslant 1.64 + \frac{5.9 - 6.49}{1.2/\sqrt{n}}\right) = 0.99$$

By Table A.1, $\qquad\qquad P(Z \leqslant 2.33) = 0.99$

and consequently $\qquad\quad P(Z \geqslant -2.33) = 0.99$

Hence, $\qquad\qquad\quad 1.64 + \dfrac{5.9 - 6.49}{1.2/\sqrt{n}} = -2.33$

and $\qquad\qquad n = \dfrac{(1.64 + 2.33)^2 (1.2)^2}{(5.9 - 6.49)^2} = 65.2$

So a sample size of 66 will meet the requirements.

Examination of the last example will show that a general formula for sample size when testing

$$H_0: \mu = \mu_0$$
$$\text{vs} \quad H_1: \mu = \mu_1$$

with significance level $100s\%$ and power t is

$$n = \frac{(z_s + z_t)^2 \, \sigma^2}{(\mu_0 - \mu_1)^2}$$

where $\quad\quad P(Z \geqslant z_s) = s, \text{ i.e., } P(Z < z_s) = 1 - s$

and $\quad\quad\quad\quad\quad P(Z < z_t) = t$

Similar algebraic manipulation leads to sample size requirements for the difference between means, i.e., for

$$H_0: \mu_1 = \mu_2$$
$$\text{vs} \quad H_1: \mu_1 - \mu_2 = d \text{ (some specified difference)}$$

and where $n_1 = n_2 = n$, then

$$n = (\sigma_1^2 + \sigma_2^2) \frac{(z_s + z_t)^2}{d^2}$$

where z_s and z_t are as before.

Also for testing the value of a proportion, i.e.

$$H_0: \pi = \pi_0$$
$$\text{vs} \quad H_1: \pi = \pi_1$$

$$\text{then } n = \left(\frac{z_s \sqrt{\pi_0(1 - \pi_0)} + z_t \sqrt{\pi_1(1 - \pi_1)}}{\pi_0 - \pi_1} \right)^2$$

Finally for testing the equality of two proportions, i.e.,

$$H_0: \pi_1 = \pi_2$$
$$\text{vs} \quad H_1: \pi_1 = p_1^0, \pi_2 = p_2^0, \text{ for } p_1^0 \neq p_2^0$$

where $n_1 = n_2 = n$, then (approximately)

$$n = \frac{(z_s \sqrt{2p_c^0(1 - p_c^0)} + z_t \sqrt{p_1^0(1 - p_1^0) + p_2^0(1 - p_2^0)})^2}{(p_1^0 - p_2^0)^2}$$

where $\quad\quad p_c^0 = \frac{n_1 p_1^0 + n_2 p_2^0}{n_1 + n_2} = \tfrac{1}{2}(p_1^0 + p_2^0)$

Notice that this last case deals with specific alternatives for both πs. If it is desired to test H_0 against the alternative that $\pi_2 - \pi_1 = d$ (some specified difference) then substitute $p_2^0 = p_1^0 + d$ in the formula for n. In fact this formula is often used to detect a relative risk (i.e., ratio of probabilities) in a cohort study (see Section 9.2.6). In this case the alternative hypothesis is $\pi_2/\pi_1 = r$ (some specified relative risk) and so p_2^0 should be substituted by rp_1^0 in the formula for n. In both cases an educated guess at the value p_1^0 would be needed. This process is described fully in Schlesselman (1974). The same paper also gives a formula for n which is appropriate in case-control studies.

Each of the four formulae for sample size given in this section deal with situations where the tests are one-sided. It is possible to alter the formulae very slightly to cover the case of a two-sided test. Two possible situations then arise: first, where the alternative hypothesis is stated as a deviation from the null hypothesis in a specified direction (as before), and second, where the magnitude but not the direction of the alternative hypothesis is specified. An example of the latter is when testing for the difference between two means where the hypotheses are

$$H_0 : \mu_1 = \mu_2$$

vs $\quad H_1 : \mu_1 - \mu_2 = \pm d$ (a difference of d in *either* direction)

In the case of a specified direction for the alternative, and a two-sided test, the only change to the previous formulae for n is in the definition of z_s which now becomes

$$P(-z_s < Z < z_s) = 1 - s \qquad \text{i.e., } P(Z < z_s) = 1 - s/2$$

In the case of a non-specific direction for the alternative, and a two-sided test, the same change needs to be made to z_s as well as a change to z_t which is now defined by

$$P(-z_t < Z < z_t) = t \qquad \text{i.e., } P(Z < z_t) = \frac{t+1}{2}$$

Example 7.18: In the health promotion programme problem, suppose that the study team decided at the outset that a difference of 0.5 mmol/l in mean serum cholesterol concentration during the 10 year programme was so important that they would not want to run a risk of more than 1 chance in 100 of missing it. To account for the possibility that cholesterol actually rises during the programme, perhaps because of external influences conflicting with those of the programme, they wish to protect against missing differences in either direction. Assuming that samples of the same size are taken 'before' and 'after', what sample size is required on each occasion if the test is to be carried out at the 5% level of significance?

The hypotheses are

$$H_0: \mu_1 = \mu_2$$

$$H_1: \mu_1 - \mu_2 = \pm 0.5$$

and the presence of the $\pm$ makes the test two-sided.

The equation for n is

$$n = (\sigma_1{}^2 + \sigma_2{}^2) \left(\frac{z_s + z_t}{d} \right)^2$$

where $\qquad P(Z < z_s) = 1 - \dfrac{s}{2} = 1 - \dfrac{0.05}{2} = 0.975$

$$P(Z < z_t) = \frac{t+1}{2} = \frac{0.99 + 1}{2} = 0.995$$

giving $z_s = 1.96$, and $z_t = 2.58$.

Thus $$n = (\sigma_1^2 + \sigma_2^2)\left(\frac{1.96 + 2.58}{0.5}\right)^2$$

Using the value 1.4 for σ_1 and σ_2 suggested in Example 7.15 gives

$$n = (2 \times 1.4^2)\left(\frac{1.96 + 2.58}{0.5}\right)^2 = 323.19$$

So a sample of 324 people is required both 'before' and 'after' the programme.

Notice that the sample sizes $n_1 = 70$ and $n_2 = 50$ actually used in Example 7.10 were insufficient to meet the requirements specified here.

7.14.3 Requirements in percentage form

So far it has been assumed that when making inferences about means sample sizes will be worked out from requirements stated in absolute terms, as in Example 7.17 where the sample size was calculated to protect against missing a real mean of 6.49. Often requirements are more easily expressed and understood in the form of a percentage: for instance it may be important to detect when the value specified in the null hypothesis is 10%, or more, out.

When requirements are stated in percentage (or proportionate) terms it will be necessary to convert back to absolute terms to be able to use the results of Sections 7.14.1 and 7.14.2. Hence, in the problem of Example 7.17, to protect against missing the fact that $H_0: \mu = 5.9$ is 10% too small, then the particular alternative hypothesis of interest is

$$5.9 + \frac{10}{100} \times 5.9 = 6.49$$

In fact this was the alternative actually used in Example 7.17, so this problem has already been solved.

Requirements which the sample size has to meet may be expressed in various different percentage forms, too many to give a comprehensive list of examples here. When converting to absolute terms some educated guesswork is sometimes needed. For instance, in the problem of Example 7.18, suppose the requirement was that a difference of 5% achieved by the programme should only have a chance of 1 in 100 of being missed. Since the initial mean is unknown the requirement cannot be exactly expressed in absolute terms. At best this initial mean can be judged by previous experience with similar situations, and so if the 'before' mean were thought to be 6.0, then the important difference, d, would be $5/100 \times 6.0 = 0.3$.

7.15 Paired data

The description of confidence intervals and tests when two means or two proportions are to be compared has so far assumed that the samples drawn from the two populations are independent. One case where this is not true is where there is dependence between pairs of data items, one member of the pair coming from each of the samples. This situation arises commonly in medical studies, for example when two types of eye drop are to be compared and both drops are given to each patient in the study, one drop in each eye. The results from left and right eyes for the same patient cannot possibly be independent as they will depend to some extent on the physiology of the individual. Another example of paired data is the 'before and after' study in which the same individuals are observed each time. In the health promotion programme example used extensively in this chapter, the data would be paired if the same individuals had had their cholesterol level measured before and after the programme. A final example of paired data is the case-control study (see Section 2.5.3) where every case is matched with a single control. The situation of multiple controls is covered by Breslow and Day (1980).

7.15.1 Confidence interval and test for comparing means

This section provides alternatives to Sections 7.5 and 7.9 when the data are paired. The method is to calculate the differences between the paired data items and then treat the differences as a single set of data. Differencing removes the dependence in the data.

If $\mu_d = \mu_1 - \mu_2$ is the difference between the two means, then a CI for the difference between μ_1 and μ_2, where data are paired, is obtained simply by finding a CI for μ_d directly. By Section 7.4.3, when the population variance is unknown, the CI is

$$\bar{x}_d \pm t_{n-1} \frac{s_d}{\sqrt{n}}$$

where $\bar{x}_d$ and s_d are the sample mean and standard deviation of the differences.

To test the null hypothesis $H_0: \mu_1 = \mu_2$ the equivalent hypothesis $H_0: \mu_d = 0$ is tested. By Section 7.8.2 the test statistic, when the population variance is unknown, is

$$\frac{\bar{x}_d - 0}{s_d/\sqrt{n}} = \frac{\bar{x}_d}{s_d/\sqrt{n}}$$

to be compared with t on $n - 1$ degrees of freedom. This test is called the *paired t test*.

Example 7.19: Patients who regularly suffer from headaches were entered into a clinical trial of two headache tablets, a new and an existing form of medication. Patients were given the two tablets, which were indistinguishable, in packets labelled 'A' and 'B'. They were told to take the tablet in packet A when their next headache occurred and the other tablet for a subsequent headache (after a specified time interval to avoid the possibility of carry-over effects). A record was kept, by the trial administrator, of which tablet was placed in which packet.

After taking a tablet patients were asked to record the amount of time that elapsed before their headache cleared. The results were:

Patient number	Time to clearing/mins. with new tablet	old tablet	Difference (mins.) new – old
1	25	21	4
2	28	18	10
3	17	42	– 25
4	20	21	– 1
5	50	41	9
6	31	38	– 7
7	21	28	– 7
8	26	15	11
9	14	16	– 2
10	12	24	– 12
11	19	18	1
12	22	29	– 7

To test whether there is a difference in clearing time between the tablets the hypotheses

$$H_0: \mu_{new} = \mu_{old} \qquad \text{which are equivalent to} \qquad H_0: \mu_d = 0$$
$$H_1: \mu_{new} \neq \mu_{old} \qquad\qquad\qquad\qquad\qquad H_1: \mu_d \neq 0$$

are tested (where the subscript d again denotes 'differences').

Now by working with the right-hand column above, $\bar{x}_d = -2.167$ and $s_d = 10.373$ and thus the test statistic

$$\frac{\bar{x}_d}{s_d/\sqrt{n}} = \frac{-2.167}{10.373/\sqrt{12}} = -0.724$$

From Table A.2, t on $n - 1 = 11$ d.f. is 1.796 for a two-sided 10% test. Since -0.724 is greater than -1.796, H_0 fails to be rejected at 10% and it is concluded that there is no evidence that either of the tablets clears headaches more quickly than the other.

Notice that it could well be that significance was not found simply because the sample size was so small (the data do slightly favour 'new'). Results from Section 7.14.2 show that 66 patients would be needed in order to detect when, an average, the new tablet does better than the old by 5 minutes with a probability of 0.95 (assuming a 5% significance test and using the sample estimate of variance).

7.15.2 Confidence interval and test for comparing proportions

This section provides alternatives to Sections 7.6.2 and 7.10.2 when the data are paired. Suppose two samples are available and these are paired in some way. Some phenomenon is observed for each sample member, and the result is recorded as either a 'success' or 'failure'. Each pair of results is then either SS or SF or FS or FF (where S = success and F = failure). The number of pairs of each type will be denoted n_{SS}, n_{SF}, n_{FS} and n_{FF} respectively. These results can be put in the form of a table:

Sample 1	Sample 2	Number of pairs
success	success	n_{SS}
success	failure	n_{SF}
failure	success	n_{FS}
failure	failure	n_{FF}
		total n

A 95% CI for the difference in the proportion of successes in the populations from which the samples were drawn is

$$p_1 - p_2 \pm 1.96 \frac{1}{n} \sqrt{n_{SF} + n_{FS} - \frac{(n_{SF} - n_{FS})^2}{n}}$$

where p_1 and p_2 are the proportions of successes in samples 1 and 2.

A test of the null hypothesis that the two population proportions are equal is given by comparing the test statistic

$$\frac{n_{SF} - n_{FS}}{\sqrt{n_{SF} + n_{FS}}}$$

with the standard normal distribution. This is called *McNemar's test* (see also Section 9.2.4). Both this test and the CI are based upon normal approximations (see Armitage and Berry (1987)) and so need reasonably large sample sizes to be acceptable procedures.

Example 7.20:   Following an outbreak of legionnaires' disease in a town a case-control study was undertaken to locate the cause. Each of 38 people with a confirmed or suspected case of legionnaires' disease was paired with a control, matched for age, sex, area of residence and social class. Each of the cases and controls was asked whether he or she had visited various parts of the town around the suspected time of the outbreak. One part of the town was a shopping centre, for which the results were:

Cases	Controls	Number of pairs
visited	visited	25
visited	not visited	11
not visited	visited	1
not visited	not visited	1
		total 38

The test statistic

$$\frac{n_{SF} - n_{FS}}{\sqrt{n_{SF} + n_{FS}}} = \frac{11 - 1}{\sqrt{11 + 1}} = 2.89$$

which has a P value of about 0.002 for a one-sided test. Hence there is evidence that cases are more likely to have visited the shopping centre than controls, and the shopping centre is a possible source of legionella bacteria.

7.16 Non-normal analyses

Throughout this chapter analyses of data have been based upon the normal distribution. This is acceptable if the original data follow a normal distribution quite closely, as can be checked by constructing diagrams, such as histograms and boxplots, of the data. Otherwise it is only acceptable if the sample size is large enough for the central limit theorem to ensure that the mean follows a normal distribution. The question remains what to do when neither of these situations pertains.

For example the length of stay data presented in Table 3.2 are certainly not even approximately normally distributed. This is clear from Fig. 3.11 (skewed histogram), Fig. 3.13 (skewed frequency polygon), Fig. 3.15 (ogive not S shaped) and Fig. 4.6 (skewed boxplot). Since the original data are so far from the normal the sample size of 56 is small enough to raise doubts about the validity of using such things as t tests with these data.

In such cases there are three possible ways of proceeding, as will now be described.

7.16.1 Using the true distribution

If the true probability distribution followed by the data is known, CIs and tests can sometimes be derived for that distribution just as was done for the normal distribution in this chapter. It is, however, unusual to be sure of the true distribution, and certainly this is the case in the length of stay example. Furthermore the procedure can be both tedious and complex.

Kendall and Stuart (1973), Section 20.9, give an example where a CI is derived from a binomial distribution.

7.16.2 Transforming the data

It is sometimes possible to carry out the same simple mathematical operation on every data value to arrive at 'transformed data' which closely follow a normal distribution. There are countless transformations that could be tried but some have frequently been found useful in special situations.

When the data are skewed the square root and logarithm transformations are most commonly successful. That is, when x represents a data value and y the corresponding transformed data value, the transformations are $y = \sqrt{x}$ or $y = \log x$ respectively. When the data are counts, such as the numbers of patients waiting in a clinic, the square root transformation is, again, often successful. When the data are proportions or percentages, such as daily percentage bed occupancies, the inverse sine transformation, $\sin y = x$ or $y = \sin^{-1} x$, is most useful. Snedecor and Cochran (1980), Section 15.10, give examples of the use of transformations. Obviously, the transformed data should be examined to ensure that the transformation has had the desired effect, that is, that the 'new' data are approximately normal. There is no guarantee that any of these transformations will work in any particular situation. Using a statistical package it is very easy to try various transformations, each time drawing histograms and such like to see if the transformation has roughly achieved normality. Some packages will also produce a *normal plot*, which is a special graph that is a perfect straight line for normal observations (see Ryan *et al.* (1985), Section 7.6). A more formal test for the fit of a normal distribution is suggested in Section 9.7.1, although in practice this is rarely used.

Although the square root transformation did give some improvement, judged by the symmetry of the histogram amongst other things, neither this nor any other simple transformation tried on the length of stay data gave particularly impressive results. So in this case another method of analysis might be preferable.

7.16.3 Non-parametric methods

Many of the methods of inference described in this chapter investigate means based on the assumption that the sample mean follows a normal distribution. When this is not appropriate non-parametric methods, which make no distributional assumptions, can be used to investigate medians. Since it was suggested in Section 4.3.4 that the median is the appropriate average for the length of stay data, and because no alternative method has been found, a non-parametric analysis could be useful for these data.

The entire scope of non-parametric statistics is vast, providing alternatives to most of statistical inference that is based on the normal distribution, including inferential correlation and regression (see Chapter 8). In situations where the normality assumption breaks down non-parametric methods give superior results, although when the assumption is correct they are inferior. One of the most common situations where non-parametric procedures are appropriate is when the data are ranks, for example patients ordered by the severity of their illness.

Although many of them are very easy to use, there are far too many non-parametric procedures to give an adequate account in this introductory text except in the particular case of the analysis of tables in Chapter 9. The philosophy behind their use is, however, exactly as for the inferential processes described in this chapter, so that the reader should be able to understand accounts of non-parametric methods given elsewhere with relative ease. Readable and comprehensive accounts of the subject are the books by Conover (1980) and Siegel (1956).

7.17 Using computer packages

Much of the business of calculating estimates and carrying out hypothesis tests, such as calculating square roots and reciprocals and looking up numbers in tables, is quite simple and yet tedious. This makes the use of a statistical computer package, which gives such things as the bounds of CIs and P values for tests, very attractive in this context. The other great advantage of using a package is that the assumptions behind the inferences can easily be checked.

There are two points of warning. First, if a package is asked to perform, say, a t test on highly skewed data, such as the length of stay data, it will not complain (although the days of expert systems which will do so may not be far away).

Second, it is all too easy to do many different analyses of the same data and take the results which best suit the user, a method appropriately known as 'ransacking the data'. This is especially true in hypothesis tests where there are sometimes many tests that can be done on the same data. Remember that a hypothesis test carried out at the 5% level of significance has a probability of 5/100 of rejecting the null hypothesis even when it is true. If, say, 5 independent 5% significance tests are carried out on the same data the chance that *at least one* rejects H_0 when it is in fact true is $1 - (95/100)^5 = 0.23$. Thus a significant result found in these circumstances should, at best, be treated with extreme caution. See Section 9.3.5 for a particular example of this problem.

Although computers are of great help in inferential statistics, few manuals, let alone the packages themselves, interpret the results or point out the limitations. It is essential for the user to understand the philosophy behind the subject, as outlined in this chapter, before embarking on any analyses.

Exercises

For these exercises you may assume normal distributions where necessary.

7.1 For the GP data in Appendix 2 answer the following questions (the answers to the exercises of Chapters 3 and 4 will provide a starting point).

(a) Find a 95% confidence interval for the mean systolic blood pressure.
(b) Find a point estimate for the proportion of smokers.
(c) Find a 99% confidence interval for the proportion of smokers.
(d) Is there any evidence that the standard deviation of systolic blood pressure is different for smokers and non-smokers?
(e) Use your answer to part (d) to find a 95% confidence interval for the difference between mean systolic blood pressure in smokers and non-smokers.
(f) Find a 95% confidence interval for the difference between the proportion of smokers with a desirable (class 0) body mass index (BMI) and the proportion of non-smokers with a desirable BMI.
(g) It is suggested that the mean systolic blood pressure for men aged 45–64 in the area served by the GP practice is likely to be 150 mm Hg. Use your answer to part (a) to test this assertion at the 5% level of significance.
(h) Similarly it is suggested that the mean diastolic blood pressure is likely to be 95 mmHg. Test this assertion. Give an approximate P value for the test.
(i) Test the hypothesis that half of the population from which the sample has been drawn are smokers against the alternative hypothesis that the true proportion is less than half. Give the P value for the test.
(j) Test the hypothesis that the mean systolic blood pressure is the same for smokers and non-smokers at the 10% level of significance.
(k) A dedicated smoker argues that people who smoke are less likely to over-eat and hence the proportion of class 2 BMI men should be less amongst smokers than amongst non-smokers. Test the null hypothesis that the proportions are the same against this alternative, specifying the P value.

7.2 In the study described in Question 3.5 the mean and standard deviation birth weights of the 3945 single births were 3302 and 519 grams respectively (compare these with the values estimated from the grouped frequency distribution in Question 4.7). In the same study 137 twins were encountered; their mean birth weight was 2445 grams with a standard deviation of 460 grams. Test the hypothesis that the two means are equal against the alternative that twins are, on average, lighter.

7.3 Edwards and Wilkins (1987) describe a study of six patients with acute cardio-respiratory failure. The heart rates (in beats/minute) of each patient before and after expansion of plasma volume were as follows.

Patient number	Before	After
1	177	102
2	155	90
3	180	110
4	147	95
5	170	115
6	150	120

Find a 90% confidence interval for the average effect of expansion of plasma volume.

7.4 A patient satisfaction survey is planned for a large acute hospital. The most crucial question on the questionnaire is:

'Are you satisfied with the amount of information given to you by your doctor?'

and it is important to be 95% confident that the true proportion who answer 'yes' to this question is found to within ± 0.05. A small pilot study has indicated that this proportion is likely to be around 0.7. Using this as a guess for the true value of the proportion determine the minimum sample size that satisfies the requirement.

7.5 In a test for the effects of alcohol ten men were randomly allocated to either group A or group B. Men in group A were given a pint of beer and men in group B a pint of orange juice. Soon after finishing their drinks each subject underwent a test of skill and judgement. Scores on the test were:

Group A (beer)	Group B (orange juice)
15	24
17	21
8	11
9	23
17	22

(a) Is there a significant difference between the average scores of the two groups at the 5% level? Looking at the figures it seems that group B tends to have bigger scores. Can you explain your answer?

(b) Experts consider a difference of 5 or more to be indicative of a real effect. Using the pooled estimate of variance from part (a), determine the sample size that would be required in each group to be 95% sure of detecting a difference of ± 5 between the two mean scores with a 5% significance test.

As the wording implies, a two-sided test is expected in this question. If you were in charge of the investigation would you use a one-sided test instead?

8
Correlation and regression

8.1 Introduction

There are many practical problems in the analysis of health data where the object is to determine the relationship between two quantitative variables. Examples are determining how the cost of running a hospital varies with size of hospital and how the number of dental cavities reported changes with the amount of fluoride in the local water supply.

Correlation is a measure of the straight-line or *linear* relationship between two variables. If there is a perfect correlation then the graph of the two variables is a straight line and the two variables always change in constant proportion to each other. Correlation has, however, limited utility because, while it can show that a linear relationship exists, it does not determine what that relationship actually is. This is the object of a linear regression analysis. Often a regression analysis is undertaken in order to predict one variable from another.

In this chapter the basic concepts of correlation and simple linear regression will be explained. Two extensions of simple linear regression will also be covered: multiple regression, where many variables are involved, and non-linear regression, where the relationship between two variables is more complicated than a simple straight line.

One thing that is essential to do before undertaking a correlation or regression analysis is to plot a graph of the observed data on the variables in question; such a plot is called a *scatter diagram* or *scattergram* (for example see Fig. 8.3). In Chapter 3 the virtues of diagrams as aids to understanding data were extolled; in exploring relationships, scattergrams often give greater insight than relatively complex correlation and regression analyses. Furthermore the scattergram can suggest which formal analysis is suitable for particular data, thus possibly avoiding useless effort or misleading results. Since the production of even a sketch graph can be quite time-consuming, the use of a computer package is advantageous. This is also very much the case when more formal analyses are performed, as will be seen later in this chapter.

8.2 Correlation

Suppose that a set of n observations $x_1, x_2, \ldots, x_n$ is available upon the first variable of interest, and a second, related, set $y_1, y_2, \ldots, y_n$ is available on the

196

second variable of interest. The object of correlation is to measure the strength of the linear relationship between the x values and their corresponding y values. For example if the linear relationship between hospital size (x) and running costs (y) can be shown to be strong then the effect of increasing size upon the running costs will be clear.

The strength of a linear relationship is measured by the *correlation coefficient*, sometimes called the *product moment* correlation coefficient. The form of the correlation coefficient will be derived via a series of arguments, beginning with an examination of the form of the scattergram for straight-line data, when the scattergram can have two possible forms, as illustrated by Fig. 8.1. In Case 1, y increases as x increases but in Case 2, y decreases as x increases. Notice that the axes have been drawn through the sample means $\bar{x}$ and $\bar{y}$ in Fig. 8.1. Although this is rarely done with real data, in this case it provides useful reference lines.

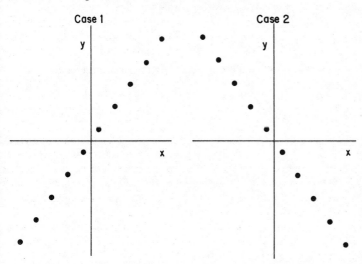

Fig. 8.1   Perfect correlations.

Case 1 is characterized by

(i) values of y above $\bar{y}$ occurring with values of x above $\bar{x}$,
(ii) values of y below $\bar{y}$ occurring with values of x below $\bar{x}$.

Case 2 is characterized by

(i) values of y above $\bar{y}$ occurring with values of x below $\bar{x}$,
(ii) values of y below $\bar{y}$ occurring with values of x above $\bar{x}$.

Now consider the summary statistic,

$$\frac{1}{n-1}\,\Sigma(x_i - \bar{x})(y_i - \bar{y})$$

which is called the sample *covariance* of x and y. That this statistic will be positive for Case 1 and negative for Case 2 can be seen from the formula, since Case 1 must give either a positive times a positive or a negative times a negative for every component of the summation, while Case 2 always gives a positive times a negative.

The covariance thus gives a means of distinguishing between Case 1, called perfect *positive* correlation and Case 2, called perfect *negative* correlation. Real-life data on pairs of x and y values might well approximate to a straight line, but are unlikely to form a perfect straight line. A measure of how near to perfect they are would be useful, but the covariance is not suitable since it depends upon the units of measurement. Thus if x and y are weight and height of patients then the covariance is bound to be bigger if height is measured in centimetres rather than metres.

To take account of this dependence a new measure, the correlation coefficient, is derived by dividing the covariance by the product of the two sample standard deviations. This new measure is thus unit free. The correlation coefficient is given the symbol r, and so

$$r = \frac{\text{covariance of x and y}}{(\text{s.d. of x})(\text{s.d. of y})} = \frac{\Sigma(x_i - \bar{x})(y_i - \bar{y})}{\sqrt{\Sigma(x_i - \bar{x})^2 \Sigma(y_i - \bar{y})^2}}$$

where the right-hand side is derived by using the formulae for covariance and

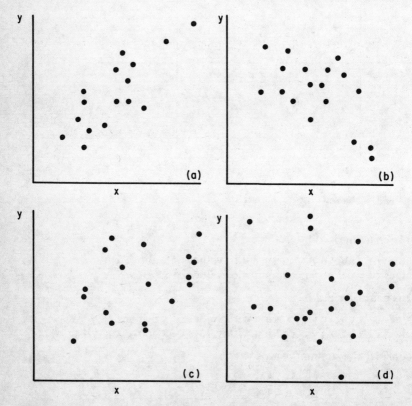

Fig. 8.2 Examples of correlations. (a) r = +0.9; (b) r = −0.8; (c) r = +0.6; (d) r = 0.0. (Reproduced from *Statistical Methods in Agriculture and Experimental Biology*, R. Mead and R.N. Curnow (1983), Chapman and Hall.)

standard deviation and then cancelling the n − 1 divisor in the top and bottom lines.

It can be proved that the correlation coefficient can only take values between −1 and +1. It is −1 when x and y are related by a perfect straight line of *negative slope* (Case 2 in Fig. 8.1) and +1 when x and y are related by a perfect straight line of *positive slope* (Case 1). The nearer to a perfect linear relationship, the nearer to either −1 or +1 the correlation coefficient becomes. A value for r of zero indicates complete absence of any linear relationship. Figure 8.2 shows a few examples.

8.2.1 Calculating the correlation coefficient

In Section 4.6.1 the term $\Sigma(x_i - \bar{x})^2$ was given the symbol S_{xx}. If similar definitions

$$S_{yy} = \Sigma(y_i - \bar{y})^2$$
$$S_{xy} = \Sigma(x_i - \bar{x})(y_i - \bar{y})$$

are made then

$$r = \frac{S_{xy}}{\sqrt{S_{xx}S_{yy}}}$$

Also in Section 4.6.1 an alternative formula for S_{xx} which is easier for hand calculations was given as

$$S_{xx} = \Sigma x_i^2 - \frac{1}{n}(\Sigma x_i)^2$$

Table 8.1 Birth data for West Midlands health districts. Note that a low-weight birth has weight less than or equal to 2.5 kg. (Data from DHSS Series LHS 27/1 and OPCS Series DH3.)

Low-weight births (%)	Perinatal mortality per 100 births
8.17	1.90
8.35	1.59
8.87	1.54
7.59	1.88
6.98	1.28
8.89	1.96
10.46	2.26
9.54	1.86
9.00	1.50
5.95	1.27
6.15	1.59
6.28	1.05
8.80	1.68
9.29	1.67
9.40	1.92
8.65	1.47
9.23	1.99
10.49	1.92
10.10	1.78
9.60	1.53

Similarly convenient formulae are

$$S_{yy} = \Sigma y_i^2 - \frac{1}{n}(\Sigma y_i)^2$$

$$S_{xy} = \Sigma x_i y_i - \frac{1}{n}(\Sigma x_i)(\Sigma y_i)$$

Statistical computer packages and some calculators with statistical functions will produce r automatically, requiring only the effort of keying in the data. If neither is available, use of the hand calculation formulae will be the most efficient.

Example 8.1: Levin (1985) describes a study of births in health districts in the West Midlands. For the sake of illustration a sample of his data has been extracted and is presented in Table 8.1. The sample data give annual low birth weight percentages and perinatal mortality rates for selected years between 1978 and 1981 for selected districts. Figure 8.3 is the scattergram of these data.

Let x represent percent low-weight births and y perinatal mortality rate. For these data n = 20 and

$$\Sigma x_i = 171.79 \qquad \Sigma y_i = 33.64$$
$$\Sigma x_i^2 = 1510.925 \qquad \Sigma y_i^2 = 58.199$$

and

$$\Sigma x_i y_i = 294.137$$

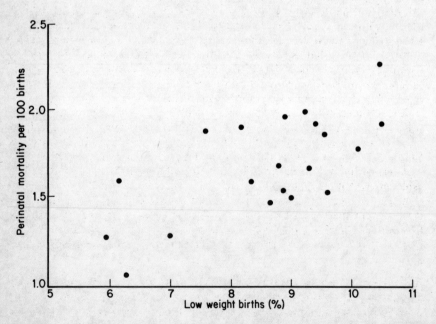

Fig. 8.3 Perinatal mortality rate against percentage of low-weight births, sample of West Midlands health districts, 1978–81. (From Levin, 1985.)

Hence $\quad S_{xx} = 1510.925 - \dfrac{1}{20}(171.79)^2 \qquad = 35.335$

$\qquad S_{yy} = \quad 58.199 - \dfrac{1}{20}(33.64)^2 \qquad = 1.617$

$\qquad S_{xy} = \quad 294.137 - \dfrac{1}{20}(171.79)(33.64) \quad = 5.186$

and $\qquad r = \dfrac{5.186}{\sqrt{35.335 \times 1.617}} = 0.69$

There is a fairly high positive correlation between the percentage of low-weight births and the perinatal mortality rate. There is evidence that one increases with the other in constant proportion.

8.2.2 Inferences about correlation

The correlation coefficient, r, calculated from a sample is an estimate of the correlation between the variables concerned for the entire population from which the sample was drawn. This is true in much the same way as the sample mean $\bar{x}$ estimates the population mean μ. Following the convention used in Chapter 7 the population correlation coefficient will be denoted by ρ (the Greek letter rho).

Confidence intervals and hypothesis tests for ρ can be calculated directly if the $x-y$ data follow a joint (*bivariate*) normal distribution. The exact form of this distribution need not be described here, but if x and y jointly follow the bivariate normal then each separately will follow a normal distribution at any fixed value of the other. If there is reason to doubt this assumption then a non-parametric analysis might be more appropriate (see, for example, Conover (1980), Section 5.4). One special case that would call for a non-parametric approach is where the data are ranks, such as two doctors' orderings of effectiveness of methods of contraception.

When the assumption of bivariate normality is valid, inferences about ρ follow in much the same way as for μ etc. in Chapter 7. The confidence interval for ρ is rather messy to calculate and is, in any case, rarely used. A simple description is given by Clarke and Cooke (1983), Section 20.8.

A test of the null hypothesis H_0: $\rho = 0$ is obtained from Table A.5. This table gives critical values of r on $n-2$ degrees of freedom at various significance levels. If ρ is found to be not significantly different from zero then the conclusion must be that there is no evidence of a linear association between the variables. The possibility should be borne in mind, however, that there may be some other relationship between the variables (see Example 8.3).

Example 8.2: For the births data of Example 8.1, $n = 20$ and $r = 0.69$. From Table A.5 the 0.1% critical value of r on $n-2 = 18$ degrees of freedom is 0.679 and hence the test is significant with a P value less than 0.001. The hypothesis of no linear relationship between perinatal mortality and low-weight births is rejected.

8.2.3　Limitations

As has been stated, correlation only measures linear relationship. This definition is really very limited in its application and is frequently misinterpreted to mean something more.

Causality

Correlation does *not* imply causality but merely association. A health spokesman, in a recent radio interview, stated that the number of deaths of babies in hospital was highly positively correlated with the number of cots available in special care baby units. He used this to demonstrate that special care baby units are not beneficial. In this example, as in many others, it is the unaccounted presence of a third variable that is producing what appears to be causality. More cots in special care baby units mean more admissions, and furthermore these are likely to be babies with a higher than average risk of death. Hence it is to be expected that the number of deaths of babies in hospital would increase in these circumstances.

Non-linear association

It may be incorrect to conclude that there is no relationship between variables simply because their correlation coefficient is near zero. There could still be a non-linear relationship.

Table 8.2　Number of notifications of cases of typhoid and paratyphoid, England and Wales 1974–84. (Data from *Annual Abstract of Statistics* (1986), Central Statistical Office.)

Year	No. of Notifications
1974	206
1975	280
1976	283
1977	296
1978	331
1979	293
1980	290
1981	254
1982	234
1983	259
1984	206

Example 8.3:　Consider the data in Table 8.2. It is clear that there is not a straight-line relationship between the number of notifications and time in Fig. 8.4. In fact, for these data r is only 0.27. On the other hand there clearly *is* a relationship of some sort; notifications rise fairly steadily to 1978 and then drop in a similar fashion. As will be seen in Section 8.5.2, a curve called a *quadratic* fits these data well.

Range of validity

The value of r over the range of observed data may be totally unrealistic when that range is extrapolated. Thus with many illnesses, increasing the dosage of

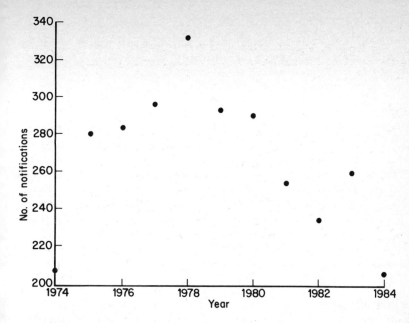

Fig. 8.4 Number of notifications of typhoid and paratyphoid, England and Wales, 1974–84. (From *Annual Abstract of Statistics*, CSO.)

a drug will improve the patient's condition according to a linear form *provided* the range of dosages is small. When the dosage gets too high the patient will suffer from side-effects of the drug which are greater than its benefits, and he will get worse instead of better.

8.2.4 Spurious correlations

The definition of the correlation coefficient is such that it approaches ± 1 as the relationship between the variables considered approaches a straight line, but unfortunately there are other circumstances in which r can be close to ± 1. An example of this is where an outlier is present in the data, that is a point on the scattergram which is far removed from all the others. As explained in Chapter 4 outliers can distort the mean and standard deviation, so it follows that they can also distort the correlation coefficient which is defined in terms of means and standard deviations. (See also Section 8.6.)

Provided that the scattergram has been drawn it will be clear when the data do not approximate to a straight line, and thus it will be clear when a correlation is spurious.

8.3 Linear regression for two variables

The relationship between two variables is illustrated by a scattergram and the degree of linear association between two variables is measured using the correlation coefficient. It is generally useful, when a linear relationship exists, to specify precisely how the two variables change together. This will allow prediction of one variable from another. For example when the equation of the straight line that relates hospital running costs to size is known, the anticipated costs for any given size of hospital can be found.

The straight line which expresses the relationship between two variables is known as the *linear regression line*. In fact it makes a difference which variable is chosen to act as the predictor and which the predicted. In many situations, such as the hospital example, it is clear which is which. In other situations a choice should be made. In essence the predictor is assumed to be fixed whilst the predicted variable is assumed to be random. The predictor variable is called the *x variable* or the *explanatory variable*, and the predicted variable is the *y variable* or the *response variable*. Some textbooks name these the independent and dependent variables respectively, but these names are misleading since, just as in correlation, there is no guarantee that x actually causes y.

Before going on to derive the regression line it may be beneficial to review the mathematical form of a straight line. Figure 8.5 shows a straight line which cuts the y axis at a, the *intercept*, and has an angle of *slope* whose tangent is b. That is for every 1 unit moved in the horizontal direction the line moves up b units in the vertical direction. The straight line illustrated has the equation

$$y = a + bx$$

Once a and b are known this equation can be used to find the value of y for any given value of x.

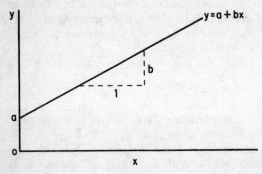

Fig. 8.5 Graph of a straight line.

8.3.1 The method of least squares

The primary object of a linear regression analysis is to find the straight line which relates the x and y variables. This straight line will be represented by

$$y = \alpha + \beta x$$

(where α and β are the Greek letters alpha and beta).

This is the line that could be determined if all possible data on x and y were available, but in most real-life situations only a sample of data is available. A straight line relating x and y for the sample data can be found. This is represented by

$$y = a + bx$$

the estimated regression line, where a and b estimate the α and β respectively of the true regression line.

Even when there is a known underlying physical mechanism which ensures that x and y are highly correlated it will usually happen that the observed data do not make a perfect line due to sampling variation. For example, height and weight are usually very highly correlated but most samples of individuals will contain people who are overweight or underweight for their height. The estimated regression line is that which best fits the data.

The easiest way to fit a line to data is to plot the scattergram and draw in the line which appears, to the eye, to go as near as possible to all the points. The reader might like to try this for Fig. 8.3. The equation of the line y = a + bx can then be determined by reading the intercept, a, and the rate of vertical to horizontal movement, b, from the line.

The problem with the 'fitting by eye' method is the subjectivity: different people will decide upon different regression lines. An objective, mathematical approach, is to be preferred. One idea would be to try to reproduce the rationale of the 'eye' method, to minimize the differences between the observed points in the scattergram and the estimated regression line. Since all these differences should be as small as possible it would be sensible to seek to minimize the overall sum of the differences.

Let the y value that is predicted by the linear regression line at the x value x_i be denoted by $\hat{y}_i$. Figure 8.6 shows both the observed y value, y_1, and the predicted y value, $\hat{y}_1$, which correspond with the first observed value of x, x_1.

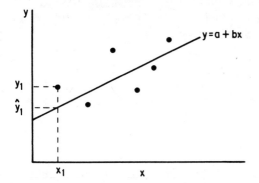

Fig. 8.6 Observed and fitted values for linear regression.

The suggested mathematical method would then be to minimize $\Sigma(y_i - \hat{y}_i)$ by moving the line around. This, however, can give strange results, since positive and negative differences can cancel each other out and make even a line which misses all the observations seem like a perfect fit. To avoid this problem the same solution that was used (in Section 4.6) to arrive at the standard deviation is employed. That is, the differences are squared so as to remove all the negatives. The result is called the method of *least squares*.

The method of least squares, then, is to find the line which minimizes

$$\Sigma(y_i - \hat{y}_i)^2$$

Since $\hat{y}_i$ lies on the regression line, $\hat{y}_i = a + bx_i$ so that

$$\Sigma(y_i - (a + bx_i))^2$$

is to be minimized with respect to a and b. The a and b that achieve this minimum are called the least squares estimates. Differential calculus may be used to prove that

$$b = \frac{S_{xy}}{S_{xx}} \qquad \text{and} \qquad a = \bar{y} - b\bar{x}$$

Example 8.4:   For the birth data of Example 8.1,

$$b = \frac{5.186}{35.335} = 0.147$$

and

$$a = \frac{33.64}{20} - 0.147 \times \frac{171.79}{20} = 0.419$$

using values previously calculated. Quoting values to two decimal places the estimated regression line is then

$$y = 0.42 + 0.15x$$

That is, perinatal mortality per hundred births can be expected to increase by 0.15 for every extra low-weight birth in a hundred. If there were no low-weight births then the expected perinatal mortality would be 0.42 in a hundred.

Of course these expectations are conditional upon the linear regression line being a true representation of the underlying physical association between the variables. This point will be considered in Section 8.3.2.

8.3.2 Goodness of fit

A linear regression line can be fitted, using least squares or otherwise, to *any* set of x–y data. On the other hand, it is certainly not the case that every pair of variables is linearly related, even approximately, so it will *not* always be sensible to fit a linear regression line. The scattergram is an important guide, but it is also useful to quantify the goodness of fit of the regression line to the observed data.

Least squares decreed that the sum of squares of the $y_i - \hat{y}_i$ should be minimized and so an obvious measure of goodness of fit is the size of $\Sigma(y_i - \hat{y}_i)^2$ once minimization has been achieved. When the regression line has been determined by this minimization, the differences $y_i - \hat{y}_i$ are known as the *residuals*. As Fig. 8.6 shows, the residuals are simply the vertical differences between the observed and the predicted, or 'fitted', y_i.

By substituting the values of the least squares estimates of a and b it can be shown that the *residual sum of squares*,

$$\Sigma(y_i - \hat{y}_i)^2 = S_{yy} - \frac{S_{xy}}{S_{xx}}$$

If the residual sum of squares is small then the linear regression line is a good fit to the data. However 'small' can only be judged in relation to the total variation in the vertical dimension, that is the variation of the ys. If the $n - 1$ divisor (degrees of freedom) for the variance is put aside for the moment, this variation in y is measured by $S_{yy} = \Sigma(y_i - \bar{y})^2$. In fact S_{yy}, the total sum of squares, is made up of two components: the variation in y which is explained by the regression line of y on x, and what is left over, the unexplained variation, which is simply the residual sum of squares.

The partition of the total sum of squares is conveniently presented in the form of an *analysis of variance table*:

Source of variation	Sum of squares	Degrees of freedom
Due to linear regression	$\dfrac{S_{xy}^2}{S_{xx}}$	1
Residual	$S_{yy} - \dfrac{S_{xy}^2}{S_{xx}}$	$n - 2$
Total	S_{yy}	$n - 1$

In this table the regression sum of squares has been calculated by subtracting the residual sum of squares from the total. The degrees of freedom for the total sum of squares are known to be $n - 1$ (the divisor for the variance) and the degrees of freedom for regression must be 1 because y was regressed on a single x variable. This leaves, by subtraction, $n - 2$ degrees of freedom for the residual.

The analysis of variance table has several uses, and most statistical packages will produce it whenever a regression line is calculated. One important use is to assess the relative sizes of the explained and unexplained components of variance. As already indicated a good fit is achieved when the residual, or unexplained sum of squares is small relative to the total variation in y. A convenient measure of goodness of fit, which will get bigger as the fit improves, is

$$r^2 = \frac{\text{regression sum of squares}}{\text{total sum of squares}} = \frac{S_{xy}^2/S_{xx}}{S_{yy}} = \frac{S_{xy}^2}{S_{xx}S_{yy}}$$

This measure is called the *coefficient of determination*. Notice that, as the symbol r^2 implies, this is simply the square of the correlation coefficient, r. Since r measures linearity it is not surprising that r^2 should provide a measure of the goodness of fit of a straight line. As r lies between -1 and $+1$ it must be that r^2 can only take values between 0 and 1.

Even when the value of r^2 is large it is still possible that the scattergram suggests that some other, non-linear, relationship exists between y and x (see Example 8.10 for instance). Then the linear regression line is certainly not an appropriate model for the data.

To see how this could happen it is useful to think of the observed data as being made up of two components

$$\text{observed data} = \text{fitted value} + \text{residual}$$

If the fit is good then the residual should be relatively small, as measured by

r^2. Also if the fit is good the 'right' model should have been used, that is, all the systematic, predictable, part of the variation should be in the fitted values. The residuals should be random, and should not present a systematic pattern. It is quite possible for the residuals to all to be small and yet, when taken together, make up a discernible and predictable pattern.

It is, therefore, essential to examine the residuals from a linear regression (or any other statistical) model. One particularly useful way of doing this is to plot the residuals against fitted values. This allows checks both that the residuals really are small compared with the fitted values, and that no patterns exist in the residuals. Often it is easier to distinguish patterns in the residual plot than in the original plot of the data (again, see Example 8.10). Furthermore, residual plots are extremely useful in multiple regression analysis (see Section 8.4). A statistical computer package should allow the residuals and fitted values to be stored automatically when the regression line is calculated. Plots can then easily be obtained.

When the x variable represents time, that is when a time series is being analysed, regular cyclic wave patterns may appear in the residual plot. This often indicates that a seasonality analysis (see Section 4.4.3) is appropriate.

When the x and y variables are each recorded serially in time another type of plot that can be useful is the plot of residuals against time. Patterns in this plot often indicate that a further x variable, not so far considered, is important in determining y. If this variable can be identified and recorded the problem can be overcome by the use of multiple regression (see Section 8.4). When patterns appear in the residual against time plot there is said to be *auto-correlation*. (See Montgomery and Peck (1982), Section 9.1, for further details on the analysis of such data.)

An excellent description of the use of plots in regression analysis is given by Anscombe (1973). A complete account of the analysis of residuals is given in Chapter 3 of Draper and Smith (1981).

Example 8.5:   The analysis of variance table for Example 8.4 is

Source of variation	Sum of squares	Degrees of freedom
Regression	0.761	1
Residual	0.856	18
Total	1.617	19

Here $r^2 = 0.761/1.617 = 0.47$. Thus 47% of the variation in perinatal mortality is explained by its regression on the percentage of low-weight births. This agrees roughly with the more complete study reported by Macfarlane *et al.* (1980). This paper mentions other factors which may account for the remaining variation such as the percentage of Caesarean sections and the expenditure on maternity services.

Notice that in Example 8.1 r was found to be 0.69. Since $0.69^2 = 0.4761$ this agrees with the value for r^2 found here, apart from rounding error.

A plot of the residuals against the fitted values from the model $y = 0.42 + 0.15x$ is given as Fig. 8.7. There are no obvious patterns in this plot to suggest an alternative type of regression model for the data.

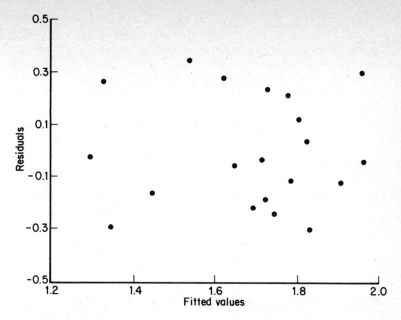

Fig. 8.7 Residual plot for linear regression fit to births example.

8.3.3 Inferences about the regression parameters

In Section 8.3.1 the least squares estimates a and b were found. These values are the coefficients in the least squares regression line of y on x,

$$y = a + bx$$

which estimates, from the observed data, the linear relationship between y and x. Recall that a estimates the true intercept, α, and b the true slope parameter, β.

In Chapter 7 it was observed that any point estimate has a limited inter-pretation. It says what is, in some way, a good estimate, but it fails to specify the accuracy of that estimate. So here, when a and b are evaluated, it is appro-priate to determine how good they are as estimates by calculating confidence intervals for the actual parameters α and β. To do this it is necessary to assume that the residuals from the linear regression fit are independent and follow a normal distribution with zero mean and constant variance σ^2. This assumption also allows hypothesis tests to be carried out.

The assumption can be tested by examining the residuals, both in the manner suggested in Section 8.3.2 and via, for example, normal plots and other diagrams (see Section 7.16.2). Experience has shown that the CIs and tests are not greatly affected unless the residuals are highly non-normal. In these cases procedures such as non-parametric methods may be used (see Section 7.16.3).

Under the stated assumptions, it may be shown that

(i) Variance of a $= \sigma^2 \left\{ \dfrac{1}{n} + \dfrac{\bar{x}^2}{S_{xx}} \right\}$

(ii) Variance of b $= \sigma^2/S_{xx}$

In fact σ^2 is unlikely to be known in any real-life situation. It is estimated by the *residual mean square*, defined to be the residual sum of squares divided by its degrees of freedom. The analysis of variance table of Section 8.3.2 is usually extended to include a column giving both this and the regression mean square, that is the regression sum of squares divided by its degrees of freedom (see Example 8.6).

If the residual mean square is given the symbol s^2 then,

(i) Estimated variance of a $= s^2 \left\{ \dfrac{1}{n} + \dfrac{\bar{x}^2}{S_{xx}} \right\}$

(ii) Estimated variance of b $= s^2/S_{xx}$

Then using the methodology of Chapter 7, confidence intervals for α and β are, respectively,

$$a \pm t_{n-2} \sqrt{\text{estimated variance of a}}$$

and

$$b \pm t_{n-2} \sqrt{\text{estimated variance of b}}$$

Also tests for $\alpha = 0$ and $\beta = 0$ are based, respectively, on comparing the test statistics

$$\frac{a}{\sqrt{\text{estimated variance of a}}} \quad \text{and} \quad \frac{b}{\sqrt{\text{estimated variance of b}}}$$

with the t distribution on $n-2$ degrees of freedom. Notice that t has $n-2$ d.f. because the residual mean square is calculated with $n-2$ d.f.

The test for $\beta = 0$ can also be performed in a different, but entirely equivalent, way. This is to calculate

$$\frac{\text{regression mean square}}{\text{residual mean square}}$$

from the analysis of variance table and compare it with the F distribution with d.f. $(1, n-2)$. One-tailed critical values from the F distribution are appropriate to this test.

Example 8.6:   The full analysis of variance table for the birth data is

Source of variation	Sum of squares	Degrees of freedom	Mean square
Regression	0.761	1	0.761
Residual	0.856	18	0.0475
Total	1.617	19	

The estimate of the residual variance is thus $s^2 = 0.0475$.
The estimated variance of a is

$$0.0475 \left\{ \frac{1}{20} + \frac{(171.79/20)^2}{35.335} \right\} = 0.102$$

and the estimated variance of b is

$$\frac{0.0475}{35.335} = 0.00134$$

Now the t distribution on n – 2 = 18 d.f. has 2.5% of its values above 2.101 (see Table A.2). Hence 95% confidence intervals for α and β are

$$0.42 \pm 0.67$$

and 0.15 ± 0.08 respectively.

That is there is 95% confidence that the true intercept lies somewhere between -0.25 and 1.09 and the true slope lies between 0.07 and 0.23.

Rather than carry out t tests, the fact that zero lies within the 95% CI for α but not for β leads to a rejection of the hypothesis that $\beta = 0$ but not to the hypothesis that $\alpha = 0$ (see Section 7.13). Nevertheless for the purpose of illustration the equivalent F test of the null hypothesis $\beta = 0$ will be given. The test statistic here is

$$\frac{\text{regression mean square}}{\text{residual mean square}} = \frac{0.761}{0.0475} = 16.02$$

The 1% critical value of F on (1, 18) d.f. is 8.29. Therefore the null hypothesis $\beta = 0$ is rejected with a P value of below 0.01; that is, there is strong evidence against the null hypothesis.

The fact that β is significantly non-zero means that the linear association with x *is* an important factor in determining y. Sometimes this is described by the phrase, 'The regression is significant'. The fact that α is not significantly different from zero says that there is no evidence to refute the assertion that perinatal deaths would cease if there were no low-weight births. (On the other hand the degree of uncertainty when predicting y at x = 0 is high as the next section will show.)

8.3.4 Predictions

A common use of the regression line is to predict y at a given value of x. For example the regression of hospital running costs on size may be used to predict the revenue expenditure for a new hospital of a planned size.

Prediction is achieved simply by substituting the required value of x into the equation of the regression line. That is the predicted value of y at $x = x_0$ is

$$a + bx_0$$

It is useful to identify the error associated with this prediction. When the assumption of normally distributed residuals is made once again, it turns out that the variance of this prediction is

$$\sigma^2 \left\{ 1 + \frac{1}{n} + \frac{(x_0 - \bar{x})^2}{S_{xx}} \right\}$$

which is estimated by substituting s^2 for σ^2 as in Section 8.3.3. This result leads to a confidence interval for the predicted y value of

$$a + bx_0 \pm t_{n-2} \sqrt{\text{estimated variance of predicted value}}$$

A similar confidence interval can be derived for the predicted mean value of y when x = x_0, that is the average value that y can be expected to take when x has the value x_0. This may be more meaningful when prediction is required for a group of people rather than a particular individual. The predicted value of $\bar{y}$ is a + bx_0 once more, and the confidence interval is

$$a + bx_0 \pm t_{n-2} \sqrt{\begin{array}{c}\text{estimated variance of the}\\\text{predicted mean value, } \bar{y}\end{array}}$$

where, this time, the estimated variance of $\bar{y}$ is

$$s^2 \left\{ \frac{1}{n} + \frac{(x_0 - \bar{x})^2}{S_{xx}} \right\}$$

Notice that the confidence interval will be narrower than that for predicted y, as would be expected (individual values are always subject to greater prediction error than are average values).

Due to the form of the variances both CIs will be smallest when $x_0 = \bar{x}$. As x_0 moves away from $\bar{x}$ both CIs will widen, in the manner illustrated by Fig. 8.8. Notice that the error associated with values of x_0 beyond the limits of the observed data are relatively large. Of course there is no reason why the linear regression line should provide an adequate description beyond the range of the observed data. Even within the range of the observed data the prediction is bound to be inadequate if the linear regression line is itself an inappropriate model of the y–x relationship.

Example 8.7:   Suppose that the perinatal mortality is to be predicted from the linear regression model when 9% of births are low-weight. That is $x_0 = 9$.

Predicted y = a + bx_0 = 0.42 + 0.15 × 9 = 1.77. Estimated variance of this predicted y

$$= s^2 \left\{ 1 + \frac{1}{n} + \frac{(x_0 - \bar{x})^2}{S_{xx}} \right\}$$

$$= 0.0475 \left\{ 1 + \frac{1}{20} + \frac{(9 - (171.79/20))^2}{35.335} \right\} = 0.050$$

Then since $t_{18} = 2.101$, the 95% CI for this prediction is

$$1.77 \pm 2.101 \sqrt{0.050}, \quad \text{i.e., } 1.77 \pm 0.47$$

Hence if the linear regression model is correct there is 95% confidence that perinatal mortality per 100 births lies between 1.30 and 2.24 when 9% of births are low-weight.

Similarly the 95% CI for predicted $\bar{y}$ at $\bar{x} = 9$ is

$$1.77 \pm 2.101 \sqrt{0.0475 \left\{ \frac{1}{20} + \frac{(9 - (171.79/20))^2}{35.335} \right\}}$$

$$= 1.77 \pm 0.11$$

Figure 8.8 shows the lower and upper limits of 95% confidence intervals for predicted $\bar{y}$ at different values of x_0.

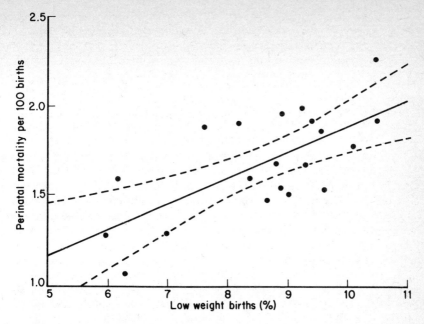

Fig. 8.8   Fitted values with confidence interval for mean predicted perinatal mortality for births example.

8.4 Multiple regression

Linear regression, as presented in Section 8.3, determines how y can be expected to change with the explanatory variable x according to a simple linear relationship. In multiple regression several explanatory variables are considered. For example, hospital running costs are unlikely to be determined solely by the size of the unit; other factors such as the age of the building, the number of personnel employed and the number of laboratory services provided are likely to have some effect on cost.

When k explanatory variables $x_1, x_2, \ldots, x_k$ have been recorded, multiple regression investigates relationships of the kind

$$y = \alpha + \beta_1 x_1 + \beta_2 x_2 + \ldots + \beta_m x_m$$

for m ⩽ k. Notice that this is a straightforward extension of the simple linear regression equation. The β are called the *partial regression coefficients* since β_i represents the effect on y of a unit change in the variable x_i keeping all other variables fixed.

The true multiple regression equation is estimated from observed data in much the same way as for simple linear regression. That is, the sum of squared residuals, $\Sigma(y_i - \hat{y}_i)^2$, is minimized so as to produce the least squares estimates a, $b_1, \ldots, b_m$ of $\alpha, \beta_1, \ldots, \beta_m$ respectively. There is no simple general formula for these estimates, but in any case the reader is advised not to undertake a multiple regression analysis without the use of a statistical computer package. Such packages will carry out the tedious arithmetic to arrive at the least squares estimates and will also evaluate goodness of fit.

8.4.1 Finding the best model

Given the computing power the problem of determining the regression equation linking y with all the explanatory variables is trivial. There still remains the question of whether all the explanatory variables are really necessary. That is, 'Do all the x variables really contribute towards explaining and predicting y?'. If not, then the model linking y with all the x variables is over complex and a simpler description of y will suffice. This is likely to happen when there are correlations amongst the explanatory variables themselves. For instance, hospital costs almost certainly do depend upon the size of the unit and the number of employees, but the number of employees is itself likely to depend upon the size of the unit, so it is unlikely that both are necessary to determine costs.

8.4.2 Comparison of fit

To be able to compare regression models with different sets of x variables, selected from the complete observed set, it will first be necessary to consider measures of goodness of fit for a multiple regression model.

Statistical packages will produce the analysis of variance table for a multiple regression model with m explanatory variables in the form:

Source of variation	Sum of squares	Degrees of freedom	Mean square
Regression	RGSS	m	RGSS/m
Residual	RSS	$n - m - 1$	$RSS/(n - m - 1)$
Total	$\Sigma(y_i - \bar{y})^2$	$n - 1$	

where the expressions for RGSS (regression sum of squares) and RSS (residual sum of squares) are different for different values of m. Notice that the regression d.f. is equal to the number of x variables included in the model.

This analysis of variance can be used in much the same way as the one from simple linear regression. In particular the coefficient of multiple determination, denoted R^2, is defined as

$$R^2 = \frac{\text{regression sum of squares}}{\text{total sum of squares}}$$

which specifies how much of the variation in the y variable has been explained by the regression of y on the m explanatory variables.

Although a large R^2 (tending towards one) will certainly suggest that the regression model provides an accurate description of the data, R^2 has the unfortunate drawback that it always increases when a new x variable is added to the model, even if the new x term has no relationship with y whatsoever. This means that while R^2 is a way of assessing how well any particular model fits the observed data, it is unsuitable for comparing goodness of fit.

In the extreme case when $n - 1$ x variable are employed this forced growth of R^2 can be seen from the analysis of variance. When m = n – 1 the residual degrees of freedom are zero, meaning that all the sum of squares are attributed to regression, and hence R^2 = 1. This would be true even if none of the x variables had any linear relationship, taken singly or as a group, with y! This is a good reason for always ensuring that the number of observations taken is appreciably bigger than the number of variables considered.

R^2 can, therefore, only be used to *reject* models or to make comparisons of different models with the *same* number of x terms. Naturally if R^2 is small for the model which includes all the x variables for which data are available, then any smaller model will be even worse in absolute terms.

To compare models with different numbers of x terms a measure which compensates for the number of terms is required. Some packages produce the so-called *adjusted R^2* statistic (see Montgomery and Peck (1982), Section 7.1.3). Using this is entirely equivalent to using the *residual mean square*, which is easier to interpret since this is an unbiased estimate of the residual variance.

Just as R^2 increases as more x terms are introduced into the model, the residual sum of squares must always decrease. However, the residual mean square compensates for this decrease by dividing RSS by its degrees of freedom, which will also decrease. The net result will be that the residual mean square (RMS) will get smaller as important x variables are introduced into the model, but once all the important x variables have been included the introduction of any of the remaining variables will increase the residual mean square.

It is frequently useful to plot the residual mean squares from all possible regressions against the number of x terms fitted on a graph, such as Fig. 8.9. The best regression model will then be somewhere near the lowest point in the diagram. The word 'somewhere' is used here because it may be sensible to choose a model with fewer x terms but only a slightly larger RMS (that is a point to the left and just above the lowest point in the diagram, if such a point exists). This simpler model would be worth considering, despite its slightly larger RMS.

Example 8.8: To illustrate the methods and problems of multiple regression consider the set of data in Table 8.3

The expenditure of a regional health authority is likely to be determined by such things as the population served, the general health of the population and the staff and facilities provided. This suggests that an appropriate model which links the variables recorded in Table 8.3 would be

$$y = \alpha + \beta_1 x_1 + \beta_2 x_2 + \beta_3 x_3 + \beta_4 x_4$$

where y = expenditure, x_1 = population, x_2 = perinatal mortality rate, x_3 = number of GPs, and x_4 = number of beds.

It is likely, however, that some of the x variables are dependent upon the others. For instance the number of GPs is likely to vary with population in a linear fashion. Consequently the full model, with all four x variables, is likely to be more complicated than is really necessary to determine the expenditure, y.

A useful first step in the multiple regression analysis of the data is to plot each of the variables against each other to obtain pictorial descriptions of the

Table 8.3 Regional health authority data. (Data from *Compendium of Health Statistics*, (1984) Office of Health Economics.)

Regional health authority	Total expenditure (£ million) 1981/82	Population (millions) 1982	Perinatal mortality* 1981	No. of GPs 1982	No. of beds (thousands) 1981
Northern	677	3.1	13.2	1569	34
Yorkshire	772	3.6	13.9	1867	29
Trent	925	4.6	11.3	2280	31
E. Anglia	394	1.9	10.4	997	13
N.W. Thames	868	3.5	10.7	2035	27
N.E. Thames	928	3.7	11.1	2015	28
S.E. Thames	879	3.6	12.1	1935	27
S.W. Thames	697	3.0	10.7	1580	25
Wessex	562	2.8	9.9	1505	19
Oxford	449	2.4	9.4	1215	13
S. Western	681	3.1	11.4	1849	24
W. Midlands	1058	5.2	12.8	2651	36
Mersey	560	2.4	12.4	1281	21
N. Western	962	4.0	12.4	2056	31

*Rate per 1000 live and stillbirths.

pairwise relationships. To save space these are omitted here, but instead the *correlation matrix* is given below. This matrix shows all the pairwise correlation coefficients for the five variables. The diagonal elements are one since these are correlation between a variable and itself, and the off-diagonal elements are repeated in an obvious way. Some computer packages will produce the entire correlation matrix directly, others require it to be built up element by element.

	y	x_1	x_2	x_3	x_4
y (expenditure)	1	0.94	0.47	0.96	0.87
x_1 (population)	0.94	1	0.45	0.98	0.85
x_2 (perinatal)	0.47	0.45	1	0.41	0.72
x_3 (GPs)	0.96	0.98	0.41	1	0.83
x_4 (beds)	0.87	0.85	0.72	0.83	1

The correlation matrix shows very high linear associations between the expenditure, population and GP variables. 'Beds' is also highly correlated with these three, but perinatal mortality is less well linearly related. All correlations are positive.

These correlations are what would be expected. Expenditure, GPs and beds are likely to be highly correlated with population size, and thus with each other. On the other hand perinatal mortality is measured as a rate and is thus already corrected for population size. Possibly the only surprising result is the fairly high positive correlation between beds and perinatal mortality, which suggests that regions with high low-age mortality have been well provided with hospital beds.

Simple linear regression can, of course, uncover the form of all these relationships, but here interest lies in using the other variables to determine expenditure. The correlation matrix indicates that there is likely to be redundancy in the full model,

$$y = \alpha + \beta_1 x_1 + \beta_2 x_2 + \beta_3 x_3 + \beta_4 x_4$$

For instance x_1 (population) and x_3 (GPs) are so highly correlated that it is extremely unlikely that when both are used to predict y (expenditure) the accuracy would be better than when only one is used. However they are not absolutely perfectly correlated, so one of the two will be slightly better.

Since there are four x variables to consider, there are

$$\binom{4}{1} + \binom{4}{2} + \binom{4}{3} + \binom{4}{4}$$

$$= 4 + 6 + 4 + 1 = 15$$

possible models for y, ignoring the model with no x terms. Each model has been fitted by computer, and the results are given in Table 8.4.

Table 8.4 All regression models fitted to RHA data (from Table 8.3). Note that a dash indicates that the term has been left out of the model.

Number of x terms	R^2	Residual mean square	Estimated coefficients				
			Intercept a	Population b_1	Perinatal b_2	GP b_3	Beds b_4
1	0.88	5 298	21.8	215.5	—	—	—
1	0.22	34 760	– 100.1	—	73.1	—	—
1	0.92	3 745	– 28.5	—	—	0.435	—
1	0.75	11 118	105.4	—	—	—	25.0
2	0.88	5 655	– 60.7	209.6	8.8	—	—
2	0.92	4 073	– 27.1	17.1	—	0.402	—
2	0.90	4 931	0.7	166.9	—	—	7.2
2	0.92	3 785	– 155.2	—	13.5	0.419	—
2	0.80	9 599	524.5	—	– 51.5	—	31.8
2	0.93	3 354	– 42.7	—	—	0.351	6.4
3	0.92	4 164	– 156.6	– 3.2	13.6	0.425	—
3	0.90	5 204	140.9	153.1	– 16.2	—	10.8
3	0.93	3 669	– 45.2	– 22.0	.—	0.389	6.7
3	0.93	3 666	5.0	—	– 5.4	0.341	7.6
4	0.93	4 053	0.9	– 21.1	– 5.2	0.378	7.9

As R^2 can give misleading results, comparison of fits will be based upon the residual mean square (RMS). Figure 8.9 shows the RMS for all possible models (although the model with x_2 alone is so much worse than all the others that it has been omitted). The best model of those that only involve one x variable is that with x_3 (GP) alone since it has the smallest RMS of all the one x variable models. Similarly the best two x variable model is that with x_3 and x_4 and the best three x variable model is that with x_2, x_3 and x_4. There is, of course, only one model with four x variables, the full model.

The best overall model for determining y will be one of these four. The full model and that with x_2, x_3 and x_4 cannot be the overall best since their RMS is bigger than the simpler model relating y to only x_3 and x_4. The model involving x_3 and x_4 has the smallest RMS overall, and so is a strong candidate

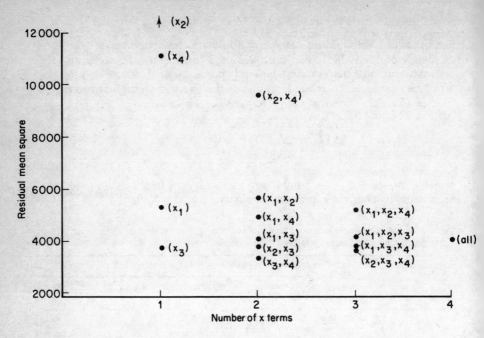

Fig. 8.9 Residual mean square for all regressions, RHA data. Note that the RMS for (x_2) is over 34 000.

for the best model, but the model with x_3 alone is not very much worse and is simpler, so it should also be considered.

The choice of best model thus lies between

$$y = -28.5 + 0.435x_3 \tag{1}$$

$$\text{and} \quad y = -42.7 + 0.351x_3 + 6.4x_4 \tag{2}$$

Model (1) says that expenditure increases by £435 000 for each extra GP. Model (2) is more complex. It says that if beds are kept constant then expenditure increases by £351 000 for each extra GP *and* if the number of GPs is kept constant then expenditure increases by £6.4 million for every extra thousand beds. If it is more likely that beds will be withdrawn, the last phrase might be better thought of as a £6.4 million saving for every thousand beds closed (when GP sizes are maintained).

The final choice between (1) and (2) rests on whether the $3745 - 3354 = 391$ extra units of RMS is considered an important difference. This could involve a practical decision regarding the trade off between lost accuracy and increased simplicity. In fact 391 is not a particularly big difference in the context of this problem. Model (2) has a RMS of 3354 on 11 d.f. (14 minus 1 minus the number of x terms fitted). This means that the average RMS per d.f. is $3354/11 = 305$. Model (1) has a RMS of 3745 on 12 d.f. The increase of 391 associated with the extra d.f. is not substantially more than the 305 expected according to model (2). So model (1) seems to be acceptable. In Example 8.9 models (1) and (2) will be formally compared through a test of significance based on the F distribution.

Notice that the determination of the 'best' model is by no means the only result of any importance of this multiple regression analysis. The correlation matrix tells a lot about the structure of the data (plots would tell more) and many of the 'rejected' regression models are potentially useful. For instance the model relating y (expenditure) to x_1 (population) alone is a good model since it explains 88% of the variation in y and, although not shown here, the residuals have no obvious patterns. For many planning purposes this model, predicting expenditure from population size, could be applied.

One final point about this analysis, which will be true of any multiple regression problem, is that even the best model is only as good as the data. Other variables which have not been considered, such as the number of nurses, prescriptions issued and notifications of diseases could also be fundamentally important in determining health expenditure.

8.4.3 Further analysis of variance

In addition to the basic analysis of variance table for multiple regression, many statistical computer packages produce a 'further' analysis of variance in which the regression sum of squares is partitioned so as to introduce the x variables one at a time, rather than all at once. Then, for a regression with m explanatory variables, the regression sum of squares is split into m different sums of squares, each representing the *extra* effect of each newly introduced x variable.

For example when there are three x variables (i.e., m = 3) the basic analysis of variance table is:

Source	SS	d.f.	Mean square
Regression	RGSS	3	RGSS/3
Residual	RSS	n − 4	RSS/(n − 4)
Total	$\Sigma(y_i - \bar{y})^2$	n − 1	

which may be partitioned into the further analysis of variance table:

Source	SS	d.f.	Mean square
Regression on x_1	$SS(x_1)$	1	$SS(x_1)/1$
Regression on x_2 given x_1	$SS(x_2 \mid x_1)$	1	$SS(x_2 \mid x_1)/1$
Regression on x_3 given x_1 and x_2	$SS(x_3 \mid x_1,x_2)$	1	$SS(x_3 \mid x_1,x_2)/1$
Residual	RSS	n − 4	RSS/(n − 4)
Total	$\Sigma(y_i - \bar{y})^2$	n − 1	

Here x_1, x_2 and x_3 have been introduced in ascending order of the subscript. Other orderings are possible; except for the residual and total (which are always the same) they would give different sums of squares. In the analysis of variance tables above

$$RGSS = SS(x_1) + SS(x_2 \mid x_1) + SS(x_3 \mid x_1,x_2)$$

From the further analysis of variance F tests can be constructed, in a similar way to that in Section 8.3.3, to see whether any of the components contributes significantly to the variation in y. These tests are valid if the residuals are independent and follow, at least approximately, a normal distribution with constant variance. For example to test whether x_1 makes a significant contribution to the determination of y, the test statistic

$$\frac{SS(x_1)/1}{RSS/(n-4)}$$ is compared with F on $(1, n-4)$ d.f.

To test whether x_2 makes a significant contribution to the determination of y over and above the contribution of x_1

$$\frac{SS(x_2|x_1)/1}{RSS/(n-4)}$$ is compared with F on $(1, n-4)$ d.f.

and so on.

This is easily generalized to the situation of m explanatory variables. All that is essentially different is that the residual degrees of freedom become $n - m - 1$, as explained in Section 8.4.2.

Example 8.9:   In the health expenditure example the choice of best model was left between the model with x_3 and the model with x_3 and x_4 (Example 8.8). Since the more complex model only differs by the addition of x_4 an F test can be used to evaluate this addition. Prior to this the residuals should be checked for independence and normality with constant variance, without which the F test is not strictly valid. In this case the residuals checked out reasonably well.

The analysis of variance table for the model relating y to x_3 and x_4 is

Source	SS	d.f.	Mean square
Regression	497 401	2	248 700
Residual	36 892	11	3 354
Total	534 293	13	

When the regression sum of squares is partitioned this becomes

Source	SS	d.f.	Mean square
Regression on x_3	489 353	1	489 353
Regression on x_4 given x_3	8 048	1	8 048
Residual	36 892	11	3 354
Total	534 293	13	

To test for a significant regression effect of x_3 on y compare $489\ 353/3354 = 145.9$ with F on $(1,11)$ d.f. Table A.4 shows that the 1% point of F on $(1,11)$ d.f. is only 9.65, so this is extremely significant. To test for a significant additional regression effect of x_4 when x_3 is already present compare $8048/3354 = 2.4$ with F on $(1,11)$ d.f. Since the 5% point of F $(1,11)$ is 4.84 this is not significant at this level.

Hence the conclusion is that x_4 is not particularly important *over and above* x_3. The model

$$y = -28.5 + 0.435x_3$$

is sufficient.

8.4.4 Automatic variable selection

One large drawback to the comparison of fit method based on the residual mean squares of all possible regression models is that the number of regression models to be evaluated is quite large even when there are few x variables (e.g., 63 models when there are 6 x variables). This may not be a problem, especially if a package which evaluates all regressions at one command is available. Nevertheless, even these packages have their limitations and, further, it is attractive to have a method for variable selection which is totally automatic.

The F tests introduced in Section 8.4.3 form the basis of a set of methods for automatic selection of variables suitable for the computer. The first of these is *forward selection*, in which new variables are introduced one at a time, each time carrying out an F test at a prescribed level of significance to decide whether or not the new variable should be allowed to enter the model.

The order in which variables are considered can make a substantial difference to the results of a forward selection analysis. For example if the variables of Example 8.8 were introduced in ascending order, that is starting with x_1, then when it comes to considering the addition of x_3 over and above x_1, x_3 will be rejected. This is because x_3 is so highly correlated with x_1 that it can have little additional explanatory power. The unfortunate fact is that the rejected variable, x_3, was found to be the most important of all when all the regression models were considered.

To alleviate this problem, forward selection starts with the variable most highly correlated with y (x_3 in the example), then considers adding the variable most highly correlated with y after regression on the previous variable has been accounted for, and so on.

The other two automatic selection methods are both variations on this theme. *Backward elimination* begins by considering the full model and tries to delete terms one at a time, once again using F tests to decide whether a variable can be dropped. *Stepwise regression* is a mixture of the forward and backward methods which continually cross-checks to see whether a variable introduced at an earlier stage may now be rejected.

Many computer packages incorporate these searching techniques. In general they can be expected to find a good model, but not necessarily the best, and so they should be thought of as quick alternatives to the method of 'all regressions' used in Example 8.8. Another disadvantage is that they do not give as much detail about the structure of the data as the earlier method. Further it should be remembered that the F tests are only strictly valid if the residuals follow a normal distribution, although even when this is not the case the F tests still measure relative importance and provide a useful cut-off criterion. A more complete description of these automatic selection methods may be found in Draper and Smith (1981), Chapter 6.

8.5 Non-linear regression

The method of regression analysis for explaining y from x presented in Section 8.3 has one large limitation: it only allows for *linear* relationships. In practice many variables are, at least approximately, related in this way, and furthermore the simplicity of the linear relationship makes it a fruitful area of study.

Of course not all relationships are linear. For instance Fig. 8.4 shows a non-linear trend over time. Fortunately the techniques of simple linear and multiple regression can be extended to deal with non-linear relationships. Sometimes this extension is achieved very easily, when the data can be transformed into the simple linear or multiple regression forms. Examples are given in Sections 8.5.1 and 8.5.2. Other times, when the data cannot be 'linearized', a more complex and time-consuming approach is necessary. This will not be described here (details are given in Draper and Smith (1981), Chapter 10) but it is essentially the method of least squares, described in Section 8.3.1, applied in a different context. The more advanced statistical computer packages contain a non-linear regression routine, the output from which should be understandable to someone who has mastered the ideas presented in this chapter. A great deal of the skill required for non-linear regression analysis is the identification of the functional relationships between y and x from inspection of the scattergram. In Sections 8.5.1–8.5.3 three curves which are often found to be reasonable models for health or demographic data will be considered. Unlike the other two, the model of Section 8.5.3 cannot be reduced to a simple linear or multiple regression form. These three examples provide only a minute selection of the non-linear models that have been fitted to health data.

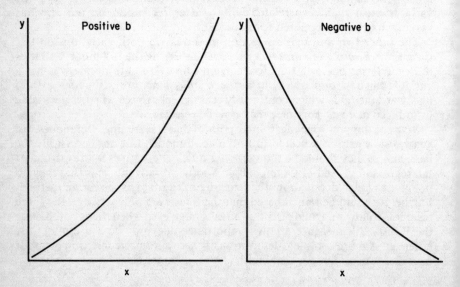

Fig. 8.10   The exponential model.

8.5.1 The exponential model

This model assumes that there is a constant *proportional* increase (or decrease) in y for every *unit* increase in x. The mathematical form of this model is $y = ke^{bx}$ where e is the mathematical constant 2.71828 (as in Section 6.6.5) and k and b are other constants. Figure 8.10 shows examples of positive and negative b.

The exponential model is often a good description when y is cost and x is time over a period of near-constant rate of inflation. In this case b is positive. The model with negative b is often useful when y is something which is decaying, such as a toxic substance in the body, and, again, x is time.

The exponential model can be transformed into a linear form by taking logarithms (to base e) of both sides of the equation. That is $y = ke^{bx}$ becomes

$$\log_e y = \log_e(ke^{bx}) = \log_e k + bx$$

So if $v = \log_e y$ and $a = \log_e k$, then the equation becomes

$$v = a + bx$$

and linear regression can be applied to the v–x data.

Most calculators have a button to produce logarithms, and many statistical packages will calculate the entire set of logarithms of y from a single command.

Example 8.10:   A study was carried out for the Audiology Department of the Royal Berkshire Hospital in Reading to investigate the relationship between number of hearing aids issued and number of staff, month by month. For simplicity a small selection of data from this study is given in Table 8.5, together with the associated scattergram, Fig. 8.11.

Table 8.5   Hearing aid data, Royal Berkshire Hospital.

Average daily no. of staff*	Number of hearing aids issued
6.08	136
6.08	128
4.54	49
5.08	74
3.74	34
4.54	54

*Whole time equivalents.

Figure 8.11 suggests that the relationship is exponential rather than linear. For the purposes of illustration a linear regression was first fitted to the data. The result was the model

$$y = -147 + 45.1x$$

where x = number of staff and y = number of aids issued. This regression model explains 95.8% of the variation in y. However a plot of the residuals against fitted values gives a systematic pattern (see Fig. 8.12). The pattern in the residuals, basically a V shape, is obvious, suggesting that the linear

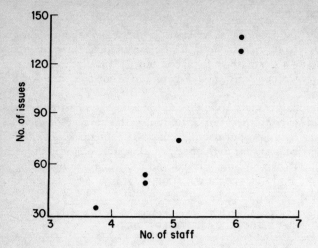

Fig. 8.11 Number of issues of hearing aids against staff for the Audiology Department, Royal Berkshire Hospital.

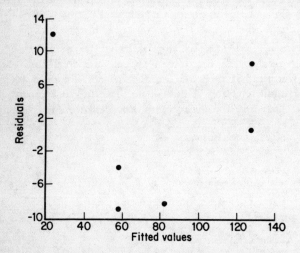

Fig. 8.12 Residual plot for linear regression fit.

model is not entirely satisfactory, despite the large value of r^2.

Taking logs of the y values and plotting them against x produces the scattergram of Fig. 8.13. With these transformed data a linear regression does seem appropriate.

Fitting a linear regression line to the transformed data produced the result

$$\log(y) = 1.29 + 0.59x$$

This model explains 99.4% of the variation in y and the residuals from the model show no obvious pattern (see Fig. 8.14).

It is interesting to transform the fitted linearized form back to the expo-

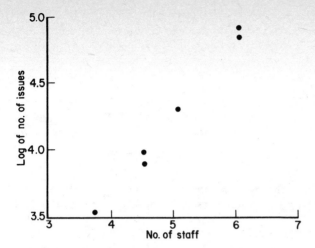

Fig. 8.13 Logarithms of number of issues of hearing aids against staff.

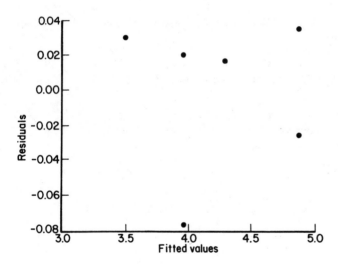

Fig. 8.14 Residual plot for exponential regression fit.

nential form $y = ke^{bx}$. Since $a = \log k$, $k = e^a = e^{1.29} = 3.63$. This gives the equivalent formulation,

$$y = 3.63e^{0.59x} = 3.63(1.80)^x$$

which can be interpreted as saying that the number of hearing aids issued is multiplied by a factor of 1.8 for every extra member of (whole-time) staff. In the context of the example it presumably means that when only a few staff are available their total time is taken up with administration, and other duties such as repairs, but extra staff have more and more time available to deal with issues of hearing aids.

Note that in this example some of the x values are repeated (i.e., 6.08 and 4.54 both appear twice). When this happens it is possible to carry out an F test for the lack of fit of the linear regression line. (See Montgomery and Peck (1982), Section 3.4 for details.)

8.5.2 Polynomial models

If y has a polynomial relationship with x then the equation linking y and x is a sum of powers of x multiplied by constants. A kth degree polynomial has the equation

$$y = a_0 + a_1x + a_2x^2 + a_3x^3 + \ldots + a_kx^k$$

The linear form, $y = a + bx$, is a 1st degree polynomial.

The number of *turning points* (changes in the direction of movement) in the graph of the polynomial will be the degree of the polynomial less one. For example the linear form ($k = 1$) has no turning points, the *quadratic* form ($k = 2$) has one turning point, the *cubic* form ($k = 3$) has two turning points, and so on. The quadratic and cubic are illustrated by Fig. 8.15.

Leaving aside the linear form, which has been already discussed, the other polynomials form a useful family of models to represent x–y relationships with turning points.

A kth degree polynomial can be transformed into a multiple regression model by defining $x_i = x^i$ for all i. That is, new variables $x_2 = x^2, x_3 = x^3, \ldots, x_k = x^k$, are calculated from the original variable, x. Then take $x_1 = x$ and the polynomial becomes

$$y = a_0 + a_1x_1 + a_2x_2 + a_3x_3 + \ldots + a_kx_k$$

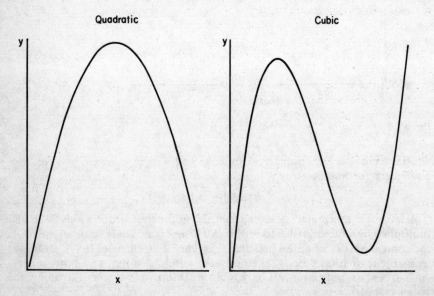

Fig. 8.15 Polynomial models.

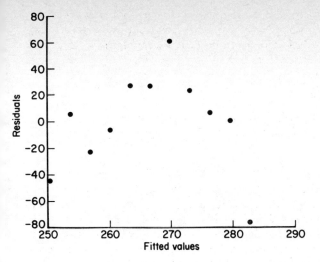

Fig. 8.16 Residual plot for linear regression fit.

the multiple regression model with intercept a_0 and partial regression coefficients $a_1, a_2, \ldots, a_k$.

Example 8.11: Consider the typhoid data of Example 8.3. Figure 8.4 suggests that a linear regression would be inappropriate due to the fact that the data have a turning point, but a quadratic regression model should fit these data reasonably well.

Nevertheless a linear regression was fitted to the data (as in Example 8.10) so as to illustrate what happens when this inappropriate model is used. The linear regression model was produced by first coding the years as 1 to 11 (e.g., 1974 becomes 1, 1975 becomes 2, etc.). The resulting model has a percentage r^2 value of only 7.5% and furthermore the residuals show a clear, quadratic, pattern (see Fig. 8.16). So the linear model can be firmly rejected.

The quadratic model was fitted by calculating a new variable which is the square of the coded time (x) and then using a multiple regression command in a statistical package. The result was

$$y = 197 + 37.6x - 3.41x^2$$

which accounts for 72.4% of the variation in the number of typhoid and paratyphoid notifications (y). The residuals from this model have no obvious pattern (Fig. 8.17). Hence the quadratic fits the data well over the short time range considered. On the other hand there is no reason to suppose that the quadratic pattern will continue in the long run – extrapolation from a small set of time series data is especially prone to error.

8.5.3 The logistic model

For sound theoretical reasons (see Mead and Curnow (1983), Chapter 10) a model which is often used to represent the growth of populations is the

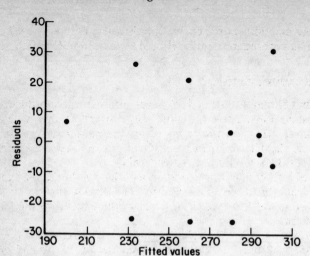

Fig. 8.17   Residual plot for quadratic regression fit.

logistic. Figure 8.18 illustrates the logistic curve, the mathematical equation of which is

$$y = \frac{s}{1 + t(r^x)}$$

for particular values of the constants r, s and t. This model has a relatively complex mathematical form, which cannot be linearized, and consequently the fitting of this model to data is beyond the scope of this book. Snedecor and Cochran (1980), Section 10.1, give an example of fitting the logistic to US

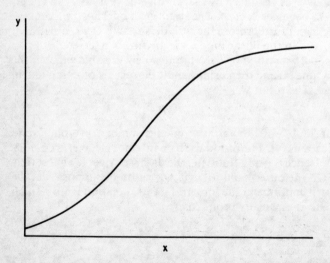

Fig. 8.18   The logistic model.

population figures. As mentioned previously, a statistical computer package with a general non-linear regression routine should be used to fit this model.

8.6 Influential observations

The regression routines of many statistical packages automatically produce a list of influential observations. These are those observations, if any, which have a disproportionate effect on the estimates of the parameters of the regression line and the fit. That is, if these observations were removed the estimates and measures of fit would alter considerably.

One obvious type of influential observation is an outlier, in the sense of an observation remote from the others in a scattergram. This observation may pull the fitted regression model towards it; if it wasn't there the fitted model would be quite different.

Methods for determining influential observations are quite complex (see Draper and Smith (1981), Section 3.12), but if a computer package which is able to detect them is available, this information can be very useful. It might be that the influential observation was in error, or it might be perfectly correct, yet unusual. Either way it could be interesting to re-fit the regression model without the influential value and compare the results. Care must be taken not to omit data points from the final analysis simply because they 'seem odd'; unless they can be found to definitely be erroneous they should, at the very least, be reported, even if they are left out of the regression analysis. The very fact that an influential value is unusual makes it of special interest.

Example 8.12:   Figure 8.19 shows length of stay against age for the data of Appendix 1. Intuitively it might be anticipated that the influential observations would include the two that are ringed, since these are considerably far from the main clump of data. These are the patients with the two longest

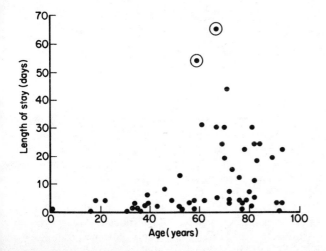

Fig. 8.19   Age against length of stay for a sample of inpatients. (Data from Appendix 1.)

lengths of stay. Another influential observation is the extreme left-hand point since this patient's 'x' value (age) is so much smaller than anybody else's. Whether or not these observations were omitted from the analysis would depend upon the questions which the analysis was designed to answer.

For the full set of 56 observations the fitted least squares linear regression line is

$$y = -1.17 + 0.207x$$

where y = length of stay and x = age. When the two ringed observations are excluded from the analysis the fitted line is

$$y = -2.75 + 0.204x$$

Hence the intercept is altered considerably. Measures of fit, predicted y etc., are also substantially changed.

8.7 Dummy variables

Up to now it has always been assumed that the variables appearing in a regression model are quantitative. Sometimes it may be that some of the explanatory variables are categorical in which case it is still possible to fit a regression equation using *dummy variables*, that is codes for the outcomes of the categorical variables. Chapter 9 deals with situations where all the variables are categorical.

Example 8.13:   In the hospital ward data of Appendix 1 the sex of the patient was recorded. Suppose that it is hypothesized that length of stay varies linearly with age, but in a different way for each sex (a not unreasonable assertion).

Let y be the length of stay and x_1 be age. Define a dummy variable x_2 which takes the value 0 if the patient is male and 1 if female. Then the multiple regression of y on x_1 and x_2 is required. When fitted the resulting model is

$$y = 11.6 + 0.238x_1 - 16.1x_2$$

That is for males ($x_2 = 0$) the linear regression of length of stay on age is

$$y = 11.6 + 0.238x_1$$

and for females ($x_2 = 1$),

$$y = -4.5 + 0.238x_1$$

When drawn on the scattergram these would appear as two parallel lines.

The further analysis of variance table for this example has a slightly larger sum of squares for sex given age than for age alone, which indicates that sex is more important than age in determining length of stay in the multiple regression model. This is not surprising when it is noticed that out of only five male patients included in the sample, two gave observations that were identified as influential in Section 8.6. Hence the sex of the patient is likely to be an important factor in determining length of stay. Indeed, it may even be sensible to fit regression lines to male and female patients completely separately, that is not forcing the lines to be parallel as here. Armitage and

Berry (1987), Section 9.4, gives an example where the fit of parallel and non-parallel lines are compared.

The technique used above easily generalizes to the situation where the categorical variable has many (more than 2) levels. For instance, source of admission has four observed values (2, 3, 5 and 6) in Appendix 1. To cope with this three dummy variables are needed, that is, x_1, x_2, and x_3 with values

x_1	x_2	x_3	
0	0	0	for source code 2
1	0	0	for source code 3
0	1	0	for source code 5
0	0	1	for source code 6

Further details on dummy variables are given by Montgomery and Peck (1982), Chapter 6.

Exercises

8.1 For the GP data in Appendix 2

   (a) Plot the systolic blood pressure against the diastolic blood pressure.
   (b) Find the correlation between the same two variables.
   (c) Calculate the least squares linear regression line of systolic blood pressure on diastolic blood pressure.
   (d) Give the analysis of variance table for this regression analysis.
   (e) Calculate the coefficient of determination.
   (f) Calculate the fitted values (i.e., calculate y at every observed value of x using the equation y = a + bx from part (c)).
   (g) Calculate the residuals (i.e., calculate the observed y minus the fitted y from part (f)).
   (h) Plot the residuals against the fitted values to check for patterns in the residuals.
   (i) Draw a boxplot and a stem and leaf diagram of the residuals to check whether they are, at least approximately, normally distributed.
   (j) Calculate a 95% confidence interval for the intercept of the regression line.
   (k) Calculate a 95% confidence interval for the slope of the regression line.
   (l) Calculate the predicted systolic blood pressure when the diastolic blood pressure takes the value 100 mmHg.
   (m) Give a 99% confidence interval for the predicted value in part (l).

   [Note: parts (f)–(i) can be tedious without the aid of a computer.]

8.2 Cummins (1983) studied the relationship between the number of grams of sodium purchased per person per week and the deaths due to cerebrovascular disease. His data (for males) are:

Year	Table salt	Total bread	15 high sodium items	Cerebrovascular disease mortality, deaths/100 000 men aged 65–74
	Sodium source (g/person a week)			
1958	10.8	7.2	26.5	673
1959	10.8	7.2	26.4	667
1960	10.0	7.0	25.6	671
1961	9.2	6.9	24.8	662
1962	9.4	6.7	25.0	672
1963	9.9	6.6	25.3	670
1964	10.4	6.4	25.6	670
1965	9.6	6.2	24.7	645
1966	9.6	5.9	24.3	629
1967	9.7	6.1	24.9	595
1968	10.6	5.9	25.4	597
1969	10.1	5.8	24.9	588
1970	11.0	5.8	26.0	573
1971	10.1	5.5	24.5	564
1972	10.8	5.3	24.8	569
1973	9.4	5.1	23.2	547
1974	11.9	5.1	25.6	532
1975	8.1	5.2	21.9	511
1976	8.1	5.1	21.8	494
1977	9.4	5.0	23.0	475
1978	8.6	4.9	22.1	476

In order to develop a predictive model multiple regression analysis may be applied to the data where mortality is the y variable and the three sodium sources are the x variables. Since 'high sodium items' include table salt and bread, high correlations amongst the x variables are likely. In fact the correlation matrix is

	y	x_1	x_2	x_3
y	1	0.349	0.935	0.764
x_1	0.349	1	0.273	0.845
x_2	0.935	0.273	1	0.726
x_3	0.764	0.845	0.726	1

where x_1 = table salt, x_2 = bread and x_3 = high sodium items.

The seven possible multiple regression models were found to have the following values of R^2 and residual mean square (RMS).

Variables in model	R^2	RMS
x_1	0.122	4467
x_2	0.875	637
x_3	0.584	2117
x_1, x_2	0.884	621
x_1, x_3	0.892	579
x_2, x_3	0.890	591
x_1, x_2, x_3	0.903	554

(a) Which x variable is, by itself, most important in predicting y?
(b) Which *pair* of x variables is 'best' in predicting y?
(c) When the full model (with all three x variables) was fitted introducing x terms in the order x_1, x_3, x_2 the further analysis of variance table was

Source	SS	d.f.
Regression on x_1	11 752	1
Regression on x_3 given x_1	74 450	1
Regression on x_2 given x_1 and x_3	999	1
Residual	9 420	17
Total	6 621	20

Use F tests from this analysis of variance, as well as the earlier table of R^2 and RMS, to decide which of the seven possible models is 'best' for predicting y.

If you have access to a statistical package with a multiple regression procedure you can check all the results given here and find the values of the partial regression coefficients in the best model.

(d) Suppose someone were to say 'I do not believe that eating salt causes heart problems. In recent years cerebrovascular mortality has been declining totally independently of sodium intake; it just happens that they have both fallen by about the same amount each year. In fact it would be more accurate to say that time has 'caused' the drop in cerebrovascular mortality!'.

Fit the simple linear regression of cerebrovascular mortality on year.

Find the r^2 and RMS for the fitted model and compare these with the corresponding values in the table given earlier. This illustrates both the problem of using secondary sources of data to explore causal relationships and the danger of ignoring the time dimension in data. Consider what you would do to examine the causal relationship further.

8.3 Using the *Welsh* net ingredient cost data from Question 4.8:

(a) plot the scattergram of cost against time;
(b) using coded units for time (1968 = 1, 1970 = 2, etc.) fit the linear regression line of cost on time;
(c) take logarithms of cost and hence fit the exponential regression model of cost on time;
(d) find r^2 for both the linear and exponential models. Which model gives the best fit? Would you expect this result from your scattergram?

If a computer package is available to produce residual plots these should be inspected for both models. In fact it turns out that both exhibit obvious patterns suggesting that there is still some unexplained systematic variation. Given more data, a time series analysis would be beneficial.

9
Contingency tables

9.1 Introduction

Health data commonly take the form of frequencies or counts, grouped according to one or more classifications. In a study on the need for domiciliary physiotherapy, for instance, patients receiving treatment in their own homes were classified according to whether they were indeed suitable for domiciliary care or should have been seen in a clinic, and by their source of referral to the domiciliary physiotherapy service. As described in Section 3.4, these data may be conveniently displayed in a table such as Table 9.1.

Table 9.1 Source of referral by suitability for a sample of domiciliary physiotherapy patients.

| Referral | Suitable for domiciliary care? | | |
	Yes	No	Total
GP	20	6	26
Consultant	97	15	112
Physiotherapist	18	0	18
Total	135	21	156

This type of table is called a contingency table and this chapter will be concerned with the analysis of such tables when the data are a sample from some larger population.

The constituents of a contingency table are the rows, columns, cells and totals. The three rows of Table 9.1 correspond to the three *levels* of the *classification variable* 'source of referral' (GP, consultant, physiotherapist), while the two columns correspond to the two levels of the classification variable 'suitability' (yes, no). Except in certain special circumstances (see Section 9.3.4) the classification variables are always treated as categorical in this chapter even though they might be derived from quantitative variables by forming class intervals such as age groups (see Section 3.3.1). The *cells* in the contingency table contain the counts associated with all possible combinations of the levels of the classification variables; since Table 9.1 has three rows and two columns it has $3 \times 2 = 6$ cells. Each contingency table has a set of *marginal* totals associated with it. Table 9.1 has three sets of marginal totals – the row totals, the column totals and the overall total. As mentioned in Section 3.4, when marginal totals are presented separately they are known as marginal tables. Note that the overall total is the sum of the row totals and the sum of the column totals.

Table 9.1 is an example of a two-way contingency table, that is a table with two classification variables. In general, contingency tables may have any number of classification variables, although the practical interpretation of tables with more than three or four variables may be difficult. Table 9.2 shows an example of a three-way contingency table, taken from the Doll and Hill (1950) case-control study on the relationship between smoking and lung cancer (see Section 2.5.3). Table 9.2 is an extract from the study (see the original paper for more information).

Table 9.2 Smoking and lung cancer. (From Doll and Hill, 1950.)

| | | Type of individual | | |
| | | --- | --- | |
Sex	Smoking Status	Case	Control	Total
Male	Smoker	647	622	1269
	Non-smoker	2	27	29
	Total	649	649	1298
Female	Smoker	41	28	69
	Non-smoker	19	32	51
	Total	60	60	120
Total	Smoker	688	650	1338
	Non-smoker	21	59	80
	Total	709	709	1418

This is a $2 \times 2 \times 2$ table (three classification variables with two levels each) and so the table contains eight cells. There are seven sets of marginal totals associated with this table: three sets of 2×2 totals (one for each variable summed over), three sets of 2 totals (one for each pair of variables summed over) and the overall total (obtained by summing over all three variables). The corresponding marginal tables are shown in Table 9.3(a)–(g); these are the marginal totals from Table 9.2.

9.1.1 Analysis

The analysis of contingency tables is aimed at investigating the relationships amongst the classification variables defining the table. Two questions which might arise from Table 9.1, for example, are, 'What proportion of domiciliary physiotherapy patients could in fact have attended a clinic?', and, 'Does the proportion of suitable patients depend upon the source of referral?'. The first question can be answered by using the methods of Section 7.6.1 to construct a confidence interval for the proportion of non-suitable patients (a 95% CI is 0.13 ± 0.05). The second question requires the comparison of patterns of numbers across the rows and columns of the table.

Table 9.3   Marginal tables for Table 9.2.

(a) Summed over sex.	Type of individual		(b) Summed over smoking status.	Type of individual	
Smoking status	Case	Control	Sex	Case	Control
Smoker	688	650	Male	649	649
Non-smoker	21	59	Female	60	60

(c) Summed over type of individual.	Smoking status		(d) Summed over sex and smoking status.	
Sex	Smoker	Non-smoker	Type of individual	
Male	1269	29	Case	709
Female	69	51	Control	709

(e) Summed over sex and type of individual		(f) Summed over type of individual and smoking status	
Smoking status		Sex	
Smoker	1338	Male	1298
Non-smoker	80	Female	120

(g) Summed over sex, smoking status and type of individual	1418

This is the kind of question that this chapter will seek to answer. A similar question associated with Table 9.2 is, 'Is there a relationship between smoking and lung cancer?'. Such questions will be framed and answered in the language of hypothesis testing. The null hypothesis will be that the variables are independent and the alternative hypothesis will be that they are not independent. This is also often stated as a test of *no interaction* versus *interaction*.

In epidemiological studies the comparison of proportions of types of people affected by some disease is more appropriately measured on a relative scale, giving rise to the concept of relative risk. In the lung cancer example the question might be asked, 'Is the risk of developing lung cancer higher in smokers relative to non-smokers?'. A discussion of such problems appears in Section 9.2.6.

It is obviously not possible to cover all the techniques available for the analysis of contingency tables in a single chapter. Interested readers will find Everitt (1977) gives a readable and comprehensive account of all the basic methods. More details relating to the use of log-linear models are provided by Fienberg (1977).

9.1.2 Assumptions

The levels of any variable in a contingency table must be mutually exclusive. This is because any observation may appear in a table only once. If, for instance, one of the variables was age group, the levels 'less than 30', 'less than 50', 'less than 70', and '70 and over' would not be appropriate. On the other hand 'less than 30', '30 but less than 50', '50 but less than 70' and '70 and over' are acceptable. The levels should also be exhaustive for the population of interest. For instance, if inferences are required for all ages, it is no use sampling only people at work, thus omitting the very young and very old. That is not to say that nil values in the sample for some levels are not acceptable – on the contrary they are often the most interesting data. It is necessary that all *potential* levels in the population are considered.

Most of the techniques for the analysis of contingency tables assume that the observations (the items counted to create the cell values) are independent. Table 9.4 contains two classification variables, 'sex of baby' (boy, girl) and 'weight at birth' (below 2500 g, 2500 g or over). Suppose that the data were drawn from pairs of boy and girl twins. It is reasonable to suppose that weights of twins are related, and this must be taken account of in any statistical analysis of these data, for example an analysis of the relationship between sex and birthweight. This is an example of paired data (see Sections 7.15 and 9.2.4). If, on the other hand the babies were born to different mothers then there is probably no reason to doubt that the observations are independent and the standard methods of analysis (Section 9.2.1, for instance) are appropriate.

Table 9.4 Sex and birthweight.

Birthweight in grams	Sex		Total
	Boy	Girl	
Less than 2500	4	5	9
2500 or over	53	52	105
Total	57	57	114

9.1.3 Basic notation

Consider a two-way table consisting of r rows and c columns. Then the cell frequency associated with row i and column j will be denoted by n_{ij}, for i = 1, 2, . . ., r and j = 1, 2, . . ., c. In Table 9.1, for example, n_{11} is the number of GP referrals suitable for domiciliary care and has a value of 20; n_{12} is the number of GP referrals not suitable for domiciliary care and has a value of 6, and so on.

The marginal totals derived from these cell values will be denoted using 'dot notation'. For two-way tables these marginal totals are:

$$n_{i.} = \text{the sum over all columns for row i} = \sum_{j=1}^{c} n_{ij}$$

$$n_{.j} = \text{the sum over all rows for column j} = \sum_{i=1}^{r} n_{ij}$$

$$n_{..} = \text{the overall total} = n$$

$$= \sum_{i=1}^{r} n_{i.} = \sum_{j=1}^{c} n_{.j} = \sum_{i=1}^{r} \sum_{j=1}^{c} n_{ij}$$

In Table 9.1

$$n_1. = 26, \quad n_2. = 112, \quad n_3. = 18$$

$$n._1 = 135, \quad n._2 = 21$$

$$n = 156$$

This notation extends naturally to three-way and higher dimensional tables. For example a four-way table would have cell values n_{ijkl} and marginal totals such as $n._{jkl} = \sum_{i=1}^{r} n_{ijkl}$.

Further notation, consistent with that given here, will be introduced later as it becomes necessary.

9.2 2 × 2 tables

This section will deal with the simplest type of contingency table, the 2×2. Sections 9.2.1–9.2.3 deal with independent data, Sections 9.2.4 and 9.2.5 with paired data and Section 9.2.6 with epidemiological data.

9.2.1 The chi-squared test for independence

An example of a contingency table with independent observations and two classification variables, each with two levels, is given by Table 9.5. This is taken from a study of non-rubella immune women who delivered babies in hospital (part of a more general investigation by Cook *et al.* (1987)). These women were classified by whether or not they received rubella vaccination after giving birth in the hospital, and by the type of ward in which they delivered.

Table 9.5   Vaccination in hospital by type of ward.

Type of ward	Vaccinated in hospital?		Total
	Yes	No	
Consultant	68	11	79
GP	18	7	25
Total	86	18	104

What is of interest in these data is whether the probability of a woman being vaccinated in hospital depends upon whether she delivered in a GP ward or a consultant ward. This may be expressed formally by stating the null and alternative hypotheses as

H_0: The probability of a woman being vaccinated in hospital is independent of the type of ward in which she delivers,

H_1: Vaccination is not independent of type of ward.

To develop a test procedure for such hypotheses consider the general case of a 2×2 table. Using an extension of the basic notation given in Section 9.1.3 let:

p_{ij} = the probability of an individual being classified into cell (i,j)

(e.g. p_{11} = probability of delivering in a consultant ward and being vaccinated in hospital),
and let

$p_{i.}$ = the probability of an individual being classified into level i of the first classification variable

(e.g., $p_{1.}$ = the probability of delivering in a consultant ward),
and

$p_{.j}$ = the probability of an individual being classified into level j of the second classification variable

(e.g., $p_{.1}$ = the probability of being vaccinated in hospital).
Clearly i and j both run from 1 to 2 in a 2×2 table.

It can be shown that the random variable representing the number of women in the entire population falling into cell (i,j) has approximately a normal distribution with mean and variance equal; let this value of the mean and variance be e_{ij}. Thus

$$\frac{n_{ij} - e_{ij}}{\sqrt{e_{ij}}}$$

comes, at least approximately, from the standard normal distribution, and so from Section 6.12.1,

$$\sum_{i=1}^{2} \sum_{j=1}^{2} \frac{(n_{ij} - e_{ij})^2}{\sqrt{e_{ij}}}$$

is an observation from a chi-squared (χ^2) distribution (approximately).

The n_{ij} values are known, since they are the observed cell values, but the e_{ij} must be derived. If the row and column classification variables are independent (H_0), then the cell values would be expected to show similar patterns in each row (and similar patterns in each column). If the variables are not independent (H_1) then the patterns would be expected to be different. On the other hand, the marginal row and column totals are scaling factors, showing the relative magnitude of each individual row and column, and provide no information about the degree of relationship between the row and column variables. The expected marginal totals, whether H_0 or H_1 is, in fact, true, will consequently be taken as fixed at their observed values. If the row (and column) margins are so fixed, then the overall total is fixed to be n, and so the expected values must satisfy the equation

$$e_{ij} = np_{ij} \ (i = 1, 2; j = 1, 2)$$

Now consider the general statement of the null and alternative hypotheses. The null hypothesis is that the row and column variables are independent. By the multiplication law for probabilities (see Section 6.4.5), under H_0, it must be that $p_{ij} = p_i.p._j$ for all i and j. Hence the hypotheses are stated mathematically as

$$H_0: p_{ij} = p_i.p._j \ (i = 1, 2; j = 1, 2)$$
$$H_1: p_{ij} \neq p_i.p._j$$

Since $n_i.$ and $n._j$ are fixed, the basic principles of probability theory in Section 6.1.1 show that

$$p_i. = \frac{n_i.}{n} \quad \text{and} \quad p._j = \frac{n._j}{n}$$

for all i and j. Consequently, when H_0 is true,

$$e_{ij} = np_{ij} = n \frac{n_i.}{n} \ \frac{n._j}{n} = \frac{n_i.n._j}{n} \ (i = 1, 2; j = 1, 2)$$

and the test statistic

$$X^2 = \sum_{i=1}^{2} \sum_{j=1}^{2} \frac{(n_{ij} - (n_i.n._j/n))^2}{n_i.n._j/n}$$

then comes from a chi-squared distribution.

Values of chi-squared are given in Table A.3, and the test for independence will entail checking the observed value of X^2 against the theoretical value of χ^2 in this table. If X^2 is bigger at a particular level of significance the null hypothesis is rejected at that level.

Now values of χ^2 depend upon the degrees of freedom and so the correct d.f. for X^2 must be ascertained. In Section 6.12.1 the d.f. for χ^2 was said to be equal to the number of independent normal distributions contributing to it. Here there are four normals involved (one for each cell of the table), so there would appear to be four d.f. The parameters (the means or expected values) of these normals have, however, been constrained by the fact that the row and column marginal totals are fixed, and this causes the normals to be interdependent. Each independent constraint causes one d.f. to be lost. Here the constraints are

$$\begin{array}{ll} e_{11} + e_{12} = n_1. & e_{11} + e_{21} = n._1 \\ e_{21} + e_{22} = n_2. & e_{12} + e_{22} = n._2 \end{array}$$

but since $n_1. + n_2. = n._1 + n._2 \ (= n)$ anyway, any three of these constraints must completely specify the fourth. That is, there are three independent constraints and hence $4 - 3 = 1$ d.f.

One final thing to notice about the test based on the X^2 statistic is that as n_{ij} and e_{ij} move further apart in *either* direction, X^2 will get bigger. This is because the difference between n_{ij} and e_{ij} is squared and also because e_{ij} must always be positive in

$$X^2 = \sum_{i=1}^{2} \sum_{j=1}^{2} \frac{(n_{ij} - e_{ij})^2}{e_{ij}}.$$

Consequently it is only large positive values of X^2 that provide evidence against H_0, and, although the test based on X^2 is always a two-sided test, it is *one-sided* critical values that must be read from tables of the chi-squared distribution, such as Table A.3.

It is useful to provide a summary of this test for independence in a 2×2 table:

(1) State the hypotheses

$$H_0: p_{ij} = p_{i\cdot}\, p_{\cdot j}$$
$$H_1: p_{ij} \neq p_{i\cdot}\, p_{\cdot j}$$

(2) Calculate the expected values when H_0 is true,

$$e_{ij} = \frac{n_{i\cdot}.n_{\cdot j}}{n}$$

(3) Calculate the value of the test statistic,

$$X^2 = \sum_{i=1}^{2} \sum_{j=1}^{2} \frac{(n_{ij} - e_{ij})^2}{e_{ij}}$$

(4) Compare the computed X^2 with the tabulated χ_1^2 (chi-squared on 1 d.f.). Reject H_0 if the probability of observing a value as large as the calculated value is less than or equal to some predetermined value (such as 0.05).

Example 9.1: Returning to Table 9.5 the test for independence is carried out as follows:

(1) $H_0: p_{ij} = p_{i\cdot}\, p_{\cdot j}$ (vaccination and type of ward are independent)
$H_1: p_{ij} \neq p_{i\cdot}\, p_{\cdot j}$ (vaccination and type of ward are not independent)

(2) $e_{11} = \dfrac{79 \times 86}{104} = 65.327 \qquad e_{12} = \dfrac{79 \times 18}{104} = 13.673$

$e_{21} = \dfrac{25 \times 86}{104} = 20.673 \qquad e_{22} = \dfrac{25 \times 18}{104} = 4.327$

(3) $X^2 = \dfrac{(68 - 65.327)^2}{65.327} + \dfrac{(11 - 13.673)^2}{13.673} + \dfrac{(18 - 20.673)^2}{20.673} + \dfrac{(7 - 4.327)^2}{4.327}$

$= 2.63$

(4) From Table A.3, $P(\chi_1^2 < 2.706) = 0.9$ and as the observed value of X^2 is below 2.706 the test fails to be rejected at the 10% level of significance (that is the P value is greater than 0.1). There is no evidence that vaccination and type of ward are related.

It is instructive to look at the observed and expected values for this example together in the form of tables:

Observed			Expected		
68	11	79	65.327	13.673	79
18	7	25	20.673	4.327	25
86	18	104	86	18	104

It is clear that the observed values and those suggested by the null hypothesis are in reasonably close agreement, so the result of the chi-squared test seems correct. Notice that the marginal totals in both tables are the same, as they must be since the $n_{i\cdot}$ and $n_{\cdot j}$ are used to calculate the expected cell values.

It is possible to interpret a test for independence in a 2×2 table as a comparison of two proportions, since independence of the row and column variables would mean that the same proportion would be expected in row 1 (say) of each column. To make the analogy clearer, Example 9.2 reworks the example of a test for the equality of two proportions, Example 7.12.

Example 9.2:   The numbers of boys and girls born in two hospitals, taken from Examples 7.7 and 7.12, are presented in Table 9.6.

Table 9.6   Sex at birth for two hospitals.

Sex	Hospital		Total
	1	2	
Female	924	66	990
Male	1076	54	1130
Total	2000	120	2120

Here $e_{11} = 934.0$, $e_{21} = 1066.0$, $e_{12} = 56.0$ and $e_{22} = 64.0$ leading to $X^2 = 3.52$. From Table A.3 this is significant at the 10% but not the 5% level, a result which is in complete agreement with Example 7.12.

In fact the agreement goes even further. The test statistic in Example 7.12 was -1.8767, and the square of this value is 3.52, the value of X^2. This should not come as a surprise! It can be seen, from Section 6.12.1, that χ^2 on 1 d.f. is the square of a single standard normal. Under the null hypothesis X^2 follows χ_1^2, and the test statistic from Example 7.12 follows a standard normal (Z) distribution.

In the special case of a 2×2 table, a simpler formula for X^2 is

$$X^2 = \frac{n \, (n_{11}n_{22} - n_{12}n_{21})^2}{n_{1.} \, n_{2.} \, n_{.1} \, n_{.2}}$$

which is easier to use for hand calculation. Using the data from Table 9.5,

$$X^2 = \frac{104 \, (68 \times 7 - 11 \times 18)^2}{79 \times 25 \times 86 \times 18} = 2.63 \text{ (as in Example 9.1)}$$

This formula does not generalize to more complex situations and so will not be used again.

9.2.2 Correction for continuity

The test statistic X^2 is only *approximately* distributed as χ^2 and, for 2×2 tables, the approximation is not always very good. One particular problem is that caused by small expected values, but this situation will be covered in Section 9.2.3. In other cases the χ^2 approximation may be improved by using a correction factor developed by Yates (1934). Yates' correction is a continuity correction, much like the continuity correction used with grouped discrete variables when constructing histograms, calculating means, etc. (see Section 3.11). It is an adjustment used to cope with the fact that X^2 depends on the discrete n_{ij} values whereas χ^2 is a continuous random variable.

Yates' correction for continuity consists of subtracting $\frac{1}{2}$ from $(n_{ij} - e_{ij})$ when it is positive, and adding $\frac{1}{2}$ to it when it is negative, in the test statistic X^2. That is, X^2 becomes

$$\sum_{i=1}^{2} \sum_{j=1}^{2} \frac{(|n_{ij} - e_{ij}| - \frac{1}{2})^2}{e_{ij}}$$

where $|n_{ij} - e_{ij}|$ means 'subtract e_{ij} from n_{ij} and make the result positive'.

Clearly Yates' correction makes the contribution of each difference (observed minus expected) less extreme by $\frac{1}{2}$. As such it reduces the value of X^2; in some cases the reduction is substantial. Re-working Example 9.1, but using Yates' correction, gives a value of 1.74 for X^2 (instead of 2.63).

There is some disagreement amongst statisticians about whether Yates' correction is really necessary and so it will not be used subsequently in this chapter. Notice that it leads to a conservative conclusion, and so errs on the side of safety, and against wrongly rejecting the null hypothesis.

9.2.3 Fisher's exact test

As mentioned in the previous section, small expected values in a 2×2 table cause special problems for the test based on the X^2 statistic. This is because X^2 is particularly sensitive to small expected values, e_{ij}, even where they are close to the corresponding observed values, n_{ij}, since the e_{ij} appear in the denominator of X^2. Small e_{ij} values thus tend to give misleadingly large values of X^2 and, bearing in mind that X^2 is only approximately distributed as χ_1^2 under H_0, this can lead to incorrect rejection of the null hypothesis. Fisher's exact test is an alternative to the chi-squared test designed to overcome this problem. Unfortunately, no such simple test is available for tables of other sizes and dimensions.

Consider a sample of size n, where the observations are somehow to be arranged in the cells of a 2×2 table, and assume that the only allowable arrangements are those for which the row and column totals are fixed. Now if these fixed values are the marginal totals observed in the sample data, it is possible to calculate the probability of obtaining any of the possible arrangements of observations in the cells which would give rise to this set of margins.

Under the null hypothesis of no interaction between classification variables,

$$\frac{n_{11}}{n_{\cdot 1}} = \frac{n_{12}}{n_{\cdot 2}} \quad \text{and} \quad \frac{n_{11}}{n_{1 \cdot}} = \frac{n_{21}}{n_{2 \cdot}}$$

that is, the proportion falling into the first row is the same for each column, and the proportion falling into the first column is the same for each row. If the null hypothesis is not true, then the proportions will be different. Fisher's exact test uses the observed departure from equal proportions to generate a test for non-independence. Since the observed difference is in a specific direction, however, the test is one-sided, unlike the chi-squared test which allows for departures in both directions, and is therefore two-sided.

It must be borne in mind that the arrangement actually observed results from only one of many possible samples which might have been drawn, and therefore, even if the null hypothesis is true, there is every chance of observing arrangements for which it is not the case that

$$\frac{n_{11}}{n_{.1}} = \frac{n_{12}}{n_{.2}} \quad \text{and} \quad \frac{n_{11}}{n_{1.}} = \frac{n_{21}}{n_{2.}}$$

exactly. If the null hypothesis is true, though, the observed arrangement is unlikely to be one that would only very rarely be observed. Using this reasoning, it is possible to calculate a P value.

If there is no relationship between the two classification variables, the probability of observing, simply by chance, the particular arrangement with $n_{11} = a$, $n_{12} = b$, $n_{21} = c$ and $n_{22} = d$, is

$$P = \frac{(a+b)!(c+d)!(a+c)!(b+d)!}{a!b!c!d!n!} \tag{1}$$

Fisher's exact test consists of finding the probability of obtaining the observed table or any *more extreme* arrangement when H_0 is true. 'More extreme' means one which increases the disparity between proportions. To find this probability the observed table and each of the more extreme tables is written down and their probabilities calculated using expression (1) above; then all these probabilities are added together to produce the P value for the test. This process will be explained in detail by working through an example.

Example 9.3:   Table 9.7 shows, for the hospital ward data in Appendix 1, patients classified by sex and by whether or not they underwent one or more operations during their stay in hospital.

Table 9.7   Sex and operation status for a sample of inpatients. (Data from Appendix 1.)

Sex	Operation(s)		Total
	No	Yes	
Male	3	2	5
Female	44	7	51
Total	47	9	56

Using expression (1), the probability of observing Table 9.7, if the null hypothesis of independence is true, is

$$P = \frac{5!51!47!9!}{3!2!44!7!56!} = 0.153$$

Now, in this observed table,

$$\frac{n_{11}}{n_{.1}} = \frac{3}{47} = 0.06 \quad \text{and} \quad \frac{n_{12}}{n_{.2}} = \frac{2}{9} = 0.22$$

How can these be made to move further apart? The smallest marginal value in the table is that associated with the row male, i.e., 5, and a little thought will serve to convince the reader that the only possible arrangements of the table which maintain the observed margins are those for which the two cells in the male row vary between 0 and 5 (i.e., $n_{11} = 5$, $n_{12} = 0$; $n_{11} = 4$, $n_{12} = 1$; . . .; $n_{11} = 0$, $n_{12} = 5$). As n_{11} increases from 3 to 4,

$$\frac{n_{11}}{n_{.1}} \quad \text{and} \quad \frac{n_{12}}{n_{.2}}$$

become 0.09 and 0.11, that is, they are *more* similar, so the table containing these values is not more extreme than the observed one. As n_{11} decreases from 3 to 2, however, the proportions become 0.04 and 0.3, and so have a greater disparity than the proportions actually observed. Hence there are three more extreme tables; these are shown below along with the probabilities of observing them, derived using expression (1). Note that these tables could also have been arrived at by comparing $\dfrac{n_{11}}{n_{1.}}$ and $\dfrac{n_{21}}{n_{2.}}$.

2	3	5		1	4	5		0	5	5
45	6	51		46	5	51		47	4	51
47	9	56		47	9	56		47	9	56

$$P = 0.0238 \qquad P = 0.00155 \qquad P = 0.000033$$

From these it is clear that the alternative hypothesis is that there is a *positive* association between males and operations. The P value for the test is

$$P = 0.153 + 0.0238 + 0.00155 + 0.000033 = 0.178$$

so the probability of obtaining the observed table or one that is more extreme is 0.178, and there is no evidence to reject the null hypothesis of independence.

Many statistical packages have the facility for performing Fisher's exact test; where calculation by hand is necessary, certain 'short-cut' methods are available (see, for example, Everitt, 1977), though these do not, in fact, help a great deal. An alternative is to use a statistical table, such as that in Fisher and Yates (1957).

Some authors suggest that a two-sided test should be performed by doubling the significance level obtained. This is exactly equivalent to changing the point at which H_0 is to be rejected, that is, if the 5% level would normally be used, reject H_0 if the P value obtained is less than 0.025. This gives a conservative test, and whether or not it is appropriate will depend upon the nature of the study. As always, special care should be taken when attempting to make decisions on the basis of borderline results.

It should be noted that Fisher's exact test has very low power for small samples, particularly where the true proportions, though different, are quite similar. In such cases there may be very little chance of detecting a departure from the hypothesis of independence.

9.2.4 McNemar's test

In Section 9.1.2 an example was given in which babies were classified by birthweight and sex, but where the observations were made on pairs of (boy, girl) twins. As mentioned, this is an example of paired (non-independent)

Table 9.8   Visits of legionnaires' cases and controls to a shopping centre.

	Cases		
Controls	Visited	Not visited	Total
Visited	25	1	26
Not visited	11	1	12
Total	36	2	38

data. A similar situation arises in a one-to-one matched case-control study where controls are matched with cases for such things as age and sex (see Section 2.5.3).

The test using X^2, described in Section 9.2.1 (and the extensions in Sections 9.2.2 and 9.2.3) is not appropriate in these situations. Instead the McNemar test, described in Section 7.15.2, should be used. In Chapter 7 this test was developed for comparing proportions; here it will be described in the language of contingency tables.

Example 9.4: Consider Example 7.20 in which 38 pairs of people (suspected cases of legionnaires' disease and their matched controls) were asked whether or not they had visited a certain shopping centre. The replies may be displayed in the form of Table 9.8. It is important to note that the observations are the number of *pairs* falling into each cell, for example there were 25 pairs in which both the case and the control had visited the shopping centre.

The object of the analysis will be to see whether cases and controls show different behaviour concerning visits to the shopping centre. Now any pair for which both the case and the control had the same behaviour is clearly of no interest, since it gives no information about the difference between cases and controls. Thus only cells (1, 2) and (2, 1) in the contingency table are of interest. If the two groups (cases and controls) were the same as regards the attribute (visiting the shopping centre) it would be expected that half of the 'interesting' pairs would have a visiting case but a non-visiting control and vice versa. That is,

$$e_{12} = \frac{n_{21} + n_{12}}{2} \quad \text{and} \quad e_{21} = \frac{n_{21} + n_{12}}{2}$$

This leads to a new X^2 statistic

$$\frac{(n_{12} - e_{12})^2}{e_{21}} + \frac{(n_{21} - e_{21})^2}{e_{21}}$$

$$= \frac{(n_{21} - n_{12})^2}{n_{21} + n_{12}}$$

after some algebraic manipulation. This is to be compared with χ^2 on 1 d.f. (X^2 has two terms and there is the constraint that $e_{21} + e_{12} = n_{21} + n_{12}$).

Notice that this X^2 is the square of the test statistic given in Section 7.15.2 (when the notation is reconciled). This is correct since, under H_0, the test statistic given there followed a standard normal (Z) distribution; here X^2

follows χ_1^2 (and $\chi_1^2 = Z^2$).

Carrying out the calculations for Table 9.8,

$$X^2 = \frac{(11 - 1)^2}{11 + 1} = 8.33$$

From Table A.3 $P(\chi_1^2 < 7.880) = 0.995$ and $P(\chi_1^2 < 10.832) = 0.999$. Hence the P value for this test is between 0.005 and 0.001. In Example 7.20 the test statistic was $2.89 = \sqrt{8.33}$, as expected, but there a one-sided test was carried out, not two-sided, as here. Consequently the P value here will be twice that found in Example 7.20, that is around 0.004.

9.2.5 Gart's test

Another situation in which the data are paired is when a clinical trial to compare two treatments is carried out in which each subject is given both treatments. Some subjects are given treatment A first and the rest are given treatment B first. This type of trial is called a *two-period cross-over trial* (see Section 2.5.4 for more details). Clearly the data are paired here because every recording on treatment A is paired with a recording on treatment B from the same subject (i.e., there is self-pairing).

Suppose that the data on the two treatments consist of a record of whether treatment A or treatment B is preferred for each subject. For example the treatment which seems most effective (to the patient or doctor) would be the one preferred. Interest would then lie in discovering whether A or B is most often preferred. There may be a second effect which influences the conclusions: when the order of taking the treatments is important. This could happen if there were a carry-over effect from the first period of treatment to the second. Gart's test allows both the effect of treatments and the order of treatments to be accounted for. The description of this test that follows uses notation (as well as methodology) from Gart (1969).

As with McNemar's test no information is obtained from like pairs; in this situation like pairs are subjects who have no preference. So, considering only unlike pairs, let

y_a = the number of subjects for which treatment A is preferred, and for which A is given first;

y_a' = the number of subjects for which treatment A is preferred, and for which B is given first;

y_b = the number of subjects for which treatment B is preferred, and for which A is given first;

y_b' = the number of subjects for which treatment B is preferred, and for which B is given first;

n = $y_a + y_b$ (the number of subjects who receive A first);

n' = $y_a' + y_b'$ (the number of subjects who receive B first).

This information may be presented in two different tables, 9.9(a) and (b).

A test for independence based on Table 9.9(a) would test the null hypothesis that the first (and second) treatment is equally likely to be preferred

Table 9.9 Data for Gart's test.

(a)

Preference	Treatment order		Total
	(A,B)	(B,A)	
1st treatment	y_a	y_b'	$y_a + y_b'$
2nd treatment	y_b	y_a'	$y_b + y_a'$
Total	n	n'	$n + n'$

(b)

Preference	Treatment order		Total
	(A,B)	(B,A)	
A	y_a	y_a'	$y_a + y_a'$
B	y_b	y_b'	$y_b + y_b'$
Total	n	n'	$n + n'$

regardless of whether A or B is taken first. That is a test that there is no difference between the treatments in terms of preference. A similar argument will show that Table 9.9(b) leads to a test of the null hypothesis that there is no order effect.

Example 9.5:   Twenty patients with rheumatoid arthritis were each given two drugs, 'X' and 'Y' in a cross-over trial. After the trial each patient was asked at which period of the trial he found himself to have the greater mobility. Nine patients were not able to detect a difference. For the remaining eleven the trial organizers checked which drug was given in which period, and Tables 9.10 (a) and (b) were produced.

Clearly the expected values under the null hypothesis of independence are small in both tables, so Fisher's exact test will be applied.

Table 9.10 Results of a trial of drugs X and Y.

(a)

Increased mobility with	Drug order		Total
	(X,Y)	(Y,X)	
1st drug	4	2	6
2nd drug	0	5	5
Total	4	7	11

(b)

Increased mobility with	Drug order		Total
	(X,Y)	(Y,X)	
X	4	5	9
Y	0	2	2
Total	4	7	11

From Table 9.10(a)

$$P = \frac{6!5!4!7!}{4!2!0!5!11!} = 0.045$$

and from Table 9.10(b)

$$P = \frac{9!2!4!7!}{4!5!0!2!11!} = 0.382$$

(note that each table was itself the extreme possibility).
So the conclusions are that there is evidence to suggest that drug X is better than drug Y for improving mobility, but there is no evidence to suggest that this outcome was affected by the order in which X and Y were administered.

9.2.6 Relative risk and the odds ratio

Sections 9.2.1–9.2.5 have dealt with various methods for testing hypotheses associated with 2 × 2 tables. In some situations it is more appropriate to estimate some parameter from the data in a 2 × 2 table. One particularly important situation for health researchers is where the relative risk or, alternatively, the odds ratio, is calculated from epidemiological data.

Consider a population of N individuals, some of whom are known to have a specific disease and some of whom are known to have been exposed to a specific factor which may be related to the disease. For example, the disease could be lung cancer and the factor could be smoking. If the numbers with the disease and the numbers who have been exposed to the factor are known for the population then these values can be displayed in a 2 × 2 table such as Table 9.11.

Table 9.11   Disease status and factor exposure.

Exposure to factor?	Disease occurs?		Total
	Yes	No	
Yes	A	B	A + B
No	C	D	C + D
Total	A + C	B + D	N

Note that capital letters have been used in Table 9.11 to reinforce the fact that these are population values, rather than the results observed in a sample.

Now $\pi_{11} = A/N$ is the proportion of people who have the disease and have been exposed to the factor and $\pi_{1.} = (A + B)/N$ is the proportion of people who have been exposed to the factor. Then the ratio $\pi_{11}/\pi_{1.} = A/(A + B)$ is the disease prevalence rate for those exposed to the factor, that is the risk of disease for those exposed. Similarly define $\pi_{21} = C/N$, $\pi_{2.} = (C + D)/N$ and then $\pi_{21}/\pi_{2.} = C/(C + D)$ is the risk of disease for those not exposed to the factor. The ratio of these two risks is given the symbol R where

$$R = \frac{A(C + D)}{C(A + B)}$$

and this is called the *relative risk*. It represents the number of times more (or less) the disease occurs in the subgroup of the population that has been exposed to the factor compared with the group not exposed. If R > 1 then a greater proportion of those exposed have the disease.

If, for some population, the values A, B, C and D were known, it would be a simple matter to calculate the relative risk. In practice, data are almost always only available for a sample, and it is from such data that the relative risk must be estimated. The method of estimation depends upon whether the data are collected from a cohort or a case-control study. Both types of study will be considered here; the reader should be familiar with the essential differences between these types of special study, as described in Sections 2.5.2 and 2.5.3. The notation and general method which follow are taken from Schlesselman (1982). Notice that although risks have been defined here in terms of *prevalence* (i.e., present numbers), they could equally well be defined in terms of *incidence* (i.e., new cases of the disease).

Consider first a *cohort* or longitudinal study in which a sample of n individuals is chosen which consists of a fraction f_1 of the A + B individuals exposed (to the factor of interest), and a fraction f_2 of the C + D individuals not exposed. These individuals are observed for some specified time period and, at the end of this period, the numbers in each of the exposure groups having the disease is recorded. It is assumed that each exposure group is sampled randomly, so that a person who will develop the disease is just as likely to be included as one who will not. The data that would be expected to result from such a study can be presented as a 2×2 table: Table 9.12.

Table 9.12   The results of a cohort study.

Exposure to factor?	Disease occurs?		Total
	Yes	No	
Yes	$a(= f_1 A)$	$b(= f_1 B)$	$a + b(= f_1(A + B))$
No	$c(= f_2 C)$	$d(= f_2 D)$	$c + d(= f_2(C + D))$
Total	$a + c$	$b + d$	n

The risks of having the disease estimated from Table 9.12 are:

$$\text{for the exposed group} \quad \frac{a}{a + b} = \frac{f_1 A}{f_1(A + B)} = \frac{A}{A + B};$$

$$\text{for the unexposed group} \frac{c}{c + d} = \frac{f_2 C}{f_2(C + D)} = \frac{C}{C + D};$$

and the estimated relative risk is

$$r = \frac{a(c + d)}{c(a + b)} = \frac{f_1 A \, f_2(C + D)}{f_2 C \, f_1(A + B)} = \frac{A(C + D)}{C(A + B)} = R$$

Now Table 9.12 presents expected, or average, values and ignores sampling variation which would occur in practice (so that a would not be $f_1 A$ exactly, etc.). Nevertheless it has been shown that r will provide a good estimate of R.

Table 9.13 Oral contraceptives and myocardial infarction.

Oral contraceptive user?	Death due to myocardial infarction?		Total
	Yes	No	
Yes	7	4793	4800
No	2	4823	4825
Total	9	9616	9625

Example 9.6: Table 9.13 shows some hypothetical data for a cohort study in which 5000 women who used oral contraceptives, and the same number of women who did not, were followed up for ten years. The number of deaths due to myocardial infarction in each group was recorded. 200 oral contraceptive users and 175 non-oral contraceptive users were lost during the follow-up period, due to migration and other causes.

The risk of death during the ten-year period was:

$$\text{for oral contraceptive users } \frac{7}{4800} = 0.001458$$

$$\text{for others } \frac{2}{4825} = 0.000415$$

and the estimated relative risk is

$$\frac{0.001458}{0.000415} = 3.5$$

That is, oral contraceptive users were three and a half times more likely to die due to myocardial infarction.

Now consider a *case-control* study in which a sample of n individuals is chosen which consists of a fraction f_3 of the A + C individuals (cases) with the disease, and a fraction f_4 of the B + D individuals (controls) without the disease. All individuals are then investigated to ascertain whether or not they were exposed (to the factor of interest). It is essential that the cases and controls are sampled randomly so that within each group both exposed and unexposed people are equally likely to be included. The data that would be expected to result from such a study can be presented in a 2×2 table: Table 9.14.

Table 9.14 The results of a case-control study.

Exposure to factor?	Disease occurs?		Total
	Yes	No	
Yes	$a(= f_3A)$	$b(= f_4B)$	$a + b(= f_3A + f_4B)$
No	$c(= f_3C)$	$d(= f_4D)$	$c + d(= f_3C + f_4D)$
Total	$a + c$	$b + d$	n

The risks of having the disease estimated from Table 9.14 are:

for the exposed group $\dfrac{a}{a + b} = \dfrac{f_3A}{f_3A + f_4B}$;

for the unexposed group $\dfrac{c}{c + d} = \dfrac{f_3C}{f_3C + f_4D}$;

and the estimated relative risk is

$$r = \frac{a(c + d)}{c(a + b)} = \frac{f_3A(f_3C + f_4D)}{f_3C(f_3A + f_4B)}.$$

So in a case-control study the observed risks are not, on average, equal to the true risks in the population, and the observed relative risk is not, on average, equal to the true relative risk. Furthermore, in most practical situations the sampling fractions f_3 and f_4 are unknown; when this is the case the relative risk *cannot* be estimated from case-control studies. Notice that in a cohort study f_1 and f_2 are also unknown, but there they cancel out in the estimates of risk. For f_3 and f_4 to be known every individual in the population with the disease must have been identified; while it might be feasible to estimate f_3 and f_4 from other sources of data on the number of cases and the population size, in many situations even this is not possible.

There is one useful parameter, however, which may be estimated from both a cohort and a case-control study and this is the *odds ratio*, usually denoted by ψ, the Greek letter psi. If some event occurs with probability π, then the ratio $\pi/(1 - \pi)$ is called the *odds* for that event. The ratio of the odds of having the disease in exposed individuals to the odds of having the disease in non-exposed individuals is the odds ratio, ψ. Reference to Table 9.11 will show that

$$\psi = \frac{A/B}{C/D} = \frac{AD}{BC}$$

Then, from either a cohort or a case-control study Tables 9.12 and 9.14 respectively will show that $\hat{\psi}$, an estimate of ψ, is given by

$$\hat{\psi} = \frac{ad}{bc}$$

since this is exactly equal to ψ when sampling variation is ignored.

Now if the incidence of the disease being studied is small, then B is approximately equal to A + B and D is approximately equal to C + D. Under these circumstances the relative risk,

$$R = \frac{A(C + D)}{C(A + B)} \text{ is approximately equal to } \frac{AD}{CB} = \psi$$

the odds ratio. This approximation is good for a wide range of problems; many diseases do have extremely small incidence rates even in relatively high-risk exposure groups (especially where death is the outcome recorded). Consequently the sample estimate, $\hat{\psi}$, is often a good estimate for the relative risk. Using the data from the cohort study in Example 9.6, for instance,

$$\hat{\psi} = \frac{7 \times 4823}{2 \times 4793} = 3.5$$

which is (at least to one decimal place) the same value as was obtained for r.

In case-control studies, where no direct estimate of R is available, $\hat{\psi}$ often provides the best possible estimate of the relative risk. It may be shown that the variance of $\log_e \hat{\psi}$ may be estimated by

$$\widehat{\text{Var}}(\log_e\hat{\psi}) = \frac{1}{a} + \frac{1}{b} + \frac{1}{c} + \frac{1}{d}$$

(at least approximately). Then, assuming that $\log_e\psi$ is normally distributed, an approximate 95% confidence interval for $\log_e\psi$ is

$$\log_e\hat{\psi} \pm 1.96\sqrt{\widehat{\text{Var}}(\log_e\hat{\psi})}$$

Taking exponentials of the lower and upper limits of this interval gives lower and upper 95% confidence limits for ψ itself. As usual, other percentage confidence intervals are obtained by substituting appropriate values from Table A.1 in place of 1.96.

One final point about case-control studies is that random sampling does not necessarily ensure comparability of the case and control groups, a problem which leads investigators to select controls by matching them with cases. If matching is one-to-one then the data should be arranged as in McNemar's test (see Section 9.2.4) and $\hat{\psi}$ is then n_{21}/n_{12}. Thus in Table 9.8 the estimated odds ratio (approximate relative risk) is 11. (See Schlesselman (1982) for further details.)

Example 9.7: The smoking and lung cancer case-control study presented in Table 9.2 allows the odds ratio to be estimated for both males and females. Table 9.15 shows the male component of these data.

Table 9.15 Male smoking and lung cancer data.

Smoker?	Has lung cancer?		Total
	Yes	No	
Yes	647	622	1269
No	2	27	29
Total	649	649	1298

The odds ratio for males is

$$\hat{\psi} = \frac{647 \times 27}{2 \times 622} = 14.0$$

Lung cancer has a low enough prevalence rate for this to be a good estimate of the relative risk. That is, male smokers are fourteen times more likely to have lung cancer than other males.

Also the following approximate results hold:

$$\widehat{\text{Var}}(\log_e\hat{\psi}) = \frac{1}{647} + \frac{1}{622} + \frac{1}{2} + \frac{1}{27} = 0.5402$$

A 99% confidence interval for $\log_e\psi$ is

$$\log_e\hat{\psi} \pm 2.58 \sqrt{\widehat{\text{Var}}(\log_e\hat{\psi})}$$

i.e., $\qquad\qquad\qquad \log_e14 \pm 2.58 \times 0.7350$

i.e., $\qquad\qquad\qquad\qquad 2.639 \pm 1.896$

with an upper limit of 4.535 and a lower limit of 0.743. Taking exponentials of both of these gives the approximate 99% confidence interval for ψ of (2.1, 93.2). Hence there is extremely strong evidence of an increased risk due to smoking.

Notice that these results ignore the number of cigarettes actually smoked, information that is given in the original paper by Doll and Hill (1950). When analysed this information suggests that the risk of lung cancer increases with the number of cigarettes smoked. Also, this analysis takes no account of the matching of controls to cases by age and sex which is reported in the original paper. If data concerning the matching were available then a matched analysis would be appropriate, and produce a more accurate result.

9.3 $r \times c$ tables

Section 9.2 dealt with the special case of 2×2 contingency tables. This section will deal with the more general case of two-way tables of any size, that is tables with r rows and c columns, where both r and c are greater than or equal to two.

9.3.1 The chi-squared test for independence

The χ^2 test for $r \times c$ tables follows exactly the same procedure as that described for 2×2 tables in Section 9.2.1. Once again the null hypothesis for independence between the two classification variables is stated as

$$H_0: p_{ij} = p_{i.}p_{.j} \qquad (i = 1, 2, \ldots, r; j = 1, 2, \ldots, c)$$

and the alternative hypothesis is

$$H_1: p_{ij} \neq p_{i.}p_{.j}$$

Under H_0, the expected value in cell (i, j) of the table is

$$e_{ij} = \frac{n_{i.}n_{.j}}{n}$$

and the test statistic is

$$X^2 = \sum_{i=1}^{r} \sum_{j=1}^{c} \frac{(n_{ij} - e_{ij})^2}{e_{ij}}$$

Under H_0, X^2 has a chi-squared distribution, but with what degrees of freedom? The expected values are calculated using the fixed row margins, $n_{i.}$, of which there are r, and the fixed column margins, $n_{.j}$, of which there are c. Due to the fact that

$$\sum_{i=1}^{r} n_{i.} = \sum_{j=1}^{c} n_{.j} \, (= n),$$

this imposes $r + c - 1$ independent constraints upon the expected values.

Since X^2 consists of rc terms (the number of cells), the degrees of freedom must be

$$rc - (r + c - 1) = (r - 1)(c - 1).$$

Notice that when r = c = 2, d.f. = 1, as stated in Section 9.2.1.

Hence the test for independence in r × c tables involves comparing X^2 with χ^2 on $(r-1)(c-1)$ d.f.

Example 9.8: Table 9.16 shows results from a hypothetical survey of patients who stayed in 4 medical wards in an acute hospital. The data are replies to the question 'Did you feel that the amount of contact with nursing staff was (a) good (b) adequate (c) insufficient?'.

Table 9.16 Contact with nursing staff by ward.

Amount of contact with nursing staff	Ward				Total
	A	B	C	D	
Good	24	27	16	23	90
Adequate	12	19	15	17	63
Insufficient	6	9	7	7	29
Total	42	55	38	47	182

The question 'Was the amount of contact with nursing staff consistent across all four wards?', may be answered by testing the hypothesis of independence between the two classification variables, amount of contact and ward. The null and alternative hypotheses are

$$H_0: p_{ij} = p_i.p_{.j} \quad (i = 1, 2, 3; j = 1, 2, 3, 4)$$
$$H_1: p_{ij} \neq p_i.p_{.j}$$

and expected values under H_0 are those shown in Table 9.17.

Table 9.17 Expected values under H_0 for Table 9.16.

Amount of contact with nursing staff	Ward				Total
	A	B	C	D	
Good	20.8	27.2	18.8	23.2	90
Adequate	14.5	19.0	13.2	16.3	63
Insufficient	6.7	8.8	6.1	7.5	29*
Total	42	55	38*	47	182

*rounding error

For example,

$$e_{11} = \frac{90 \times 42}{182} = 20.8 \quad \text{and} \quad e_{23} = \frac{63 \times 38}{182} = 13.2.$$

$$\text{Then} \quad X^2 = \frac{(24 - 20.8)^2}{20.8} + \frac{(27 - 27.2)^2}{27.2} + \ldots + \frac{(7 - 7.5)^2}{7.5} = 1.91$$

and the degrees of freedom are $(3 - 1)(4 - 1) = 2 \times 3 = 6$.

From Table A.3 $P(\chi^2_6 < 10.645) = 0.90$ and thus the P value for this test is well above 0.1. There is no reason to reject the null hypothesis that the amount of contact is independent of ward.

It is important to underline the fact that this is only a test of independence between the classification variables. The fact that H_0 was not rejected in Example 9.8 does not necessarily mean that the table contains no information of interest. Across all wards, about 16% of patients felt that they had had insufficient contact with the nursing staff. Further analysis, perhaps by age group, length of stay or diagnosis might be useful.

9.3.2 Pooling levels

One thing to be remembered when using X^2 is that it is only *approximately* distributed as χ^2, as indicated in Section 9.2.2. The approximation may be improved by using Yates' correction in 2×2 tables, but no such correction is available for $r \times c$ tables. As with 2×2 tables, a special problem is that X^2 is particularly sensitive to small expected values which tend to inflate it. Unfortunately there are no hard and fast rules about how large the expected values, e_{ij}, must be before the test is reliable, and there is a great deal of disagreement amongst statisticians on this point. The most conservative rule states that no expected value should be less than 5, while a very generous one is that no expected value should be less than 1. A possible compromise is given by Cochran (1954) who states that if no more than 20% of expected values are less than 5, then the minimum expected value may be 1.

In 2×2 tables the problem of small expected values is overcome by using Fisher's exact test (see Section 9.2.3) but no such simple test is available for $r \times c$ tables. One possible way to avoid the problem is to pool levels of one or both of the classification variables, but such a procedure is only valid if particular care is taken about how the pooling is done. Take Table 9.1 as an example. There are two small expected values, $e_{12} = 3.5$ and $e_{32} = 2.4$. These represent one third of all the cells in the table, and so it is doubtful whether the chi-squared test should be used on these data as presented in Table 9.1. It may seem that the obvious thing to do, in order to produce large enough expected values and thus solve the problem, is to combine the two rows associated with the small expected values (i.e., rows 1 and 3). Table 9.18

Table 9.18   Modified version of Table 9.1.

Source of referral	Suitable for domiciliary care?		
	Yes	No	Total
GP/physiotherapist	38	6	44
Consultant	97	15	112
Total	135	21	156

shows the effect of such pooling. For this table the smallest expected value is 5.9, so there should be no problem with using the chi-squared test.

There are, however, three serious problems with this approach:

(1) *Levels have been pooled on the basis of observed data.* It is perfectly acceptable to pool levels if the decision is taken before the data are seen, perhaps because two levels have some common meaning. Pooling levels on the basis of observed data in order to produce larger expected values may have the effect of producing spuriously small P values in the chi-squared test.
(2) *The pooled levels have dubious meaning.* What does 'GP or physiotherapist' mean in Table 9.18? Levels pooled on the basis of observed data often generate new levels that are, at best, difficult to interpret and, at worst, meaningless.
(3) *The observed proportions have been changed unacceptably.* In Table 9.1, 6/26 (23%) of GP referrals were inappropriate whilst 0/18 (0%) of physiotherapy referrals were inappropriate. In Table 9.18, 6/44 (14%) of the combined group is inappropriate, and the temptation is to interpret this to mean 14% of each of the constituent groups is inappropriate. This problem is particularly important when only the reduced table is presented in a report, so that the reader does not have access to the original data.

Further drawbacks of pooling are given by Everitt (1977), but it should be stated that, where it is done correctly, pooling is acceptable. In some cases it may even aid interpretation, although it does inevitably lead to a loss of information. Where it is possible to also include the original table in a report this problem is overcome.

Unfortunately there is no easy way of analysing Table 9.1; while a researcher might be tempted to make some inferences about independence from these data, they should not be regarded as conclusive. A larger sample would probably solve the problem.

9.3.3 Residuals

In a 2×2 table non-independence of the classification variables is easy to interpret in terms of the difference between proportions. In $r \times c$ tables interpretation is not so straightforward. What is needed is some way of isolating those cells in which the observed value differs considerably from the expected value, and attaching some meaning to these differences.

One simple way of doing this is to present tables of observed (n_{ij}) and expected (e_{ij}) values side-by-side, as was done at the end of Example 9.1. Alternatively the two tables can be differenced, that is a single table of $n_{ij} - e_{ij}$ values can be presented. Recall that the e_{ij} values are the expected values when H_0 is true. Assuming H_0 to be true imposes a *model*, in this case the model of independence between the row and column classification variables. Just as with any other statistical model (e.g., regression, time series) the differences between the values actually observed and those predicted by the model are called the *residuals*. Hence $n_{ij} - e_{ij}$ is the residual for cell (i, j).

The cells which are important for causing lack of independence (that is, showing up the inadequacy of the model) are those with large residuals. There is a problem in interpreting 'large', however, because when e_{ij} is small a small residual may still indicate an important difference between n_{ij} and e_{ij} whereas when e_{ij} is big a small residual is unlikely to be important. Consequently it is more usual to look at the *standardized residuals*,

$$r_{ij} = \frac{n_{ij} - e_{ij}}{\sqrt{e_{ij}}} \qquad (i = 1, 2, \ldots, r; j = 1, 2, \ldots, c)$$

which are clearly the square roots of the components of X^2.

In fact it can be shown that $Var(r_{ij})$ can be estimated by

$$\hat{v}_{ij} = (1 - (n_{i.}/n))(1 - (n_{.j}/n))$$

and the *adjusted residuals*, d_{ij}, where

$$d_{ij} = \frac{r_{ij}}{\sqrt{\hat{v}_{ij}}}$$

are, approximately, values from a standard normal distribution, for all i and j when the classification variables are independent. Thus the d_{ij} can be compared with Table A.1 to identify 'significant' residuals.

Table 9.19   Source of admission by specialty for one month's discharges and deaths.

Specialty	Source of admission			Total
	Emergency	Booked/elective	Other	
General surgery	241	383	38	662
General medicine	317	839	74	1230
Paediatrics	102	228	18	348
Total	660	1450	130	2240

Example 9.9:   Table 9.19 shows the number of discharges and deaths for three specialties which occurred in one month in a large acute hospital, classified by specialty and source of admission. The value of X^2 from this table is 24.03 on $(3 - 1)(3 - 1) = 4$ d.f. which is extremely significant ($P < 0.0005$). The standardized residuals and adjusted residuals are given below:

Standardized residuals

	Source		
	3.29	−2.20	−0.068
Speciality	−2.39	1.52	0.31
	−0.053	0.18	−0.49

Adjusted residuals

	Source		
	4.67*	−4.41*	−0.08
Speciality	−4.23*	3.80*	0.48
	−0.07	0.33	−0.55

* significant at .01%

In both tables a negative value represents an observed value which was less than expected under the null hypothesis of independence. The asterisked adjusted residuals are the only ones to have anything but moderate departure from zero. In practical terms the significant values show that the number of General Surgery patients admitted as emergencies and the number of General Medicine patients who were booked or elective admissions are both higher than would be expected if source of admission were independent of specialty. The inverse numbers are lower than would be expected.

9.3.4 Trends in 2 × c tables

In many cases one of the two classification variables that make up the table takes only two levels. Without loss of generality, assume that this dichotomous variable labels the rows; hence the table is 2 × c for c greater than or equal to 2. The levels of this dichotomous variable can always be thought of as 'success' and 'failure' for suitable definitions of these terms. In this sense the analysis of a 2 × c table may be thought of as a generalization of the analysis of two proportions (in Section 7.10.2).

A test for the null hypothesis of independence in a 2 × c table is, of course, just a special case of such tests for r × c tables, as described in Section 9.3.1. One shortcoming with this test is that the alternative of no independence is completely general. Sometimes it can be more useful to consider a more specific alternative. When the levels of the column variable are ordered in some way it is often useful to consider testing the null hypothesis of independence against the specific alternative hypothesis of a linear (straight line) trend in the proportions falling in the first (or second) row. Ordering can occur because the column variable is quantitative (for example, length of stay groups) or ordinal (for example where the levels are 'small', 'medium' and 'large'). The ability to identify a linear trend is especially useful when the purpose is to demonstrate possible causality (i.e., the more cause, the more effect). The test for a linear trend involves calculating a test statistic, denoted $X^2_{(t)}$, and comparing this with χ^2 on 1 d.f. A significant result implies that a linear trend in the proportions exists. $X^2_{(t)}$ is actually part of the overall test statistic, X^2, which has $(r-1)(c-1) = c-1$ d.f. for a 2 × c table. The difference, $X^2 - X^2_{(t)}$, with $c-2$ d.f. provides a test for a significant non-linear component in the proportions.

$X^2_{(t)}$ is derived using a form of least squares similar to that described in Section 8.3.1. Armitage and Berry (1987) give the result

$$X^2_{(t)} = \frac{n(n \sum_{j=1}^{c} s_j n_{1j} - n_1 . \sum_{j=1}^{c} s_j n_{.j})^2}{n_1 . n_2 . (n \sum_{j=1}^{c} s_j^2 n_{.j} - (\sum_{j=1}^{c} s_j n_{.j})^2)}$$

where s_j (for $j = 1, 2, \ldots, c$) are scores representing the order of magnitude of the levels of the column variable. For instance, the scores for length of stay groups might be the mid-range of each class interval (as in Example 4.5) and for 'small', 'medium' and 'large' the scores might be -1, 0 and 1.

Example 9.10: A record of all rubella immunity tests carried out on women by the Public Health Laboratory in Reading in 1983 is given in Table 9.20. An

Table 9.20   Rubella immunity by age of woman, Reading, 1983.

Rubella immune?	Age group					Total
	15–19	20–24	25–29	30–34	over 34	
No	70	117	159	120	63	529
Yes	1293	2390	2955	1598	883	9119
Total	1363	2507	3114	1718	946	9648
Percentage 'No'	5.1	4.7	5.1	7.0	6.7	

extra line has been added to the table to stress the fact that it is linear trends in *proportions* or percentages which $X^2_{(t)}$ may be used to find. In this case there does seem to be an upward movement in the percentages, especially during the middle years of age that are recorded.

The overall X^2 statistic for these data has the value 14.40 on 4 d.f. This is significant at the 1% level $(0.01 > P > 0.005)$.

To calculate $X^2_{(t)}$ the scores, s_j, for the age groups need to be decided. For the first four age groups the mid-range is the obvious score. For the last group an educated guess has to be made for the upper limit; here it seems unlikely that many women above childbearing age would have a rubella immunity test, so 45 years seems a reasonable approximation to the maximum age. This gives scores of 17.5, 22.5, 27.5, 32.5 and 40 for age groups. In fact it will be easier to work with the scores −4, −2, 0, 2 and 5 which still reflect the numerical ordering of the groups, and the relative sizes of their mid-points. All the calculations required are presented in the following table:

j	s_j	n_{1j}	$n_{\cdot j}$	$s_j n_{1j}$	$s_j n_{\cdot j}$	$s_j^2 n_{\cdot j}$
1	−4	70	1363	−280	−5452	21 808
2	−2	117	2507	−234	−5014	10 028
3	0	159	3114	0	0	0
4	2	120	1718	240	3436	6 872
5	5	63	946	315	4730	23 650
		529	9648	41	−2300	62358

Hence
$$X^2_{(t)} = \frac{9648\,(9648 \times 41 + 529 \times 2300)^2}{529 \times 9119\,(9648 \times 62\,358 - 2300^2)} = 8.72$$

Comparing this with χ^2_1 shows that the P value is between 0.005 and 0.001, so the conclusion is that there is no evidence to reject the hypothesis that a linear trend exists in the proportions.

Finally $X^2 - X^2_{(t)} = 14.40 - 8.72 = 5.68$ on 3 d.f. This value is not significant even at the 10% level, so there is no reason to suppose that a non-linear component is important for these data.

Similar methods are available for general $r \times c$ tables; see, for example, Everitt (1977), Section 3.6.

9.3.5 Partitioning $2 \times c$ tables

In some situations one of the columns of a $2 \times c$ table corresponds to a level which is, in some way, special. In this situation it may be desirable to compare each of the other columns individually against this 'reference column'. This could happen where the first column gives data for an existing method of treatment and the remaining columns are for a set of different new treatments. The 2 × c table would then be partitioned into c – 1 2 × 2 tables of the form:

$$\begin{array}{cc|cc|c|cc} n_{11} & n_{12} & n_{11} & n_{13} & \cdots & n_{11} & n_{1c} \\ n_{21} & n_{22} & n_{21} & n_{23} & \cdots & n_{21} & n_{2c} \end{array}$$

Notice that the choice of column 1 to act as a reference column is purely arbitrary; in general any column could be used, provided that this has been decided before the data were collected.

The overall 2 × c table of the form:

$$\begin{array}{cccc} n_{11} & n_{12} & \cdots & n_{1c} \\ n_{21} & n_{22} & \cdots & n_{2c} \end{array}$$

has an X^2 statistic with c – 1 d.f. which provides an overall test of independence among the entire set of proportions observed in row 1 (or row 2). Each 2 × 2 table has an X^2 statistic with 1 d.f. which provides a test for independence between pairs of proportions. The sum of the 2 × 2 statistics does *not*, however, equal the 2 × c statistic. In fact this is an example of 'ransacking the data', a cautionary note against which was given in Section 7.17. As described there, when repeated significance tests are carried out on the same set of data the nominal significance level needs to be adjusted. In this particular case, Everitt (1977) states that when a constituent 2 × 2 table has a nominal P value (from Table A.3) of p then the true P value is $p' = 2p(c - 1)$.

Example 9.11: In a clinical trial of three anaesthetics A, B and C (one of which, A, has been in use for many years) patients were randomly assigned to one of the anaesthetics for their operation. Table 9.21 records the number of cases of serious nausea reported by the patients.

Table 9.21 Nausea for three anaesthetics.

Serious nausea?	Anaesthetic			Total
	A	B	C	
Yes	14	6	1	21
No	86	44	49	179
Total	100	50	50	200

For these data, $X^2 = 5.27$ with 2 d.f., which is not significant at the 5% level. It does, however, seem that anaesthetic C has performed well, and it might be interesting to compare this alone with the standard anaesthetic, A. The reduced 2 × 2 table is

$$\begin{array}{cc|c} 14 & 1 & 15 \\ 86 & 49 & 135 \\ \hline 100 & 50 & 150 \end{array}$$

for which $X^2 = 5.33$ (without a continuity correction) on 1 d.f. For this test the nominal P value from Table A.3 lies between 0.025 and 0.01. In fact the precise P value is 0.021 (found from a statistical computer package). Since this result comes from a partition of a bigger table with c = 3, the true value is 4 times this, that is P = 0.084. Notice that, if 5% significance tests are used this correction actually changes the conclusion made. Since the true P value is greater than 0.05 the null hypothesis fails to be rejected; there is insufficient evidence to conclude that anaesthetic C causes a different proportion of people to experience serious nausea than anaesthetic A.

9.4 Using packages to analyse two-way tables

Statistical computer packages are very useful for constructing contingency tables from raw data, and for carrying out the tedious arithmetic associated with the chi-squared test of independence. This is only the most basic analysis and, as Sections 9.2 and 9.3 have shown, in many situations will not be appropriate. Some packages will carry out Fisher's exact test and/or calculate standardized or adjusted residuals; such things as Gart's test, relative risk or trend analysis are not usually included. In fact, with perhaps a couple of exceptions, all the calculations described in this chapter are easy, and provided even a basic type of calculator is used, they should not cause any particular difficulties (the same cannot be said of what follows in Sections 9.5 and 9.6). The art is in knowing which technique to use for which kind of table.

Two things to watch out for when using packages are, first, packages which calculate X^2 even when the expected values are small and, second, packages which will not allow the table to be read directly. The first problem can lead to dubious inferences, whilst the second is a nuisance. To illustrate the second problem consider Table 9.16 again. If this table appeared in a report, a reader might like to carry out a test for independence upon it. It would be simple just to type in the cell values, in a 3×4 tabular array, and ask for a χ^2 test. Some packages *will* allow this. Others, as mentioned, will not allow such direct input. Instead these packages require the raw data to be input, including values for the levels of the classification variables (sometimes called *factors*). Thus Table 9.16 could be read into such packages as:

1	1	24	or, better still as,	G	A	24
1	2	27		G	B	27
1	3	16		G	C	16
1	4	23		G	D	23
2	1	12		A	A	12
2	2	19		A	B	19
2	3	15		A	C	15
2	4	17		A	D	17
3	1	6		I	A	6
3	2	9		I	B	9
3	3	7		I	C	7
3	4	7		I	D	7

One final point about statistical packages is that at least one package in popular use produces so-called *measures of association* for contingency tables, which attempt to assess the strength of the relationship between the classification variables (rather like the correlation coefficient for quantitative variables). These are described in Section 3.7 of Everitt's (1977) book, but in general it is far more illuminating to look at the residuals, as described in Section 9.3.3.

9.5 Log-linear models

As was discussed in Section 9.3.3, the test based on the X^2 statistic is a test of the goodness of fit of the model of independence between the row and the column classification variables in an r × c table. In a two-way table there are only two models to choose between: the model of independence of classification variables (the null hypothesis) and the model of non-independence (the alternative hypothesis).

In higher-dimensional tables the situation is not so straightforward. In a three-way table such as Table 9.22, for instance, there are nine possible models that could be fitted, including complete independence of all three variables in any combination, independence of the three taken together but non-independence of one or more of the possible pairs, and complete non-independence. It is quite possible to use the methods of the previous sections to specify models and obtain expressions for the resulting expected values, but this is cumbersome even for three-way tables, and rapidly becomes prohibitive for tables of higher dimensions.

Fortunately a simple system of specifying such models has been developed, and this will be described in this section. It is not, however, a method that is suitable for hand computation and (as with multiple regression) requires the use of a statistical computer package. No attempt will be made here to describe the somewhat complex method used by packages to fit the models (see McCullagh and Nelder (1983) for details). Any package that advertises itself as being able to fit *generalized linear models* will be able to carry out the computations required. In the present context the models to be fitted are called log-linear models, for reasons that will be made clear in Section 9.5.1. Assuming that some method of fitting log-linear models is available, Section 9.5.3 will show how to identify, from amongst all those available, the model that is, in some sense, best. Before that, Section 9.5.1 will discuss the specification of a log-linear model in the simple case of a two-way table, and Section 9.5.2 will show how to generalize this to higher-dimensional tables.

Given a table of any dimension it is always possible to reduce it to a series of marginal two-way tables; for example the three-way table, Table 9.2 was earlier reduced to three two-way tables: Table 9.3(a)–(c). So, given that a simple method for testing independence is already available for two-way tables from earlier sections, the reader may well ask why it is necessary to bother with log-linear modelling of multi-way tables. Example 9.12 will illustrate the sort of problem that can arise when only the marginal tables are analysed and therefore justify the need for the log-linear approach.

Example 9.12: Table 9.22 shows the incidence of heart attack in a hypothetical random sample of 11 100 men classified by height and weight. Height

Table 9.22   Height, weight and heart attack status for a sample of men.

					Weight				
		Low			Medium			High	
Status	Height	Low	Medium	High	Low	Medium	High	Low	Medium	High
Had heart attack		290	10	5	20	365	30	5	10	280
No heart attack		2710	190	145	320	3635	170	5	40	2870

and weight were each grouped by some predetermined formula into 'low',
'medium' and 'high'.

There are three two-way marginal tables which may be formed from these
data – height × heart attack, weight × heart attack and height × weight. In
determining risk factors for heart attack only the first two are of any real
interest. These two marginal tables, and their associated X^2 values (on 2 d.f.)
are given as Tables 9.23 and 9.24.

Table 9.23   Height by heart attack status (from Table 9.22). $X^2 = 0.398$.

Had heart attack?	Low	Medium	High	Total
		Height		
Yes	315	385	315	1 015
No	3035	3865	3185	10 085
Total	3350	4250	3500	11 100

Table 9.24   Weight by heart attack status (from Table 9.22). $X^2 = 0.015$.

Had heart attack?	Low	Medium	High	Total
		Weight		
Yes	305	415	295	1 015
No	3045	4125	2915	10 085
Total	3350	4540	3210	11 100

Since the X^2 values are so small, the conclusion from these two analyses is
that both height and weight are independent of heart attack status. That is
clearly nonsense! Inspection of Table 9.22 shows that, whilst in the low-
weight group fewer than 10% of low-height men had suffered heart attacks,
in the high-weight group 50% of the low-height men had suffered heart
attacks. There is obviously a *joint* effect of height and weight on the chance
of a heart attack (precisely as would be expected – men with a high weight-
for-height are generally at greater risk). The marginal two-way analyses
cannot pick up such subtle effects, and in this example (as in many others)
they can lead to erroneous conclusions.

9.5.1 Two-way tables

Consider again the model of independence in two-way tables. It has been shown that this leads to expected values,

$$e_{ij} = n \frac{n_{i\cdot}}{n} \frac{n_{\cdot j}}{n}$$

Taking logarithms of this gives

$$\log e_{ij} = -\log n + \log n_{i\cdot} + \log n_{\cdot j}$$

which has the form of a linear model (rather like those of Chapter 8) on the log scale, hence the term 'log-linear model'. To avoid writing 'log' each time it will be simpler to re-express this model as

$$\ell_{ij} = u + u_{1(i)} + u_{2(j)} \qquad (1)$$

for $i = 1, 2, \ldots, r$ (the row variable) and $j = 1, 2, \ldots, c$ (the column variable) where

$$\ell_{ij} = \log e_{ij}$$

$$u = \frac{1}{rc} \sum_{i=1}^{r} \sum_{j=1}^{c} \ell_{ij}$$

$$u_{1(i)} = \frac{1}{c} \sum_{j=1}^{c} \ell_{ij} - u \quad (i = 1, 2, \ldots, r)$$

$$u_{2(j)} = \frac{1}{r} \sum_{i=1}^{r} \ell_{ij} - u \quad (j = 1, 2, \ldots, c)$$

and

$$\sum_{i=1}^{r} u_{1(i)} = \sum_{j=1}^{c} u_{2(j)} = 0$$

These precise definitions of the u terms need not concern the reader; u is simply the mean of the logs of the expected cell values, and so is a scaling factor. $u_{1(i)}$ represents the extra effect of row i and $u_{2(j)}$ the extra effect of column j. Thus model (1) says that the log of the expected value in cell (i, j) is made up of an overall mean plus a term for the effect of row i plus a term for the effect of column j.

Consider the alternative model for two-way tables, that is the model of non-independence. Now there is another effect that contributes to each expected value, an effect which is particular to the cell being considered, and in this case the log-linear model is

$$\ell_{ij} = u + u_{1(i)} + u_{2(j)} + u_{12(ij)} \qquad (2)$$

for $i = 1, 2, \ldots, r; j = 1, 2, \ldots, c$ and where

$$\sum_{i=1}^{r} u_{12(ij)} = \sum_{j=1}^{c} u_{12(ij)} = 0$$

Notice that when $u_{12(ij)} = 0$ for every value of i and j, model (2) reduces to model (1), and so a test for independence is also a test for $u_{12(ij)} = 0$ in all cases. The $u_{12(ij)}$ are sometimes called the *interaction* terms for variables 1 and 2.

The chi-squared test for independence discussed in Sections 9.2 and 9.3 tests the null hypothesis that model (1) fits against the alternative hypothesis that model (2) fits. This can also be done by log-linear modelling, with the advantage that the method easily generalizes to higher-dimensional tables. In log-linear modelling the basic approach is to fit each model (as was done in a different context in Section 8.4.2) and then to compare the goodness of fit to decide which model is 'best'.

Some measure of goodness of fit is obviously needed (and the question of what is 'best' will be left to Section 9.5.3). On the original scale X^2 is the measure used, whilst on the log scale the measure

$$G^2 = 2\Sigma(\text{observed}) \log \left(\frac{\text{observed}}{\text{expected}} \right)$$

which is called the *deviance*, is used. Just as with X^2, G^2 should be compared with the chi-squared distribution to test for significance (i.e., lack of fit). The degrees of freedom for G^2 are, as with X^2,

number of cells – number of independent parameters fitted.

Here the parameters are the u terms with d.f.,

1 for u

$r - 1$ for the r $u_{1(i)}$ terms (since $\sum_{i=1}^{r} u_{1(i)} = 0$)

$c - 1$ for the c $u_{2(j)}$ terms (since $\sum_{j=1}^{c} u_{2(j)} = 0$)

$rc - r - c + 1$ for the rc $u_{12(ij)}$ terms (since $\sum_{i=1}^{r} u_{12(ij)} = \sum_{j=1}^{c} u_{12(ij)} = 0$)

Provided that a computer package is used the formulae for G^2 and d.f. need not concern the reader; the computer will automatically calculate them. It is nevertheless interesting to see that, in the case of a two-way table, for model (1) d.f. $= rc - 1 - (r - 1) - (c - 1) = (r - 1)(c - 1)$, and for model (2) d.f. $= rc - 1 - (r - 1) - (c - 1) - (rc - r - c + 1) = 0$. Looking back to Section 9.3.1 will show that the d.f. for model (1) agree with those for X^2. Why should the d.f. for model (2) be zero? Remember that model (2) includes a term $u_{12(ij)}$ which is different for each (i, j) combination, i.e., for each cell of the table. Here the expected values depend not only on the marginal row and column effects, but also on the effect of the cell itself. The only information on the effect of any one cell is the observed value in that cell, and consequently the expected values are simply the observed values in every cell. So everything is fixed under model (2) and there are no degrees of freedom. Model (2) is often called the *full model* for a two-way table.

The implication of this is that the full model always fits the data perfectly: why then test other models, such as model (1)? The answer is that if the extra terms, $u_{12(ij)}$, do not contribute substantially to the fit, then they can be left out, resulting in a simpler, more parsimonious model which is easier to interpret. Exactly the same approach is used in multiple regression.

In order to test whether $u_{12(ij)}$ is necessary for at least some (i, j) combination, model (1) must be fitted to the data and its G^2 value and d.f. obtained. If G^2 exceeds the value of χ^2 on the appropriate d.f. then model (1) is rejected in favour of model (2). The conclusion is then that the row and column variables are not independent, that is $u_{12(ij)} \neq 0$ for some i and j.

Example 9.13:   When the log-linear model of independence (model 1) was fitted to Table 9.16 the deviance, G^2, found from a statistical computer package was 1.92 on 6 d.f. This agrees almost exactly with the value of X^2 in Example 9.8, and so the conclusion will be the same, contact is independent of ward.

9.5.2 Multi-way tables

Following the notation of Section 9.5.1, the full log-linear model for a three-way table is

$$\ell_{ijk} = u + u_{1(i)} + u_{2(j)} + u_{3(k)} + u_{12(ij)} + u_{13(ik)} + u_{23(jk)} + u_{123(ijk)}$$

where the subscripts 1, 2 and 3 denote the three classification variables with levels $i(= 1, 2, \ldots, r)$, $j(= 1, 2, \ldots, c)$ and $k(= 1, 2, \ldots, t)$ respectively. The form of the full model for four and even higher dimensions should be clear by analogy. For the sake of clarity, the i, j and k subscripts will be omitted in what follows, although it must be remembered that when, for example, u_{123} is mentioned this really refers to a set of rct *different* parameters.

As stated earlier, there are nine possible different models that can be fitted for a three-way table, and these are:

Model	Terms assumed zero	d.f. for deviance
No independence		
(1) $\ell = u + u_1 + u_2 + u_3$ $+ u_{12} + u_{13} + u_{23} + u_{123}$		0
3-way independence		
(2) $\ell = u + u_1 + u_2 + u_3$ $+ u_{12} + u_{13} + u_{23}$	u_{123}	$(r-1)(c-1)(t-1)$
Partial 2-way independence		
(3) $\ell = u + u_1 + u_2 + u_3$ $+ u_{12} + u_{13}$	u_{23}, u_{123}	$(r-1)(c-1)(t-1)$ $+ (c-1)(t-1)$
(4) $\ell = u + u_1 + u_2 + u_3$ $+ u_{12} + u_{23}$	u_{13}, u_{123}	$(r-1)(c-1)(t-1)$ $+ (r-1)(t-1)$
(5) $\ell = u + u_1 + u_2 + u_3$ $+ u_{13} + u_{23}$	u_{12}, u_{123}	$(r-1)(c-1)(t-1)$ $+ (r-1)(c-1)$
(6) $\ell = u + u_1 + u_2 + u_3$ $+ u_{12}$	u_{13}, u_{23}, u_{123}	$(r-1)(c-1)(t-1)$ $+ (r-1)(t-1)$ $+ (c-1)(t-1)$
(7) $\ell = u + u_1 + u_2 + u_3$ $+ u_{13}$	u_{12}, u_{23}, u_{123}	$(r-1)(c-1)(t-1)$ $+ (r-1)(c-1)$ $+ (c-1)(t-1)$
(8) $\ell = u + u_1 + u_2 + u_3$ $+ u_{23}$	u_{12}, u_{13}, u_{123}	$(r-1)(c-1)(t-1)$ $+ (r-1)(c-1)$ $+ (r-1)(t-1)$

Model	Terms assumed zero	d.f. for deviance
Complete independence		
(9) $\ell = u + u_1 + u_2 + u_3$	$u_{12}, u_{13}, u_{23}, u_{123}$	$(r-1)(c-1)(t-1)$
		$+ (r-1)(c-1)$
		$+ (r-1)(t-1)$
		$+ (c-1)(t-1)$

An appropriate statistical computer package can be used to fit models (2)–(9) and calculate the deviance each time. Model (1) does not need to be fitted since it is the full model which always has a deviance of zero. Section 9.5.3 will discuss how to find which model is 'best' by inspecting the deviances and differences between deviances.

9.5.3 Choosing a model

In Sections 9.5.1 and 9.5.2 it was feasible to write out all the possible log-linear models that can be fitted to two- or three-way tables. For higher dimensions this is not an attractive proposition; even for only four-way tables there are 80 possible models! Although, for practical reasons, some of these models may be inadmissable for particular sets of data (see Section 9.5.4) there is, nevertheless, a need for a rational strategy for finding the model which best fits the data. This strategy begins with the identification of all the models that provide an adequate fit to the data (i.e., those models with a non-significant deviance). Next, a term which does not contribute significantly to the fit is identified by comparing a pair of models which are identical except for the term in question. Where each of two models provides an adequate fit, and yet one model has all the terms of the other but also one extra term, then it may be that the extra term is, in fact, not significantly different from zero. This may be tested by calculating the difference between the deviances of the two models, which approximately follows a χ^2 distribution with d.f. given by the difference between the d.f. of the two models, if the extra term is not significantly different from zero.

In some ways this process is similar to multiple regression, since in both cases the aim is to identify the model which is an adequate fit to the data without containing any unnecessary terms. It also suffers from the same drawback: the solution obtained may well depend upon the order in which the terms are tested. Even when it is possible to fit all possible models (that is with a table of no more than three, or perhaps four, dimensions), there is no simple solution to this problem, and the investigator may have to use his own judgement.

There are, nevertheless, two rules that should always be followed in log-linear model selection. Rule one is that the one-factor u terms, u_1, u_2, etc. (sometimes called the *main effects*) must be included in any model, as must the overall mean, u. The second rule is that when a particular term is included in a model, then so must all the terms marginal to it. For example, when u_{123} is included then u_{12}, u_{13} and u_{23} must also be included. The justification for these rules is beyond the scope of this text, although they are essentially a generalization of the principle already used in two-way tables where marginal totals were kept fixed. Precise details are given in Bishop *et al.* (1975). These rules were followed to obtain the nine models for a three-way table in Section 9.5.2.

Applying these rules to model selection shows that model fitting should be hierarchical. That is, the most complete model should be fitted first, then the second most complete and so on down to the smallest model, which is the one with only main effects and the overall mean term. It will be assumed for now that it is practically possible to fit all possible models; some suggestions of what to do when this is not the case are given after Example 9.15.

Assuming, then, that all models can be fitted, model selection proceeds hierarchically as follows:

(i) Calculate the deviance for a model. Compare it with χ^2 on the appropriate d.f. If the deviance is significant (at a predetermined level) then reject this model, if not go to (ii).

(ii) Calculate the deviance for a model with one fewer term. If this deviance is significant then reject this model, if not calculate the difference between the deviances of this model and that at the previous step. Check the difference against χ^2 with d.f. given by the difference between the d.f. of the two models being compared. If this difference is significant then reject this model, otherwise go to (iii).

(iii) Calculate the deviance for a model with one fewer term, as above . . . etc.

The procedure continues until dropping any term leads to a rejected model; the last model which was not rejected is identified as the best fit.

This would work very easily if there were only one model at each stage of the hierarchy (that is, only one model with a given number of u terms) but this is generally not the case. Consequently there are likely to be many hierarchical paths through all possible models (Fig. 9.1 gives an example), and this may mean that judgement has to be made in choosing between models that are 'best' through different paths. Examples 9.14 and 9.15 illustrate the model selection procedure.

Example 9.14: Log-linear models were fitted to Table 9.22. First, the full model has $G^2 = 0$ and d.f. $= 0$. Next the model with one fewer term was fitted. By the rules for fitting log-linear models the term dropped must be the highest order term, u_{123}, representing the 3-way 'interaction'. For this model $G^2 = 40.0$ with 4 d.f. Comparing this with Table A.3 shows that the P value is below 0.0005, so the deviance, G^2, is highly significant and this model is rejected. The full model (of no independence) is the best fit.

Notice that this conclusion agrees with the observation made in Example 9.12: there *is* an important joint effect of height and weight upon heart attack, that is heart attack is not independent of height and weight taken *together*.

Example 9.15: In order to investigate the importance of weight and height on the chance of having an accident involving broken bones, a random sample of 1010 females was interviewed to ascertain whether they had ever suffered one or more broken bones due to an accident. The subjects were classified by height and weight as in Table 9.22. The resulting data are given in Table 9.25.

Table 9.25 Height, weight and broken bone status for a sample of women.

	Weight								
	Low			Medium			High		
Status \ Height	Low	Medium	High	Low	Medium	High	Low	Medium	High
Had broken bones	65	7	5	9	100	6	1	5	100
No broken bones	135	13	10	21	300	14	5	10	204

Models (1)–(9) of Section 9.5.2. (all possible models) were fitted, taking broken bone status as classification variable 1, height as classification variable 2 and weight as classification variable 3. A statistical package gave the following results (note that deviances were given to five digits of accuracy).

Model no.	Model fitted	Deviance G^2	d.f. for deviance
(1)	$\ell = u + u_1 + u_2 + u_3 + u_{12} + u_{13} + u_{23} + u_{123}$	0	0
(2)	$\ell = u + u_1 + u_2 + u_3 + u_{12} + u_{13} + u_{23}$	1.0896	4
(3)	$\ell = u + u_1 + u_2 + u_3 + u_{12} + u_{13}$	1347.7	8
(4)	$\ell = u + u_1 + u_2 + u_3 + u_{12} + u_{23}$	2.2596	6
(5)	$\ell = u + u_1 + u_2 + u_3 + u_{13} + u_{23}$	1.4157	6
(6)	$\ell = u + u_1 + u_2 + u_3 + u_{12}$	1353.9	10
(7)	$\ell = u + u_1 + u_2 + u_3 + u_{13}$	1353.0	10
(8)	$\ell = u + u_1 + u_2 + u_3 + u_{23}$	7.5464	8
(9)	$\ell = u + u_1 + u_2 + u_3$	1359.2	12

Model (1) obviously fits the data. By comparing the deviances with χ^2 on the appropriate d.f., all deviances but those for models (2), (4), (5) and (8) gave significant results (5% significance tests are used in this example). Hence each of these five models will provide an adequate fit to the data. To help decide which is 'best' Fig. 9.1 shows the hierarchical paths through the set of models.

Consider first, in Fig. 9.1, the path (1) to (2). Both models are acceptable and the difference in their deviances is $1.0896 - 0$ on $4 - 0$ d.f. which is not significant. This implies that there is no significant difference between the two and so, as model (2) contains fewer terms, model (1) is rejected.

Moving downwards from model (2), the path to model (3) can be ignored since (3) has already been rejected. Consider the path (2) to (4). The difference between the deviances of models (4) and (2) is $2.2596 - 1.0896 = 1.1700$ on $6 - 4 = 2$ d.f. This is not significant, so model (4) is to be preferred to model (2). Next consider the path (2) to (5). The difference in deviances here is 0.3261 on 2 d.f. which again is not significant, and so model (5) is preferred to model (2). Clearly model (2) can now be rejected.

Consider moving downwards from (4). Only the path to (8) is of interest.

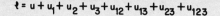

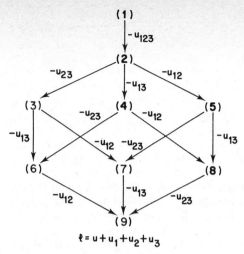

Fig. 9.1 The hierarchy of models for three-way tables. Acceptable models for Example 9.15 are in bold type.

The difference in deviances is 5.2868 on 2 d.f. which is not quite significant at 5%. So model (8) is just preferred to model (4).

Moving downwards from (5), again only the path to (8) is of interest. The difference in deviances is 6.1307 on 2 d.f. which is just significant at 5% and thus model (8) cannot be preferred to model (5).

So, by two different paths, models (8) and (5) have been identified as 'best'. Which is the overall best? Going by strict 5% significance, model (5) must be chosen since it has been shown that when u_{13} is dropped from (5) to make model (8) the fit gets significantly worse. Notice, however, that the result of the last 5% significance test was quite marginal; there does not really seem to be a lot of difference between (5) and (8), and so it might be argued that (8) could be taken as best by virtue of its simplicity.

Nevertheless, taking (5) as the best model, how can it be interpreted? It says that terms u_{13} and u_{23} are needed in the model, so that there is a relationship between broken bones (variable 1) and weight (variable 3) and also between height (variable 2) and weight. The latter would certainly be anticipated; the former is an interesting result of the study (for these hypothetical data), since it says that there is a difference in the propensity to suffer broken bones for people of different weights. Notice that model (5) omits u_{123} and u_{12} so there is no joint effect of height and weight, and no effect of height alone, on the chance of having an accident that causes broken bones.

Having identified the best model it would be useful to obtain the expected (or 'fitted') values for this model, and then to calculate the adjusted residuals. These should be produced by the computer package. The residuals can then be analysed to identify the cells responsible for the non-independence effects in a manner similar to that used in Section 9.3.3, although here the interpretation is not so straightforward.

For three-way tables it is not too onerous to specify and fit all the possible models. As the dimensionality of the table increases this becomes more and more tedious, and the need for a 'shortcut' method is felt. As with multiple regression (see Section 8.4.4) automatic methods of model selection, which identify models that are very good if not the best, have been suggested (see Section 5.3 of Fienberg (1977) for details). Given a d-dimensional table, the simplest shortcut method is to fit the model with all (d – 1)-way terms, then the model with all (d – 2)-way terms, . . . and finally the model with all 1-way terms (main effects), and restrict further investigation to the stage preceding the one at which the deviance became significant. For Example 9.15 this would have meant fitting models (2) and (9), and then investigating the models 'between' (2) and (9), that is, the models with mixtures of 2-way u terms (u_{12}, u_{13} and u_{23}).

Finally, a word about small expected cell values. The log-linear approach deals well with situations where some expected values are small and, at least in theory, will even cope with zero values. There are, however, limits to what even log-linear models can do, the exact limits being dependent upon the computing method used. The authors have personal experience of a health researcher attempting to fit log-linear models to a table with over 500 cells but fewer than 300 observations. This is surely nonsense; even if the computer can be forced to test the goodness of fit of models to such data, the power of the tests is likely to be extremely poor. If it is sensible, the solution would be to pool levels as in Section 9.3.2. Failing this, it may be necessary either to assume that higher order effects are non-significant (but there must be a good reason for believing this) or simply to abandon the analysis until a larger sample is available. Such problems are unlikely to occur in a study that has been properly designed.

9.5.4 Sampling constraints

Sometimes zero values in tables are caused not by sampling variation but by necessity, such as in a table of sex by specialty on discharge, in which zero values are sure to occur in the (male, gynaecology) and (male, obstetrics) cells. The fact that zeros must occur in such cells (for observed and expected values) leads to these being termed *fixed zeros*. When a fixed zero occurs a degree of freedom is lost for the cell itself and for any margin in which it occurs. It is unlikely that the computer package used will make these adjustments automatically. Exactly the same procedure would, of course, be used if any number other than zero were fixed anywhere in the table.

Another form of sampling constraint is where the tables contain fixed n-way margins for n greater than or equal to 2. This will often occur in case-control studies where a predetermined number of controls is selected in, say, each age × sex group, so as to make the size of each group equal to the size of the corresponding 'case' group. The case/control × age × sex margin is then fixed by the design of the study (as are the subsidiary case/control × age, case/control × sex and age × sex margins, of course). Where such fixed margins occur the corresponding u terms *must* be present in any model fitted to the data. This is because when such a u term is included the expected margin will be equal to the observed margin; this need not be true when the u term is excluded and the result would be nonsensical.

Example 9.16: Consider the data shown in Table 9.2. To select this sample Doll and Hill (1950) fixed the number of men (649) and women (60) in the control group to be equal to the number of men and women cases respectively, so case/control × sex is certainly a fixed margin. Furthermore, since males and females were not sampled randomly, but only within case/control groups, the data can give no information about the relationship between sex and any further factor (here, smoking status), unless possibly something is known about the relative proportions of male and female cases and controls in the entire population. So the sex × smoking status margin is also fixed.

Taking smoking status as variable 1, case/control status as variable 2 and sex as variable 3, this means that the terms u_{23} and u_{13} must be included in any model, along with u, u_1, u_2 and u_3 as always, and the range of models that can be fitted is restricted to models (1), (2) and (5) of Section 9.5.2.

These models, their deviances and d.f. are:

Model	G^2	d.f.
$\ell = u + u_1 + u_2 + u_3 + u_{12} + u_{13} + u_{23} + u_{123}$	0	0
$\ell = u + u_1 + u_2 + u_3 + u_{12} + u_{13} + u_{23}$	5.737	1
$\ell = u + u_1 + u_2 + u_3 + u_{13} + u_{23}$	31.95	2

Comparing the deviances with χ^2 shows that the full model is the only acceptable one. Hence the u_{123} term is necessary; the chance of developing lung cancer differs across the various sex/smoking groups. As well as implying that smoking is associated with lung cancer, this shows that the relative risks for smokers versus non-smokers will be different for the two sexes. When considering these results the notes at the end of Example 9.7 should again be taken into account.

9.5.5 Further aspects

A three-way interaction such as that found to be significant in Example 9.16 may be difficult to interpret, and four and higher dimension interactions may have very little practical meaning at all. Fortunately, it often happens that higher order interactions are non-significant, but what can be done when, as in the previous example, they must be included? A simple way to obtain further understanding of the relationships in the table is to *condition* on the levels of one or more variables involved in the interaction, that is to consider separately each of the tables obtained by fixing a level or set of levels. In Table 9.2, for instance, the 2×2 table corresponding to each sex could be analysed separately. The male table has been given already as Table 9.15, for which the estimated relative risk was 14.0.

The log-linear approach can be used to embrace all the material of Sections 9.2 and 9.3 in a general framework. The text by Fienberg (1977) provides an excellent introduction to these more specialized topics.

9.6 Logistic regression

In log-linear models all of the classification variables are given equal status, but in many situations one of the variables can be identified as being a response which might be predicted by the others (just as the y variable is so identified in regression). In the Doll and Hill (1950) data shown in Table 9.2, for example, case/control (lung cancer/not lung cancer) is clearly the response variable that, it is hypothesized, may be determined by smoking and sex group.

If the response variable is dichotomous then logistic regression may be used, instead of log-linear modelling, to fit models to the data. Dichotomous response variables are very common in health problems, for instance life and death or, as in the Doll and Hill (1950) study, case and control group. As in other situations the two outcomes may be thought of, in general, as 'success' and 'failure'.

The basic idea behind logistic regression analysis is that the proportion of successes for each combination of levels of the non-response (or *design*) variables should be identified. So in Table 9.2 the proportion of female smokers with lung cancer (here taken as a success) is $41/69 = 0.59$, the proportion of female non-smokers with lung cancer is $19/51 = 0.37$, etc. Any effect of the smoking and sex variables would cause differences in the proportion of lung cancer cases for different combinations of levels of the smoking and sex variables. Logistic regression looks for such differences by first calculating a transformation of the proportions known as the *logit*. For a proportion, p, the logit is

$$L = \log_e\left(\frac{p}{1-p}\right)$$

This transformation is merely for mathematical convenience. Linear models, known as logistic models, are then fitted to the logits.

Logistic models may be fitted by the same statistical packages that fit log-linear models (they are both types of generalized linear models). The method of model fitting proceeds exactly as for log-linear models. Hence the full logistic model for a three-way table where variables 1 and 2 are the two design variables is

$$L = w + w_1 + w_2 + w_{12}$$

where the 'w' terms are comparable to the 'u' terms of log-linear models. The precise relationship will be discussed following an illustrative example.

Example 9.17:　A logistic regression analysis was carried out on the Doll and Hill (1950) data of Table 9.2. As in Example 9.16 let '1' be smoking status, '2' be case/control status (the response variable) and '3' be sex. Then, since the proportion of males and females in the control group was fixed to be the same as in the case group the term w_3 *must* appear in any model. This gives three possible logistic models that can be fitted to the data. These models, their deviances and degrees of freedom (obtained from a computer package) are:

Model	G^2	d.f.
$L = w + w_1 + w_3 + w_{13}$	0	0
$L = w + w_1 + w_3$	5.737	1
$L = w + w_3$	31.95	2

Notice that these deviances and d.f. agree exactly with the log-linear analysis of Example 9.16, so again, only the full model provides an adequate fit. For a further analysis of Doll and Hill's data see the article by Wilson (1985).

Comparison of Examples 9.16 and 9.17 shows that the w terms replace u terms with the same subscripts except for the deletion of a '2' subscript (for the response variable). Thus w_1 replaces u_{12}. The only exception to this is that the overall mean w replaces the overall mean plus main effects $u + u_1 + u_2 + u_3$. This makes perfect sense; in a log-linear model u_{12} denotes the effect of variable 1 on variable 2 (or vice versa) since it is the 1×2 'interaction' term. In the logistic model w_1 denotes the effect of variable 1 on the response variable, variable 2, so they represent the same thing.

In general a log-linear model can always be found that corresponds exactly with a logistic model for any contingency table. Notice, however, that there will always be more log-linear models that can be tried on any table than there are logistic models (unless, as in Examples 9.16 and 9.17, sampling constraints exist). This is because terms like u_{13} in the context of Example 9.16 have no analogy in Example 9.17. In other words logistic regression cannot be used to test for independence among all combinations of the variables. If this is not the purpose of the analysis (as with the Doll and Hill (1950) study), then there is no particular advantage in using one approach rather than the other, provided that the data are of the correct form for logistic regression (one or more design variable and one dichotomous response variable). It should be borne in mind, however, that even when the logistic approach is used, model fitting should always be hierarchical and any sampling constraints must be accounted for.

One advantage that logistic regression has over log-linear modelling is that it is quite possible to have some design variables that are quantitative as well as the qualitative variables that are the concern of this chapter (indeed they may all be quantitative). Thus in identifying risk factors for heart attacks, Shaper *et al.* (1986) used a logistic regression analysis where heart attack was the response variable and the design variables included categorical variables such as parental history of heart trouble and evidence of angina (both yes/no) and quantitative variables such as blood pressure and serum total cholesterol.

A more complete account of logistic regression is given in Fienberg (1977).

9.7 Tests of goodness of fit for frequency data

In this final section of the chapter, a simple method of determining the adequacy of fit of expected frequencies, generated by some hypothesis, to observed frequencies will be described. This is not strictly a method concerned with contingency tables but, since it relies upon the X^2 statistic introduced in Sections 9.2 and 9.3, it is most appropriate to include it here. In this context an expected set of frequencies (the 'e' values) is suggested according to some theory or model, data are collected (the 'n' values) and the two are compared for goodness of fit using

$$X^2 = \Sigma \frac{(n - e)^2}{e}$$

As usual the null hypothesis (that the theory is a good fit to the data) is rejected whenever X^2 exceeds the tabulated value of chi-squared. As before, the degrees of freedom are the number of cells in the tables less the number of constraints imposed upon the expected frequencies.

Example 9.18:   It has been suggested that the total number of admissions to a certain Accident and Emergency Department is the same on each day of the week. To test this assertion data were collected for 10 weeks, during which time 175 patients were admitted. The distribution of these admissions by day of the week was

Day	Sunday	Monday	Tuesday	Wednesday	Thursday	Friday	Saturday
No. of admissions	17	22	21	23	25	21	46

If the assertion is correct the same number of admissions would be expected on each day of the week, that is $175/7 = 25$ patients. Looking at the above figures it appears that Saturday is quite different from the rest (with almost twice as many admissions) and Sunday's total is rather low. So it may be anticipated that the theory of an equal spread of admissions will be rejected; the X^2 statistic provides a formal test of whether there is a real difference, or whether the observed differences are merely due to chance. The relevant calculations are:

Day of the week	n	e	$\dfrac{(n-e)^2}{e}$
Sunday	17	25	2.56
Monday	22	25	0.36
Tuesday	21	25	0.64
Wednesday	23	25	0.16
Thursday	25	25	0.00
Friday	21	25	0.64
Saturday	46	25	17.64
Total	175	175	22.00

So $X^2 = 22.00$. The number of cells in the table is 7 and one constraint has been imposed upon the data (that the sum of the expected frequencies should be 175, that is, equal to the sum of the observed frequencies). Clearly any other total for the expected values would be nonsense. Hence the degrees of freedom are $7 - 1 = 6$. The value of chi-squared on 6 d.f. is 22.464 at the 0.1% level and thus the result is very nearly significant at this extreme level. There is considerable evidence that admissions are not equally spread amongst days of the week. It appears that whilst weekdays are similar, Saturdays are extremely busy and Sundays are quieter than usual.

9.7.1 Fitting with probability distributions

In Example 9.18 the goodness of fit test compared observed frequencies with a given set of expected frequencies. With a slight modification the same procedure can be used to test whether observed data follow a particular probability distribution.

In Section 6.6.3, for instance, it was suggested that the number of babies born to a family up to and including the first girl would have a geometric distribution. To test this assertion many different couples would be observed and the number of children born until the first girl recorded for each. These recordings, presented in the form of a frequency distribution, would be compared with expected frequencies generated according to a geometric distribution. Now to generate such expected frequencies the parameter (π, the probability of a girl birth) of the geometric distribution needs to be specified. In most real-life situations the parameter of the fitted probability distribution is unknown and so it will need to be estimated. This is usually achieved by making the mean of the probability distribution equal to the mean of the observed data. In the case of the geometric the mean is $1/\pi$ (see Section 6.9.1) and so $1/\pi$ is put equal to $\bar{x}$; hence π is estimated by $1/\bar{x}$.

The process of estimating a parameter of a probability distribution introduces one extra constraint upon the expected frequencies in addition to the constraint that the sum of expected frequencies should equal the sum of the observed frequencies. When two parameters have to be estimated (in the case of fitting a normal distribution, for example) sample and theoretical variances are equated as well as sample and theoretical means. Yet another degree of freedom is then lost. In general the degrees of freedom for the X^2 statistic are

$$\left(\begin{array}{c}\text{number of cells}\\\text{in the table}\end{array}\right) - \left(\begin{array}{c}\text{number of estimated}\\\text{parameters of the}\\\text{probability}\\\text{distribution}\end{array}\right) - 1$$

Example 9.19:   Data were collected from another Accident and Emergency Department (but also over 70 days) to check whether the number of admissions each day follows a Poisson distribution (an assertion used in various places in Chapter 6). The observed frequency distribution was:

Number of admissions	0	1	2	3	4	5	6	Total
Number of days	9	8	23	11	10	6	3	70

Poisson probabilities (see Section 6.6.5) take the form

$$P(x) = \frac{e^{-\lambda}\lambda^x}{x!} \qquad (x = 0, 1, 2, \ldots)$$

where $P(x)$ is the probability of observing outcome x (the number of admissions per day in this example). The only unknown parameter is λ, and according to Section 6.9.1 the mean of the Poisson is itself λ so that λ will be estimated by $\bar{x}$. Here,

$$\bar{x} = \frac{(0 \times 9) + (1 \times 8) + \ldots + (6 \times 3)}{70} = 2.5$$

Thus the Poisson distribution to be fitted is

$$P(x) = \frac{e^{-2.5}2.5^x}{x!}$$

Now the expected number of occurrences of the outcome x must be $70P(x)$ since there were 70 trials (that is observations were taken on 70 days). The formula for expected frequencies is, therefore,

$$e_x = 70e^{-2.5}\frac{2.5^x}{x!} = 5.7459\frac{2.5^x}{x!}$$

For example the expected number of days out of 70 on which there are 3 admissions is

$$e_3 = 5.7459\frac{2.5^3}{3!} = 14.963$$

Continuing in this way leads to the figures:

x	n	e	$\dfrac{(n-e)^2}{e}$
0	9	5.746	1.843
1	8	14.365	2.820
2	23	17.956	1.417
3	11	14.963	1.050
4	10	9.352	0.045
5	6 ⎫	4.676 ⎫	
6	3 ⎬ 9	1.948 ⎬ 7.618	0.251
over 6	0 ⎭	0.994 ⎭	
Total	70	70	7.426

Note 1: The 'over 6' category is included because Poisson outcomes have no upper limit (remember that all potential levels have to be included – see Section 9.1.2). The expected frequency 'over 6' is obtained by subtraction.
Note 2: The final three 'cells' of the table have been combined to avoid the problem of small expected frequencies (see Section 9.3.2). When fitting distributions in this way there is no problem with pooling cells from consecutive categories simply to avoid small frequencies.

The number of cells in the final version of the table (after grouping) is 6. Degrees of freedom are consequently $6 - 1 - 1 = 4$. The 10% critical value of chi-squared is 7.779 on 4 d.f. so that the P value for this test is above 0.1. There is no evidence to refute the assertion that admissions to the Accident and Emergency Department follow a Poisson distribution. By the conditions that give rise to the Poisson (see Section 6.6.5) this suggests that patients tend to arrive randomly, independently and uniformly in time.

Except for the way in which an unknown parameter is estimated (and possibly the degrees of freedom) the methodology of Example 9.19 is applicable to other probability distributions. When the distribution to be fitted is continuous, such as the normal, the data must be aggregated into a grouped frequency distribution, and the expected frequencies found from the probabilities of values falling within the class intervals.

Exercises

9.1 In a study on workload in neonatal units, Field *et al.* (1985) reported the numbers of live and still births in Nottingham health district during two periods (1 January–31 December 1977 and 1 April 1983–31 March 1984). The results are shown below. Do these data suggest that the proportion of still births is different in the two periods?

	Period		
Type of birth	01/01/77–31/12/77	01/04/83–31/03/84	Total
Live	6819	7602	14 421
Still	51	45	96
Total	6870	7647	14 517

9.2 Over a certain weekend, the medical and surgical wards which were on take in a hospital admitted 17 patients, of which 8 were emergencies and 9 were booked or from the elective admission list. The following table shows the number of each type of admission to each type of ward. Did the proportion of emergency admissions differ significantly in medical and surgical wards?

	Type of admission		
Type of ward	Emergency	Booked/elective admission list	Total
Medical	2	7	9
Surgical	6	2	8
Total	8	9	17

9.3 Thirty randomly chosen hospital employees took part in a trial at the hospital cafeteria to compare a new recipe for beef stew, incorporating recent recommendations for healthy eating, with the one normally used. Each person was first given the usual dish and then the new recipe. They were asked to rate each of the two dishes simply as either 'Acceptable' or 'Not acceptable'. The results of the trial were:

	New recipe	
Normal recipe	Acceptable	Not acceptable
Acceptable	15	2
Not acceptable	8	5

(a) Is there any evidence to suggest that the new recipe would not be acceptable as a replacement for the existing one?
(b) Do you have any criticism of this study?

9.4 In a trial of two drugs for hypertension, patients were each given a new drug (Blandolol) and a well-tried standard formulation (Hyperlol). Twenty-five patients completed the study in which each patient was given the two drugs in random order. One of the records made of the outcome of the drug was a statement, by the patient, about which drug he or she felt better when using. 15 patients expressed no preference, but the data for the remaining 10 were:

Felt better with	Drug order		Total
	Blandolol–Hyperlol	Hyperlol–Blandolol	
1st drug used	3	1	4
2nd drug used	1	5	6
Total	4	6	10

Use Gart's test to ascertain whether there is:

(a) a treatment (drug) effect;
(b) an order of treatment effect.

9.5   In a hypothetical study to compare the incidence of suicide, 60 000 people living in an 'inner city' area and 190 000 people living in the suburbs of the same city were followed up for 10 years to discover how many committed suicide (ICD codes E950–E959, ninth revision). The results were:

Habitat	Death by suicide?		Total
	Yes	No	
Inner city	102	59 898	60 000
Suburbs	121	189 879	190 000
Total	223	249 777	250 000

(a) Estimate the risk of death by suicide, during the ten-year period, for inner city dwellers.
(b) Estimate the risk of death by suicide, during the same period, for suburb dwellers.
(c) Estimate the relative risk (inner city to suburbs).
(d) Estimate the odds ratio and compare with your answer to part (c).

9.6   For the Doll and Hill (1950) case-control data on *females* in Table 9.2:

(a) calculate the sample odds ratio (the approximate relative risk);
(b) calculate a 99% confidence interval for the odds ratio.

9.7   In a hypothetical survey of types of care for mentally handicapped children, four districts were asked to report on the type of care received by each such patient in the district. The results are presented in the following table.

Type of care	District				Total
	A	B	C	D	
Hospital	108	60	73	206	447
Community home or hostel	6	21	12	35	74
Patient's home	204	85	24	70	383
Total	318	166	109	311	904

(a) Do the districts differ in their type of care for mentally handicapped children?
(b) If there are differences calculate the adjusted residuals to see where they occur.

9.8   In a study on a new prognostic indicator, Ellis *et al.* (1985) reported the intensity of staining of tissue for 109 women for whom the oestrogen receptor state was either positive or negative. These findings are shown in the following table:

Oestrogen receptor state	Staining intensity				Total
	+	+ +	+ + +	+ + + +	
Positive	18	14	19	9	60
Negative	14	11	21	3	49
Total	32	25	40	12	109

(a) Do the data suggest that oestrogen receptor state and staining intensity are related?

(b) Are the proportions positive and negative linearly related to staining intensity?

9.9 A random sample of 600 patients who had had a particular operation were selected from all patients who had had the operation in two districts in one year. Of these, only patients under 55 years old were considered, resulting in 583 patients in the study. These were then classified into two age groups: child (0–14 years) and adult (15–54 years) and by whether or not the operation had been done as day surgery. The resulting numbers were:

	Age group			
	Child		Adult	
District	A	B	A	B
Day-surgery				
No	90	118	86	120
Yes	22	61	19	57

A statistical package was used to fit a set of hierarchical models; the models fitted and their associated deviances are shown below.

Model No.	Model fitted	Deviance
1	$\ell = u + u_1 + u_2 + u_3 + u_{12} + u_{13} + u_{23}$	0.0015892
2	$\ell = u + u_1 + u_2 + u_3 + u_{12} + u_{13}$	0.14511
3	$\ell = u + u_1 + u_2 + u_3 + u_{13} + u_{23}$	14.215
4	$\ell = u + u_1 + u_2 + u_3 + u_{12} + u_{23}$	0.22597
5	$\ell = u + u_1 + u_2 + u_3 + u_{23}$	14.392
6	$\ell = u + u_1 + u_2 + u_3 + u_{13}$	14.311
7	$\ell = u + u_1 + u_2 + u_3 + u_{12}$	0.32168
8	$\ell = u + u_1 + u_2 + u_3$	14.488

(Variable 1 = day surgery status, variable 2 = district, variable 3 = age group.)

Which is the best model, and what practical interpretation does it have?

9.10 Fifty women with a specific fertility problem were each given a drug to redress their hormonal imbalance. Subsequently all fifty became pregnant (although not all went on to give birth). The number of months between the end of the drug treatment and the onset of pregnancy were:

No. of months	1	2	3	4	5	6	7	8	9	10	11	12
No. of women	29	8	3	2	1	0	1	1	1	1	2	1

If the chance of pregnancy in each month is constant and unaffected by the outcome of earlier attempts, then a geometric distribution should fit these data. Carry out a test to see whether this is indeed the case.

10
Surveys

10.1 Introduction

Methods of data collection were discussed in Chapter 2 where it was suggested that the survey is the most commonly used method for obtaining health data when the relevant information is not available from routine sources. A brief introduction to the conduct of surveys was given in Section 2.5.1. In this chapter further aspects of the design of surveys will be described in conjunction with methods of analysis.

In the main the methods of analysis are those described in Chapters 3 to 9. For instance survey outcomes are often presented in the form of tables and diagrams, and, when the survey is a sample survey rather than a census, inferences which relate the sample results to the overall population are often required. There are two particular aspects of sample surveys that require special attention in the analysis. First, the population from which the sample has been drawn may not be infinitely big, as is assumed in the 'standard' methods of inference described in Chapter 7. When, for example, sampling consultants in a regional health authority (RHA) the population of consultants is certainly of a finite, in fact small, size. Second, the sampling method used may not be a straightforward selection of individuals from a population but may, instead, involve selection at different levels or within subgroups. When sampling consultants in England, a number of RHAs may first be selected and then consultants sampled from within each of the chosen RHAs (a 'multi-stage sample' – see Section 10.9.5).

The effect that these two special situations have on the basic methods of data analysis presented earlier in this book are shown in Section 10.9. Since these effects are slight it has seemed sensible to place this chapter at the end of the book, after the basic methods have been covered, enabling this to be a compact, comprehensive introduction to surveys. Some of the material covered here is, however, applicable to other methods of data collection. In particular questionnaire design for surveys is essentially no different from questionnaire design for other special studies (cohort, case-control, etc.).

Furthermore other special studies may well involve sampling, for example when the controls are sampled in a case-control study, and consequently the types of error encountered in sample survey results may also be encountered in other special studies.

10.2 Methods of questioning

Surveys seek to extract data from the subjects selected for study. The data may be acquired by observation, measurement or by obtaining responses to questions, the method used depending upon the type of information required. Individual data items are then usually entered onto a *questionnaire*, a collection of pieces of paper containing whatever questions are to be asked together with blank spaces for the data items. In the case of observations and measurements these 'questions' may be no more than the headings to blank columns or rows. A questionnaire may be completed either by the selected subject (the *respondent*) himself, or by a trained survey employee (an *enumerator*) on the respondent's behalf.

When a respondent is expected to complete the questionnaire himself the questionnaire is often sent and returned by post. This is relatively cheap, but not always very productive, since many people fail to respond. Sending an enumerator out to ask questions is expensive, in terms of salary and travelling costs, but a well-trained and experienced enumerator can usually persuade the respondent to answer the questions. This is particularly important when the questions are of a sensitive nature, as often happens in health studies. If a postal questionnaire is used it is essential that it is accompanied by a covering message which explains the importance of obtaining accurate and truthful answers from the respondent, if possible showing how the information could eventually benefit the respondent himself.

The rate of response to both self-fill and enumerator questionnaires can be improved, and costs reduced, if the respondent can be 'captured' on health premises. A questionnaire about the standard of nursing care, for instance, could be given to the patient in hospital rather than mailing it to him after discharge. The danger with this method, however, is that the act of questioning may influence the services received, such as when the nurses try harder because they know that they are being assessed. Furthermore the respondent may feel that it is in his best interests to approve of the nursing care whilst he is still reliant upon it.

Quite often a questionnaire about health will include questions about complex medical or behavioural ideas. In such cases the respondent may not understand or be able to complete the questionnaire without the assistance of an enumerator. It may well be that the enumerator, in such circumstances, must have medical, nursing or social-work qualifications. Such highly-qualified enumerators will certainly be needed if the survey involves professional assessments of health or measurements of such things as blood pressure.

If enumerators are used, care must be taken to ensure that they do not distort the data by allowing, however unconsciously, their own prejudices to colour the responses they obtain. Similarly it is not unknown for enumerators to fabricate results. Comprehensive training and critical selection of enumerators is necessary to avoid such problems, and it is a good idea to check the quality of data as they are collected.

Whether the questionnaire is self-filled or completed by an enumerator it is essential to assure the respondent of the confidentiality of his own data in order to obtain his co-operation, that is to increase accuracy and reduce non-response. In a postal survey the confidentiality aspect should be advertised in the covering message.

10.3 The content of questionnaires

A good questionnaire is one which extracts accurate responses to the questions which the survey seeks to answer. A questionnaire can go wrong either by getting inaccurate responses, such as when a person lies about his alcohol consumption, or by getting accurate responses to the wrong question, such as when the question, 'Do you regularly use drugs?' is interpreted, by the respondent, to encompass only illegal drugs whereas the survey organizers wished to know about the use of medicinal drugs also. To produce a good questionnaire, it is essential for the designer to put himself in the mind of the respondent, to imagine how a particular question might be interpreted. Since all respondents will not be alike this may mean imagining the responses of different types of respondent, such as different ethnic groups, people with different levels of education, the sexes and so on.

When enumerators are used they will interpret both the question and the answer, so there are extra levels of complexity. A question may then be inadequate because of the way it is asked or the way in which its answer is recorded. The survey organizers must ensure that each enumerator knows what is expected of each question (such as 'drugs' includes medicinal drugs) and any prompts or additional explanations that are allowed.

Most surveys will generate their own special problems of questionnaire content but a few general points on good practice can be given. For further details see Chapter 13 of Moser and Kalton (1971).

(1) *The questions should be unambiguous*

The questions should have a unique meaning that is understood by *all* respondents. In the question, 'How many people share your household?', for instance, the term 'household' can be interpreted in different ways; people living under the same roof sharing expenses such as heating and lighting but not food, might or might not be considered as members of a common household. Ambiguities arise when the scope of the question is not made explicit, such as in the drug question mentioned earlier where medicines may or may not be included, or in the household question where a further ambiguity arises because it is not clear whether the respondent should include himself in his answer. Another common type of ambiguity is that caused by the use of such adjectives as 'large' in 'Is the tumour large?' which will obtain subjective answers depending upon the respondent's concept of tumour sizes. An objective, factual, question such as, 'What size is the tumour?' is preferable.

(2) *The questions should be understandable to all the potential respondents*

Simple language is best; technical terms and words not in everyday use should be avoided or, where really necessary, explained fully. Complex language can irritate and confuse the respondent, and is likely to lead, especially in self-fill questionnaires, to non-response, and to guessing, especially when enumerators are used. Since few people are prepared to appear ignorant it cannot be assumed that respondents will necessarily ask for help from the enumerators. Examples of complex language would be the use of medical terminology not familiar to the layman, or a question such as 'How many siblings do you have?' which should be replaced with the universally understood 'How many brothers and sisters by the same mother or father do you have?'.

(3) *All respondents should be able to answer each question accurately*

If not the answers may, again, be guessed or left blank or, at best, answered to the limits of the respondent's ability. In a door-to-door smoking survey, to give an example, housewives might be expected to answer questions about family smoking habits when their knowledge of their teenage son's or daughter's smoking (at school, perhaps) is incomplete. When the question involves an element of recall the respondent's accuracy can be poor, especially if the time scale is large. Few people could, for instance, answer the question 'How many headaches have you had in the last 6 months?' with confidence.

(4) *The questions should all be relevant*

Long forms are demoralizing for respondents, enumerators and the people who will process the data after collection, leading to carelessness and lack of enthusiasm which in turn results in loss of quality and an increase in non-response. To avoid these problems it is a good idea to draw up such things as dummy tables for the required results of the study before the questionnaire is compiled. This may help to avoid the tendency to include questions which 'may be useful later'. Equally the temptation to dovetail different study questionnaires together, or include a question on a colleague's behalf, should be avoided in order to maintain the quality of the results.

(5) *Leading questions should be avoided*

Most respondents like to please, or, if confused, will seek the answer which is offered them. So the practice of leading a respondent into a particular answer should be avoided. 'Would you agree that soluble aspirin cures headaches more quickly?' and 'You don't think that insoluble aspirin is better, do you?' are both examples of leading questions (especially bad when asked of a patient by his doctor). A neutral question such as, 'Which type of aspirin do you prefer?' is preferable. Another type of leading question is that which explicitly suggests one of a possible range of answers, as in, 'Did you like any features of the hospital meals, such as the wide choice of vegetables?' This can be expected to provoke the response, 'wide choice of vegetables', more often than if it were, like everything else, left for the patient to consider for himself.

(6) *The questions should not upset or distress any respondent*

Otherwise non-response can be expected. Sometimes, due to the nature of the enquiry, intimate questions need to be asked, for instance on sexual habits or drug abuse. One way round this problem is to put the question in the third person, but probably the most important factor is the rapport established by the enumerator; as mentioned earlier it is advisable to use enumerators, perhaps professionally qualified, when intimate questions are asked. In general 'problem' questions, if they cannot be omitted entirely, are best placed at the end of the form so that the respondent has, by then, warmed to his task, and also so that the earlier answers are already captured in case of refusal to respond.

(7) *The respondent should not be expected to make calculations*

Calculations unnecessarily introduce a possible source of error, especially as they place an extra burden on the respondent. The question 'How much did you spend last week on cigarettes?' requires the number smoked to be multiplied by the average cost (perhaps involving different brands). The cost can easily be found by the study team and a computer used to do the multiplication leaving the respondent to only count his cigarette consumption. Two other faults, incidentally, could be found with the spending question: there is an ambiguity in whether cigarettes bought but smoked by someone else should be included, and the recall time of one week is too long to produce accurate answers in some cases.

(8) *The respondent should not be expected to invent a classification scheme*

If classifications are to be made then the range of possibilities should be given in the questionnaire. The question 'What do you think of the standard of nursing care you have received?' may extract the answer 'Good' from one respondent and 'Very good' from another. There is no guarantee that they both use the same scale of measurement; the first may not allow 'Very good', for example, his classification meaning 'Good' as opposed to 'Bad' or 'Average'. In this case it would not make sense to compare the answers of different respondents or to make such statements as 'Only 5% of patients said the nursing care was very good'.

10.4 The design of questionnaires

In general the simpler the questionnaire the easier it is to follow and hence the more accurate the data obtained. The complexity of the questionnaire is, however, necessarily dependent upon the complexity of the individual questions and the relationship between the questions. For instance, in a survey of nursing activity, where the number of hours spent on various tasks is recorded, the questionnaire can be very simple, perhaps no more than a table with blank spaces for the results. In other cases, particularly where personal opinions are sought, where some answers are a choice from a set of alternatives, or where questions may be skipped because they are not relevant to a particular respondent, the questionnaire is inevitably more complex. It would then take the form of a series of verbatim questions listed down the page with boxes to be marked or spaces filled in for the answers. When question skipping is possible the pathways must be clearly marked on the questionnaire, either by using such constructions as, 'If . . . then go to question X, otherwise go to question Y' or by using arrows to mark the pathways as in Fig. 10.1.

---

11. Do you have private health insurance? | Yes | No |

12. Who pays for this insurance?
    (Please tick the appropriate box)

    The company for which you work?

    The company for which a member of your family
    works?

    You or your family?

    Other (please state below)

    ....

13. During the last five years have you spent at
    least one night as a patient in a hospital bed? | Yes | No |

Fig. 10.1 Excerpt from a questionnaire.

It is important to ensure that instructions for completion of the questionnaire are clearly given. As well as the pathways this includes such things as the units of measurement (otherwise, for example, metres and yards might be mixed up) and the degree of rounding for numerical answers. Also the convention for indicating the choice when a set of answer boxes is offered (as in Fig. 10.1) must be stated. If not the respondent could use an X, say on the female box in the question, 'What is your sex?'. Does this mean that X marks the correct response (respondent is female) or X cancels the alternative response (respondent is male)? It is safest to state the rules in the introduction to the questionnaire and because of the problem with an X it is best to ask for the correct box to be ringed or ticked.

Adequate space must be provided for each response, the paper used should be of good quality and a sensible size and the printing should be readable by all respondents (and enumerators) without strain. Small print should be avoided. The questionnaire should contain space for the respondent's name, enumerator's name, hospital, date of interview, etc. where such things are appropriate, although names will usually be recorded as codes. Besides the obvious uses to which such information might be put, it can be useful when checking the data, for instance, to identify a hospital which has consistently misrecorded a particular answer. The questions should appear in a logical order, all questions of the same type appearing together with one question leading to the next where possible.

With the more complex questionnaires the layout of the questions needs to be decided. To make the questionnaire easy to follow during filling in and later transfer of the data (perhaps to a computer), it is advantageous to have a consistent format throughout. Each question and the boxes or blank space for its answer should be separated in a standard way, perhaps with questions to the left two-thirds of the page and answers to the right. This reduces the chance of missing a question or mixing up questions and answers as well as reducing fatigue, all of which lead to errors in the data. For similar reasons questions of the same type should receive the same style of answer, so that if one question has boxes for Bad/Satisfactory/Good responses then, if appropriate, so should the next.

10.4.1 Coding

Once the questionnaire has been completed it will need to be interpreted to extract useful information. Almost always this requires the data from the complete set of questionnaires to first be transferred to some convenient storage medium, such as a computer. The stored data are known as a *database* (see Section 2.6). To save time during keying in of data, to save storage space, and to make the analysis easier, the data would normally be summarized as a series of codes, so that, for example, the answer 'Male' becomes M and 'Female' becomes F.

If the questions are all *closed*, that is the entire range of possible answers is provided as a set of boxes, then coding is a simple operation. The sex question is a simple example. If any question is *open*, that is a blank line or space is left for the answer, then coding for this question may be impossible and the answer may simply be read for background information but left off the database. Open questions are sometimes advantageous because they allow the respondent to express himself, but they are clearly difficult to analyse. Closed questions have their own disadvantages: they can force people into a category to which they do not really belong and do not allow for qualification as in, 'Yes I prefer . . ., but only if . . .'. The first of these problems can be eased if an 'Other' category of response is included, preferably with a 'Please state' open rider, as in Fig. 10.1. This is always a good idea if there is a chance that all possible responses have not been covered by the given closed alternatives.

Provided that the questionnaire has been well designed it should rarely be necessary to code as a separate operation to keying in; indeed there is less chance of error if the two are done simultaneously. The only exception is where it is possible, after all the data are available, to identify a small range of types of response to an open question or the 'Please state' open rider to an 'Other' response. Then a separate coding operation is required to translate the written answers into codes. Undoubtedly the coding-keying operation is less error prone if a database package is used (see Section 2.6.1) and this is recommended if the database is of at least a moderate size.

The codes adopted should be meaningful (as in the sex question – M and F are better than 1 and 2) and should be documented, preferably on the questionnaire itself or the screen if a database package is used. Otherwise miscoding is liable to occur and, furthermore, when referring back to the data in later months the codes may well have been forgotten.

It is frequently important to be able to identify a true non-response to a question since it can identify faults in the questionnaire or interviewing technique, which can, perhaps, be rectified by call-back, or at least accounted for. Non-response can be identified if the answer 'Don't know' receives a code, just as with any other answer. Often a special code, such as 9, is used for 'Don't know' so that if someone knows the month, October, and year, 1986, but not the day of a particular event, the date is coded as 99/10/86. The answer 'No' should always receive a code to distinguish it from non-response or 'Don't know'; it is not sufficient to say, for example, 'Please tick if you have ever had any of the following diseases'.

10.5 Errors in survey results

There are three types of error that can occur in survey results: sampling, bias and processing error.

Sampling error is the difference between the characteristics of the chosen sample and the characteristics of the population as a whole. When sampling is undertaken sampling error is unavoidable; it is present in all sample surveys but absent from all censuses. As with bias and processing error, sampling error cannot be measured directly because this would require complete information on the population of interest; if this were available there would be no need for the survey! Unlike bias and processing error there exists a simple way of measuring the likely degree of sampling error when estimating a population parameter such as a mean or a proportion: the *standard error*. Recall from Section 7.2 that the standard error is a measure of the average difference, over many samples, between the values taken by a sample estimate and a fixed value. In the absence of bias this fixed value is the population parameter that is to be estimated. So, loosely speaking, the standard error is the average sampling error. Alternatively sampling error can be measured by the difference between the upper and lower limits of a confidence interval, such as the 95% confidence interval for the population mean (see Example 10.4). As was seen in Chapter 7, sampling error can be reduced by increasing sample size. Sometimes it can also be reduced by improving the sample design (see Section 10.9).

Bias error is a general term for a large number of different errors that arise from lack of care at the design or data collection stages of the survey, and which tend to push the survey results in a particular direction. Without bias error the sample means, for instance, collected from various sample surveys of the same population would vary about the true mean, but when bias error is present in any particular sample survey the sample mean automatically starts out with a predilection to take a value above (or alternatively below) the true mean. Many instances of bias error are caused by quantifying the wrong population; that is the survey is meant to extract responses from a specified 'target' population but, for some reason, responses are extracted from a different population. Frequently this alternative population is a special subset of the target population. Examples of this, and other kinds of bias error, follow in Section 10.5.1. Because bias error can arise in such a large number of ways, it is difficult to be certain that it has been totally avoided, and it is frequently the largest source of hidden error in survey results. Since, with fixed resources, more care can be taken per response when the sample size is smaller, bias error tends to increase as sample size increases (the opposite to sampling error).

Processing error arises through lack of care at the coding, keying and analysis stages of the survey and includes such things as miscoding and mistakes when counting up the entries in a frequency table. As with bias error, the chance of an error occurring at the processing stage will increase with the amount of work to be done, and thus processing error tends to increase with sample size. Processing error can, and should, be completely removed by careful checking and sensible use of computer facilities, even in a large-scale survey.

Example 10.1:   In a survey of personal health and hygiene, individuals in a town were selected by taking every hundredth name from the town's telephone directory. Each respondent was asked the question, 'How many times have you visited your dentist for a routine check-up in the last 12 months?', and the average number of visits was calculated from the sample data.

Bias error occurs in the result because the population sampled, those with a telephone, is not the same as the target population, all individuals in the town. People without a telephone are likely to be relatively poor, and it is reasonable to suggest that poor people are less likely to attend for dental check-ups because of their worry that potentially expensive corrective work may be diagnosed. Hence the average number of visits calculated from the sample is an over-estimate – it is biased towards the more affluent people.

Sampling error occurs in the result because the average number of visits for the sample is almost certainly not the same as the average number of visits for the whole population (that is the whole population sampled, the town-dwellers who own telephones). Processing error would also occur if an arithmetic mistake was made in calculating the average number of visits from the sample data.

10.5.1 Examples of bias error

Bias can arise from a wide variety of sources, some of which have already been instanced in this chapter. It is impossible to give an exhaustive list of sources, since many will be particular to certain types of survey. Instead some examples are given here. In order that the interpretation of these examples may be made in a more general context they have been grouped into four sources of error. Further examples, although not related to health management and research, are given by Huff (1954).

(1) *Bias when selecting*

The results of a sample survey are supposed to represent the entire target population. If the method of selection is biased in favour of some particular subgroup then the results will be biased in the same way.

Consider a survey recently carried out in a health district whose object was to determine the healthiness of the life-style of the population by choosing a sample of people in the district and asking them questions about the amount of exercise they took, how much they smoked and drank and the constituents of their diet. To obtain a representative sample cheaply and easily the Community Medicine Department conducting the survey proposed to sample individuals by first sampling GP practices and then sampling individuals within the chosen practices. This seemed a sensible strategy because some GP practices in the district held registers of patients classified by age and sex and this list would provide a suitable medium (called a *sampling frame*) from which a sample with a representative age/sex profile could be drawn.

In principle the method outlined is sound. The problem was that not all GP practices kept an age/sex register and so the plan was not to consider for selection practices without registers. The results from the survey would then almost certainly be biased, since practices with registers are, in some sense, more progressive and hence more likely to be active in health education. Indeed those practices with computerized age/sex registers often use them to identify 'problem patients' such as those with a high body mass index and those who have not recently had a check-up or attended a screening clinic. Patients from such practices have thus probably received more indoctrination about healthy living than the rest of the district. The survey results would be biased in favour of those with a healthy life-style and would not give an accurate representation of the district in general. In the end, the researchers agreed to take a random sample (see Section 10.9.1) of GP practices.

Bias through selection can also occur in situations where patients are expected to present themselves for selection. Suppose, for example, that a survey of road traffic accidents (RTAs) were organized by having nurses at an Accident and Emergency Department fill out a questionnaire for every tenth RTA patient admitted over a specified time period. Many RTA victims will not go to hospital at all, perhaps because they were only slightly injured, because their illness takes some time to appear (e.g., concussion) or due to death. In some cases this may not matter, for example if the survey aims to estimate the average amount of nursing time used per admitted RTA patient. In other cases the fact that only a certain type of RTA victim is included will bias the results, for instance if the survey seeks to estimate the average cost to the NHS of each RTA. Clearly the RTAs that lead to hospitalization will tend to be most costly and thus the result will be an over-estimate.

Bias when selecting may occur when a non-probability sampling scheme is used (see Section 10.8).

(2) *Bias when defining the population*

If the population itself is mis-specified in some important way then any sample taken from it cannot truly represent the target population. A case of this kind appears in Example 10.1 where the telephone directory was used incorrectly as a sampling frame.

A further example would be where the voting register is used as a sampling frame in a survey of the use of contraceptives. This register excludes teenagers below the age of 18 and it is likely that young teenagers using contraceptives have somewhat different attitudes and habits than older people. Hence the survey would be biased in favour of older people's views and use of contraceptives.

(3) *Bias through non-response*

Non-response not only reduces the sample size and wastes effort but may also induce bias. Some examples of how and why non-response arises have been mentioned earlier, particularly in postal surveys where the questions are sensitive in their nature and where the respondent is concerned about the confidentiality of his replies or the purpose of the survey.

Non-response produces bias in survey results if the non-respondents are, in some important way, different from the rest of the target population. If, for example, a postal questionnaire asked the potential respondents for their view on the government's recent health policies it is likely that only those with a strong opinion would reply. In fact, it is quite possible that only those with a strong view opposing the policies would reply, in which case the survey results are biased against the government.

Non-response can sometimes occur at a group ('cluster') rather than individual level. This would happen, for instance, where a health authority wrote to a number of factory managers asking for permission to carry out an occupational health survey amongst their labour force. In factories where health and safety at work is conscientiously monitored the response is likely to be favourable; where it is not the response is likely to be unfavourable. Bias in the results is then obvious.

A third, and final, example of non-response bias is a situation where enumerators are involved. Consider a door-to-door survey of health status in which community nurses go into people's homes to measure height, weight, blood pressure, etc. of every member of the household. Suppose this was done in the daytime on weekdays. Then most full-time workers would be missed. On the other hand, suppose it was done on Saturdays. Then there is a danger that regular participants in sports would be under-represented in the sample. In both situations the results would be biased because those missed are generally likely to be different in terms of the variables measured. Although it might be expensive, the only sure way to avoid such bias is to call back on non-respondents, repeatedly if necessary. Clearly in a door-to-door survey it is most sensible to call back on a different day of the week and at a different time.

(4) *Bias in the response*

Sometimes the answers extracted from respondents are not accurate. This may not be important in the final analysis, for instance when calculating an average, if negative and positive errors cancel out. In some cases, however, the inaccuracy is systematic from respondent to respondent, that is, each respondent tends to be inaccurate in the same direction. The final survey results will then be biased in that direction. Perhaps the clearest example of this phenomenon is where the observations are measurements, such as height of patients, and the errors are due to rounding (see Section 1.5). If patient's height is always measured to the nearest centimetre the measurements are individually inaccurate, but when pooled (to calculate, say, the mean) over many patients the errors will tend to cancel out. On the other hand if every height is rounded down to the nearest centimetre below the exact height the survey results are biased downwards.

Response bias will be likely if the respondent feels that some personal advantage is to be had from giving a particular answer. For instance age is frequently understated. To take another example, few patients would be prepared to say that their doctor's treatment was entirely useless, even when that is what they feel, if they are concerned that their opinion would be communicated to their doctor. Finally, consider a health economics survey where the chronically ill were asked about the monetary costs of their disability. If the respondents believed that it was likely to lead to an award of increased government benefit they might well be tempted to overstate (perhaps by rounding upwards) their personal expenditure. Such problems can largely be overcome by stressing the confidentiality of the response and the precise purpose of the survey.

Other examples of response bias already mentioned in this chapter are leading questions, which push the respondent into a particular response, and influential enumerators. Enumerators can induce bias in various ways, such as by phrasing questions so that they are effectively 'leading' or by openly expressing their own views, a particular problem when the enumerators are doctors because of the respect they command. In some cases the enumerator's own personal characteristics can influence the response such as when female respondents are reluctant to give complete accounts of their health experience to male enumerators, and this could well bias the final results of the survey.

Finally, another type of response bias that, again, has been touched upon earlier is the so-called 'prestige error' where the respondent gives an answer which makes him appear more knowledgeable. Such error is likely to arise if the question, 'Did you know that X is the Minister of Health?' is asked. A 'testing question', such as, 'Who is the Minister of Health?', would be better. Barnett (1974) gives an example of prestige error in a survey about cervical cancer where a woman was asked 'Is your husband circumcized?' to which the answer was, 'Yes doctor, very'.

10.6 Pilot surveys

Before the survey proper is begun it is *essential* to conduct a pretest or pilot survey. At its most basic this involves trying out the questionnaire on a few volunteer subjects. At its most complete it involves a thorough testing of all aspects of the survey including sample selection and data processing using a relatively small number of respondents. There are a number of advantages to having a thorough pilot survey which certainly outweigh the extra costs.

(1) *Testing of questionnaires* In Sections 10.3 and 10.4 an account was given of some problems that can occur in questionnaires. It is very easy for the survey designer to overlook some of these problems, particularly

ambiguities in the questions, since he will already know what is expected. The best way to uncover such problems before the survey begins is to try out the questionnaire beforehand. The pilot survey can also help to determine the response categories for closed questions.

Before the 1981 population census of the UK the questionnaire was tested on a sample. It was found that the response rate was very low amongst ethnic minorities due to the inclusion of a question on ethnic origin. Consequently this question was omitted in the census proper (see OPCS Monitors CEN 80/2 and CEN 80/3).

(2) *Identifying bias* If checks are maintained during the pilot survey it may be possible to uncover bias in the method chosen to conduct the survey. This can help to eliminate bias error, or at least quantify the expected bias error, in the survey proper.

(3) *Estimating cost and duration* The cost and time taken per unit response can be calculated from the pilot survey. For instance, the cost might be the average number of postal requests to a response times the postal charge, or the average cost of an enumerator's time per interview. Knowledge of average cost and time leads to an estimate of total cost and time for a survey proper of any particular size. If the expected cost of the planned survey exceeds the available budget, or the expected duration is too long, then it will be necessary to make changes in the plans, such as reducing the size of the sample or the length of the questionnaire.

(4) *Estimating variability* To calculate the sample size needed to keep sampling error within acceptable bounds it is essential to have some idea of the variability of the survey material (see Sections 7.14 and 10.9.3). It will almost always be necessary to obtain this from a pilot survey.

(5) *Testing manpower and equipment* The pilot survey can be used to test the techniques of enumerators, keyboard operators and other personnel, and also to test how well they interrelate (particularly important with a large field-force). Knowledge of costs and duration per response enables estimation of the number of personnel necessary to complete the planned survey within a specified time. Equipment, such as computer facilities and stationery, can be similarly costed.

10.6.1 Post-survey tests

In large-scale surveys a sub-sample of the respondents is sometimes selected, after the survey proper is complete, and then requestioned during a thorough interview in order to estimate the bias error in the survey results. Another useful form of post-survey checking, in some circumstances, is to sample the enumerators to ascertain the completeness of their coverage in the field.

10.7 Censuses

One of the three types of error specified in Section 10.5, sampling error, occurs only in sample surveys. It seems, therefore, attractive to remove this error entirely by carrying out a census; that is questioning every member of the population. There are two important drawbacks to a census which can be best appreciated by considering the census as a 100% sample.

The first drawback is obvious – the larger the sample the more expensive the survey in terms of time and money, and the census is the largest survey of all. Should the budget run out there is a real danger that the survey could be abandoned without producing any results. On the other hand, if the time taken in data collection and processing becomes too long the results will have become irrelevant. These problems are illustrated by the most famous census of all, the national census of population, often simply called 'the census'. The 1981 census of England and Wales received a budget of £38 million at 1978 prices (OPCS Monitor CEN 80/1) and employed 103 100 enumerators and 8610 Census Officers, Assistant Census Officers and Supervisors (*Census 1981 Preliminary Report*). The time from planning to publishing results in a census of this scale takes many years.

The second drawback arises because, as stated in Section 10.5, sampling error decreases with increasing sample size but unfortunately bias error tends to increase at the same time. In a census, sampling error has reached its minimum (zero), but bias error is potentially at its maximum. In some cases the expected bias error in a census is so large that it exceeds the expected bias plus sampling error in a sample survey of moderate size. High bias can be expected because of the difficulty of taking sufficient care each time when a large number of respondents is involved. It will not, for instance, be feasible to monitor the work of each enumerator thoroughly to check for unprofessional conduct such as influencing the respondent or fabricating answers. This problem is made worse if, because of the sheer size of the fieldforce, it is not possible to give detailed training to each enumerator. Even the bias errors normally associated with sampling procedures can be expected to be important in a census. The target population may not be 100% sampled, for example, because an enumerator is too lazy to bother visiting relatively inaccessible places or to call back where non-response occurred. In a postal census there will almost certainly be non-response to cause the coverage to be incomplete.

For these reasons sample surveys are much more common than censuses. Even so censuses are unavoidable if data are required at a very low level of disaggregation. This, of course, is the motivation for the census of population. Naturally if the population is quite small the problems involved with censuses will largely disappear.

Sometimes it is sensible to combine a census with a sample survey by giving a standard questionnaire to every member of the population, but giving extra questions at the same time to a sample. The extra questions would be those for which information at a low level of disaggregation is not so crucial, so that it would be a waste of resources to solicit a 100% response.

10.8 Non-probability sampling schemes

The remainder of this chapter will assume that a sample survey is to be carried out, and consider how the sample should be selected. Each member of the population is called a *sampling unit*, since they are all potentially available for sampling. In most of the examples used in this chapter the sampling units are people but, as will be seen in some examples in Section 10.9, the sampling units could equally well be health authorities, hospitals, days of the week, etc. The basic problem is to draw a sample which is, as far as practical constraints allow, a true representation in miniature of the population itself.

The simplest sampling scheme of all is *accessibility sampling* wherein only the most convenient sampling units are selected. Such a scheme is not advised since it leads to bias error through 'bias when selecting'. When, for instance, a doctor carries out a survey of drinking and driving amongst his immediate friends and colleagues he is choosing sampling units with a higher than average exposure to health education. This probably makes them not typical of the population as a whole in relation to drinking and driving. Another example is where volunteer subjects comprise the sample. The chance of bias error here is large because volunteers often have some special characteristic, such as a higher than average knowledge of the subject of the survey or a lower than average income (where payments are made). Other examples of accessibility sampling leading to bias error can be seen in Section 10.5.1.

An improvement on accessibility sampling is *quota sampling* where an enumerator is told to pick a sample of a certain size and structure but the choice of the actual sampling units selected is left to the enumerator. This is a widely used method of sampling because it is cheap, quick and does not require a sampling frame from which to draw the sample. It is most commonly seen in the high street where enumerators stop people to ask questions about voting intentions, shopping habits, and so on. The structure of the sample is arranged so as to spread the sample over the different sub-groups of the population where these subgroups are thought to have important differences as far as the survey material is concerned. It would not, for example, be satisfactory in most market research surveys to have a sample that was, due to enumerator preference, virtually all made up of young women because their buying habits are likely to be different from others. Consequently enumerators are each given daily quotas to interview which specify that so many respondents should be, for example, of a particular sex and so many in a particular age group, over many age/sex groups. The aim would be to make the sample similar, with respect to age and sex, to the population overall (as recorded in the population census).

Despite the controls on quotas, quota sampling is extremely susceptible to bias error because the enumerator still has the final choice of whom to interview and whom to reject. Many enumerators would prefer to stop and question people who look helpful, which can lead, for example, to survey results biased towards the middle class. Another problem is that enumerators often exclude extreme subjects within the quota controls. Thus, when asked to pick men aged 15–24 years old, enumerators may tend to choose men aged around 20 to be sure of satisfying the age criterion. Such problems are much less likely to occur when the enumerator is well-trained and experienced, and for this reason it is advisable to employ the services of a reputable market research organization if a quota sampling scheme is envisaged.

A further disadvantage with quota sampling is common to all non-probability sampling schemes: sampling error cannot be specified. As stated in Section 10.5, the most basic measure of sampling error is the standard error, a measure of the variation, from sample to sample, of the values taken by a sample estimate. Unless the probabilities associated with different samples are known no formula can be derived for the standard error. Since the standard error is used to calculate confidence intervals these are not available either; indeed none of the methods of statistical inference described in Chapter 7 are appropriate.

10.9 Probability sampling schemes

In a probability sampling scheme every sampling unit has (before the sample is drawn) a known non-zero probability of entering the sample. With these schemes sampling error can be specified and bias when selecting is eliminated, and so they are to be preferred. Various probability sampling schemes have been devised; in this section the five basic schemes are explained. Most probability schemes used in practice are either one of these five, or a combination of two or more.

Besides tables and diagrams, the most usual requirement of a survey is that it should produce estimates of means (such as the mean blood pressure of residents in a health district) or proportions (such as the proportion of residents who have suffered heart attacks). Chapter 7 has already considered estimation and other topics in statistical inference with respect to means and proportions. In Chapter 7, however, two things were assumed. First, that the sample is drawn as a simple random sample (see Section 10.9.1), and, second, that the population is infinite. These conditions will be relaxed here, with the consequence that the formulae for unbiased estimation of means, proportions and their standard errors will change. The revised formulae will be specified and these should be used in place of those previously stated (see Example 10.4).

One further population characteristic that is occasionally required is a total, such as where a sample of hospital admissions is to be used to estimate the total number of cross-boundary flows in a year. Given an estimate for a mean or a proportion the corresponding total is estimated simply by multiplying the given estimate by the size of the population, and similarly for the standard errors. Thus if the proportion of patients who live outside the district providing care in the sample is p and the total number of patients admitted in the year is N, then the total number of cross-boundary flows is estimated by Np.

Throughout this section the population size will be denoted by N and the sample size by n.

10.9.1 Simple random samples

When the sample of size n from a population of size N is drawn so that every possible different sample of size n has an equal chance of selection it is called a simple random sample. A consequence of this method of selection is that every member of the population has an equal probability of being included in the sample. Intuitively this seems a sensible way of being fair to everybody, that is avoiding bias, when selecting the sample. Indeed this is the most basic probability sampling scheme, and the one that is assumed in previous chapters.

To draw a simple random sample some method of randomization is required. One possibility is to take a collection of N identical balls, write the name of (or a code for) a different individual on each one, place the balls in an urn and then pick them out one at a time, without replacement, until n balls have been taken. Since this can be a tedious process, tables of random digits are often used instead. To use them every member of the population (sampling unit) must first be given a unique number. Numbers are then selected from the random number table until n different numbers are obtained, and those whose numbers are drawn make up the sample. Some statistical computer packages and calculators will produce random numbers of the kind that appear in random number tables.

Example 10.2: Consider a hypothetical regional health authority with 15 districts. The initials and sex of the General Managers of each district are:

P.S. (m)	W.H. (f)	A.M. (m)	N.N. (f)	I.T. (f)
L.A. (f)	M.H. (f)	N.D. (m)	S.D. (m)	G.H. (m)
G.J. (m)	J.S. (m)	J.R. (m)	G.S. (f)	M.J. (m)

(in brackets m denotes male and f denotes female)

Suppose that a questionnaire is to be given to a sample of 5 District General Managers (DGMs) in the region. The list above provides a sampling frame for the population of DGMs.

To select a simple random sample it is first necessary to number the sampling units:

1 P.S.	4 W.H.	7 A.M.	10 N.N.	13 I.T.
2 L.A.	5 M.H.	8 N.D.	11 S.D.	14 G.H.
3 G.J.	6 J.S.	9 J.R.	12 G.S.	15 M.J.

Then 5 random numbers between 1 and 15 need to be selected. This will be achieved using Table A.6.

Since there are 15 sampling units it clearly would be no use simply taking single digit random numbers (between 0 and 9), since 10–15 would then have no chance of selection. Therefore the digits in Table A.6 must be taken in pairs, for instance, 1 0 2 7 5 3 is read as 10, 27, 53. This produces random numbers between 00 and 99. These numbers can be selected from Table A.6 until 5 different numbers between 1 (or, strictly, 01) and 15 have been obtained (00 and any number above 15 are ignored). If a number that has already been selected comes up, its second and subsequent occurrences are ignored.

If, however, the numbers are really to be a random selection they should be selected from the table in random fashion. A direction of movement through the table (left to right across, downwards, diagonally, etc.) should first be chosen arbitrarily. Then a starting point in the table should be chosen in some haphazard fashion, such as by inserting a pin with eyes closed. It is *not* correct to always use the same starting position and direction in the same random number table every time random numbers are selected.

Suppose that the direction chosen is left to right across the table, and the starting point chosen is row 4 column 27. The following 2-digit random numbers are then selected from Table A.6:

44	31	89	16	91	47	75	**03**	20	54	20	70	56
77	59	95	60	19	75	29	94	**11**	23	59	30	**14**
47	64	17	18	43	97	37	66	55	86	**08**	74	50
43	43	23	29	16	24	**15**						

The numbers between 1 and 15 are underlined; every other number is outside this range and is thus ignored. Hence the random numbers selected are:

$$3, 11, 14, 8 \text{ and } 15$$

which implies that the simple random sample of DGMs is

$$\text{G.J., S.D., G.H., N.D. and M.J.}$$

When a quantitative variable, x, is recorded for each sampled individual, the sample mean, $\bar{x}$, where

$$\bar{x} = \frac{\Sigma x}{n}$$

is an unbiased estimate of the corresponding population mean. Similarly the sample variance, s^2, where

$$s^2 = \frac{1}{n-1} \Sigma (x - \bar{x})^2$$

is an unbiased estimate of the corresponding population variance. Then the variance of $\bar{x}$ is estimated by

$$\frac{(1-f)s^2}{n}$$

where $f = n/N$, and so the estimated standard error of the sample mean (the most basic measure of sampling error when estimating a mean) is the square root of this,

$$s \sqrt{\frac{(1-f)}{n}}$$

Although $\bar{x}$ and s^2 are the same as in Chapter 7 and elsewhere, the formula for the standard error is not. A factor $1 - f$, called the *finite population correction factor*, has been introduced. This is necessary to ensure unbiased estimation when the population being studied is finite. Notice that f is the sampling fraction n/N. When N is large compared with n then f will tend to zero. Hence with an infinite population the correction factor, $1 - f$, becomes 1, which explains its omission in Chapter 7. Even when N is certainly finite, say the size of a town's population, f will often be negligible. Thus when selecting a sample of 500 from a town of 100 000 inhabitants the correction factor is 0.995 which is unlikely to make much difference to the estimated standard error unless s^2 is very small. The correction factor is, however, important when the fraction sampled is large.

When a proportion is to be estimated the sample proportion, p, is used as in Section 7.6. Again the difference is that the formula for the estimated standard error changes. Here the estimated standard error of p is

$$\sqrt{(1-f)\ \frac{p(1-p)}{n-1}}$$

Example 10.3:   In the survey described in Example 10.2 two of the variables recorded on the questionnaire are the length of service with the NHS and whether or not the DGM has a medical qualification. Although this information would be unknown to the survey organizers, suppose, for the sake of illustration, that the values for these variables for all DGMs are as follows:

Initials (sex)	Length of service (years)	Medical qualification?
P.S. (m)	28.2	yes
L.A. (f)	24.4	yes
G.J. (m)	32.7	no
W.H. (f)	20.2	yes
M.H. (f)	24.4	yes
J.S. (m)	30.8	no
A.M. (m)	33.4	no
N.D. (m)	24.6	yes
J.R. (m)	1.0	no
N.N. (f)	29.6	yes
S.D. (m)	27.2	yes
G.S. (f)	26.3	yes
I.T. (f)	25.0	yes
G.H. (m)	0.8	no
M.J. (m)	31.1	no
Total	359.7	yes = 9
		no = 6

The mean length of service and proportion of DGMs with medical qualification are then $359.7/15 = 24.0$ years and $9/15 = 0.6$ respectively for the entire population of DGMs.

Now consider how these two parameters, that are really unknown, would be estimated from the sample G.J., S.D., G.H., N.D. and M.J. selected in Example 10.2. The sample mean $\bar{x}$ is

$$\frac{32.7 + 27.2 + 0.8 + 24.6 + 31.1}{5} = 23.3 \text{ years}$$

and the sample proportion of medically qualified DGMs is $2/5 = 0.4$.

Since, artificially, the true population mean and proportion are known here the sampling error can be measured exactly as $24.0 - 23.3 = 0.7$ years when estimating the mean and $0.6 - 0.4 = 0.2$ when estimating the proportion. In a real-life sampling situation the true values would be unknown; the sampling error would then be represented by the estimated standard error. Using the formulae given earlier these turn out to be 4.73 years and 0.2 for the mean and proportion respectively. As mentioned in Section 10.5, each of these can be interpreted as estimating an average sampling error over many different samples of size 5 from the 15 (thus, by chance, the sample actually drawn appears to be better than average for estimating the mean and average for estimating the proportion).

When the population is finite the finite population correction factor should be used in any of the formulae involving the standard error in Chapter 7.

Example 10.4:   A 95% confidence interval for the mean is given by

$$\bar{x} \pm t_{n-1} \text{(estimated standard error of } \bar{x})$$

where 2.5% of the Student's t distribution on n – 1 d.f. exceeds t_{n-1}, provided that the usual normality assumption holds (see Section 7.4.3). Then a 95% confidence interval for the mean length of service in Example 10.3 would be

$$23.3 \pm 2.776 \times 4.73$$

i.e., $$23.3 \pm 13.1 \text{ years}$$

(using Table A.2). However, the normality assumption does appear to be dubious in this case.

10.9.2 Systematic samples

Sometimes it may be convenient to sample by choosing every ith unit in a list or queue; the resulting sample is called a *systematic sample*. To obtain a sample of size n from a population of size N, i should have the value N/n. For example in the survey described in Example 10.2, N = 15 and n = 5 so that if systematic sampling were used, every 3rd DGM in the list should be selected. That is the sample will consist of units 1, 4, 7, 10 and 13, or 2, 5, 8, 11 and 14, or 3, 6, 9, 12 and 15, depending on where the starting point is chosen. In order to ensure randomness the starting point should always be chosen at random. This is achieved simply by selecting a random number between 1 and N/n (between 1 and 3 in the example). Notice that N/n may not be a whole number; the best that can then be done is to take i (the sampling interval) to be the nearest whole number to N/n.

One great advantage of the systematic method is its lack of requirement of a sampling frame. As implied in the introduction, a queue could be used instead. By 'queue' here is meant a steady stream of sampling units, such as patients leaving a hospital. If it were wished to sample patients continually over a year then it would be impossible to draw a simple random sample since no list of patients exists *before* the year begins. A systematic sample would be perfectly possible, for instance every 10th discharge or death from the hospital could be selected, and the patient's personal details, diagnosis, etc., recorded. Indeed this was a recommended method of sampling to obtain the 10% sample for the early years of the *Hospital In-Patient Enquiry* (DHSS *et al.*, 1974). Other advantages are the ease of administering the systematic scheme, and its need for only one random number selection.

The disadvantage with systematic samples is that the result, or at least the precision of the result, can be adversely affected by systematic patterns in the sampling frame or queue from which the sample is drawn. Suppose, for instance, that every 20th patient consulting a GP were sampled in a survey of GP workloads which recorded, amongst other things, time taken per consultation. Now if the GP were actually booked to see 40 patients each day it would happen that the sampled patients always received a consultation at about the same period into a half-day surgery. It is quite possible that consultation times vary during a surgery, perhaps getting shorter towards the end of the day and just before lunch.

To take another example, consider a survey of the effects of pollution on the inhabitants of a street which samples every 10th house number. It is quite possible, due to the position of industrial buildings, that one side of the street suffers more from pollution than does the other, but every 10th house number falls on the same side of the street.

When there are *no* systematic patterns in the list or queue the systematic sample will behave like a simple random sample, since in this case the ordering is random. When there are systematic patterns it cannot be treated as a simple random sample, and for details of how to carry out data analysis in such situations, the reader is referred to Chapter 8 of Cochran (1977). Cochran shows that, notwithstanding the type of example given in the last paragraph, systematic sampling is actually better than simple random sampling when certain special systematic patterns pertain.

10.9.3 Stratified samples

In many situations it is useful to consider the overall population as an amalgam of several different subgroups (or *strata*) of sampling units, where the strata are themselves of separate interest or importance. It is, for example, common when surveying human beings to want to quantify the attributes or opinions of the different sexes and different age groups separately.

A stratified sample consists of a collection of simple random samples (s.r.s.s), one from each stratum of the population. Thus, if it is important to consider four distinct age groups as well as sex, then s.r.s.s should be drawn from each of the eight strata to produce a sample stratified by age and sex. This is similar to the idea behind quota sampling (see Section 10.8), except that here the random or probabalistic element enables sampling error to be quantified whilst reducing the chance of bias. The disadvantage is that the stratified sample requires a sampling frame.

Example 10.5:   Consider selecting a sample stratified by sex from the DGM population listed in Example 10.2. The first step would be to number males and females separately:

	Males			Females	
1 P.S.	4 A.M.	7 S.D.	1 L.A.	4 N.N.	
2 G.J.	5 N.D.	8 G.H.	2 W.H.	5 G.S.	
3 J.S.	6 J.R.	9 M.J.	3 M.H.	6 I.T.	

Suppose that a sample of overall size 5 is to be drawn (as before). Since 9 out of the 15 are men it seems reasonable, in the absence of any other information, to have the same proportion of men in the sample (so-called *proportional stratification*). Hence a simple random sample of $9/15 \times 5 = 3$ men will be drawn. By a similar argument (or simply by subtraction) a second s.r.s. of 2 women will be drawn.

To draw the s.r.s. of men 3 random numbers between 1 and 9 need to be selected. This is achieved by taking single digits from Table A.6. Starting at column 16 of row 1 and moving right to left, the numbers selected are 4, 5, 0 (rejected), 5 (repeated, so rejected) and 1 corresponding to A.M., N.D. and P.S. To draw the s.r.s. of females 2 random numbers between 1 and 6 need to be selected. Continuing where the last selection left off in Table A.6 the random digits are 7 (rejected), 3 and 2 corresponding to M.H. and W.H. Hence the complete stratified sample is A.M., N.D. and P.S. (males) and M.H. and W.H. (females).

Stratified sampling has two important advantages over simple random sampling. First, when the different strata are important in their own right, it can ensure that a reasonably large sample is drawn from each stratum. In a s.r.s. it is quite possible for at least one stratum to be poorly represented in the sample, leading to poor precision in estimation. Indeed it is possible for one (or more) strata to be missed entirely which makes sample-based comparison impossible (this happened in Example 10.2 where, by chance, the sampling units selected were all male).

The second advantage of stratified samples arises when the variables that define the strata are related to the item, or items, measured in the survey. In this case the sampling error of a stratified sample is less than that of a s.r.s. of the same overall size. For example, most health outcomes are related to the individual's age and sex, and so these are sensible stratification variables in human health surveys. Also in the street pollution survey described in Section 10.9.2, it would be sensible to stratify by side of street since different pollution levels are expected on each side. Given a fixed overall sample size, the most precise estimates are obtained when the strata are internally as homogeneous as possible. When the strata are internally similar, each individual stratum represents a certain type of individual. Since samples are then taken from all strata this ensures that the overall sample represents all sectors of the population. Consider sampling admissions to Accident and Emergency Departments. Different workloads might be expected at different times of the day and week. Stratification by day and time of day would ensure that each type of workload was sampled. Consider also sampling acute hospitals, where it might be sensible to stratify by the number of beds (perhaps 'few', 'average', and 'many', suitably defined). Since hospitals of similar size will tend to have similar facilities, different to those of a different size, it would seem sensible to seek to spread the survey across the size groups.

The disadvantage of stratified sampling, compared with simple random sampling, is the requirement of information on the stratification variable(s) *before* the sample is drawn. In some situations this is not available. Moser and Kalton (1971) give the example of selecting individuals from the electoral register and being unable to stratify by age because the age of the voter is not recorded on the register.

From stratified samples means and proportions will usually be estimated for each individual stratum, as well as for the population as a whole. The ith stratum mean is estimated by $\bar{x}_i$, the mean of the s.r.s. that has been drawn from the ith stratum. Its standard error is estimated by

$$s_i \sqrt{\frac{(1 - f_i)}{n_i}}$$

where s_i is the standard deviation of the s.r.s. from the ith stratum and f_i is the ith sampling fraction, n_i/N_i, for ith stratum size N_i and ith stratum sample size n_i. Notice that with proportional stratification f_i is the same for all i (in Example 10.5 a sampling fraction of one third was used for both male and female strata). Proportional stratification is *not* necessary; indeed it may not even be the best way of dividing the sample amongst the strata. If, for instance, one stratum was known to be more variable than the others (information that could come from a pilot survey) then it would be sensible to use a larger sampling fraction in this stratum. (See Moser and Kalton (1971), Section 5.3 for further details.)

The unbiased estimate of the population mean, denoted $\bar{x}_{st}$, is then a weighted average of the individual stratum sample means where the weights are the stratum sizes, i.e.,

$$\bar{x}_{st} = \frac{1}{N} \sum_{i=1}^{k} N_i \bar{x}_i$$

where k is the number of strata and

$$N = \sum_{i=1}^{k} N_i, \text{ the overall population size.}$$

The estimated variance of $\bar{x}_{st}$ is similarly a weighted average of stratum sample variances, leading to an estimated standard error of

$$\frac{1}{N} \sqrt{\sum_{i=1}^{k} N_i^2 (1 - f_i) s_i^2/n_i}$$

There are similar estimates for proportions. For the ith stratum use p_i, the proportion in the ith stratum sample, with estimated standard error

$$\sqrt{(1 - f_i) \frac{p_i (1 - p_i)}{n_i - 1}}$$

For the overall proportion use

$$\frac{1}{N} \sum_{i=1}^{k} N_i p_i$$

with estimated standard error

$$\frac{1}{N} \sqrt{\sum_{i=1}^{k} N_i^2 (1 - f_i) \frac{p_i (1 - p_i)}{n_i - 1}}$$

Example 10.6: Using the sample drawn in Example 10.5 and the information given in Example 10.3, the sample data are:

	Male			Female	
	Service	Qualification?		Service	Qualification?
A.M.	33.4	no	M.H.	24.4	yes
N.D.	24.6	yes	W.H.	20.2	yes
P.S.	28.2	yes	Total	44.6	yes = 2
Total	86.2	yes = 2			no = 0
		no = 1			

Let the subscript M denote males and F denote females.

$$\bar{x}_M = \frac{86.2}{3} = 28.7 \text{ years}$$

with estimated standard error

$$s_M \sqrt{\frac{(1 - f_M)}{n_M}}$$

$$= 4.424 \sqrt{\frac{(1 - \frac{1}{3})}{3}} = 2.1 \text{ years}$$

Similarly $\bar{x}_F = 22.3$ years with estimated standard error 1.7 years. Then,

$$\bar{x}_{st} = \frac{(9 \times 28.7) + (6 \times 22.3)}{15} = 26.1 \text{ years}$$

with estimated standard error

$$\frac{1}{N} \sqrt{N_M^2 (1 - f_M) s_M^2/n_M + N_F^2 (1 - f_F) s_F^2/n_F}$$

$$= \frac{1}{15} \sqrt{81 \times (2.1)^2 + 36 \times (1.7)^2}$$

$$= 1.4 \text{ years}$$

So the average length of service for all DGMs in the region is estimated to be 26.1 years, but male DGMs are estimated to have an average of 28.7 years and females 22.3 years.

Notice that the standard error of $\bar{x}_{st}$ is much lower than that of $\bar{x}$ found in Example 10.3. This suggests that the stratification has been successful in improving the precision of estimation, although the reason for stratification here was to be able to make meaningful comparisons between the sexes. It must, however, always be remembered that standard error is a measure of average error over many samples, and in practice only one such sample would ever be drawn. In fact comparison with Example 10.3 will show that the particular $\bar{x}_{st}$ obtained here is actually further from the true mean than the particular $\bar{x}$ obtained earlier.

Notice also that in any calculation of mean and standard error the two outliers in the population are likely to have considerable influence. These outliers are two men who were appointed straight to the position of DGM from outside the health service. One outlier turned up in the simple random sample, but neither appeared in the stratified sample (by chance) and this probably accounts for the smaller standard error in the latter. The reader might like to consider whether the mean is the most appropriate measure of average length of service given the practical knowledge that a few unusually small values are likely.

Now consider estimation of the proportion who are medically qualified.

$$p_M = \tfrac{2}{3} \text{ or } 0.667 \text{ with estimated standard error}$$

$$\sqrt{(1 - f_M) \frac{p_M (1 - p_M)}{n_M - 1}}$$

$$= \sqrt{\tfrac{2}{3} \times \tfrac{2}{3} \times \tfrac{1}{3}/2} = 0.272$$

$$p_F = \tfrac{2}{2} = 1 \text{ with estimated standard error}$$
$$\sqrt{\tfrac{2}{3} \times 1 \times 0/1} = 0$$

The overall proportion is estimated by

$$\frac{1}{15}(9 \times \tfrac{2}{3} + 6 \times 1) = 0.8$$

with estimated standard error

$$\frac{1}{15}\sqrt{81 \times (0.272)^2 + 36 \times (0)^2} = 0.163$$

So the proportion of DGMs with medical qualifications is estimated to be 0.8, although the proportion of males is estimated as 0.667 and females as 1.

10.9.4 Cluster samples

In many situations it is useful to consider the overall population as a two-stage hierarchy wherein each individual belongs to a unique group (or *cluster*) at the higher stage. A cluster sample is produced by taking a probability sample of clusters and then taking all the individuals within the chosen clusters. For instance in a nationwide survey of GPs, the GPs could be considered as clustered into areas managed by individual Family Practitioner Committees (FPCs). A cluster sample survey would then proceed by sampling a few FPCs (the clusters) and then giving the questionnaire to every GP 'within' every chosen FPC. The sample of FPCs could, perhaps, be drawn using simple random sampling (but see Section 10.9.5); notice that it is the clusters and not the individual sampling units (GPs) that are sampled. Another example is a survey of personal health of the individuals in a town where the households are the clusters.

There are two advantages of cluster sampling compared with simple random sampling. First, fieldwork is reduced in situations where enumerators are used. A simple random sample is likely to spread the sample over the entire population whereas in a cluster sample it is grouped (i.e., clustered) in some convenient way. Thus in the GP survey a s.r.s. would probably involve sending enumerators to all parts of the country, whereas a cluster sample only requires travel to certain, randomly selected, areas, saving both money and time.

The second advantage is that a sampling frame is only needed for the clusters, not for the individuals at the lower stage of the hierarchy. Taking the personal health survey example consider how difficult it would be to obtain an up-to-date list of *all* people in a town, regardless of age. On the other hand, a list of households is relatively easily available, for example from rating records or, at least approximately, from maps showing individual dwellings (taking account of obvious cases of multiple occupancy such as high-rise flats).

The disadvantage with cluster samples is that they generally give less precise results than do simple random samples of the same overall size. Cluster sampling is most accurate when the clusters are internally as heterogeneous as possible in terms of the survey subject matter. Most clusterings that give the two advantages just quoted are, unfortunately, internally homogeneous. People living in the same household, for instance, are exposed to the same sanitary conditions and tend to share the same preferences and prejudices. Notice that, whilst clusters are better if they are internally heterogeneous, strata are better if they are internally homogeneous (see Section 10.9.3). This difference is explained by the method of sampling: only *some* clusters are sampled, whereas samples are taken from *all* strata. If the few clusters chosen are to represent the whole population accurately then they should reflect the diversity of the whole, a thing they cannot do if they are very similar internally.

Generally cluster sampling is used for its convenience and low cost rather than for its accuracy. For some extra cost, improved precision in estimation can be obtained by sub-sampling within the chosen clusters (see the next section). Frequently it is beneficial to stratify the clusters, for example FPCs could be stratified by zones (say urban–rural, north–south, etc.) of the country, and then to sample the clusters using stratified sampling. Notice that what is a cluster in one survey might be an individual sampling unit in another. Thus in this section the household has been a cluster but in a household expenditure survey the expenditure on a range of goods by the household is recorded, making the household the individual sampling unit. In a true cluster sample it is quite possible to want to analyse data at both the cluster and individual level.

10.9.5 Multi-stage samples

In a cluster sample every single individual in a chosen cluster is included in the sample. If, instead, the individuals are themselves sampled from within each chosen cluster the method of sampling is called *two-stage*. The clusters are then known as *primary*, or *first-stage sampling units* and the individuals within the clusters as *second-stage* sampling units. Hence FPCs would form the primary sampling units and the GPs within the FPCs would form the secondary sampling units. At the first stage of sampling a few FPCs would be chosen, as before, but now at the second stage a number of GPs would be sampled from within each chosen FPC. Similarly the sampling in Examples 10.2 or 10.5 could be second-stage sampling within a particular regional health authority, this being one of the regions selected at the first stage of sampling.

Clearly there is no reason why this hierarchical sampling should be restricted to two stages. For example, a nationwide sample of nurses could be drawn by selecting regional health authorities at the first stage, districts within the chosen regions at the second stage, hospitals within the chosen districts at the third stage and nurses within the hospitals at the final stage of sampling. Similarly individual people could be sampled through the sequence county-borough-ward-household-person. In general such samples are called *multi-stage*; in fact cluster sampling is a special case of multi-stage sampling where a 100% sample of second-stage units is selected. Many government surveys are of this type, such as the *General Household Survey* (see Section 2.4.5) which uses postal sectors (defined by the first part of the postal code) as the primary sampling units, and then samples households within the chosen postal sectors.

In multi-stage surveys there is considerable scope when choosing the sample at each stage. Perhaps the simplest scheme is to take simple random samples at each stage, but even then there is the question of which sampling fraction to use (e.g., 1 in 100 at the first stage, 1 in 50 at the second, 1 in 20 at the third?). More commonly primary sampling units are chosen with probability proportional to size; for instance a regional health authority with 14 districts is given twice the probability of selection compared with a region with only 7 districts when districts form the second stage sampling units. With a fixed overall size of sample it is possible to allocate the sample in many ways. That is with, say, a sample size of 100 in a two-stage survey it would be possible to sample 10 clusters and then sample 10 second-stage units from within each chosen cluster, or 5 clusters with 20 in each or 2 clusters with 50 in each, and so on. Generally the fewer the number of primary sampling units selected the lower the cost, but also the lower the precision of the estimates obtained. In particular circumstances a suitable balance between cost and precision can be struck. The question of sampling frames may also influence the decision of how to allocate the sample, that is when it is expensive to create sampling frames within clusters it may be necessary to select only a few clusters.

The analysis of multi-stage (and cluster) samples depends upon the method of sampling used and other considerations. Moser and Kalton (1971) give a clear description of the major issues involved, but as in this book, a complete description of the methods of analysis lies beyond their scope. Formulae for estimation are given, over several chapters, by Cochran (1977), but the reader is advised to consult a statistician before undertaking a complex sample survey of the hierarchical kind.

Exercises

10.1 In 1981/82 the total expenditure of the 14 English regional health authorities (RHAs) were (from Table 8.3, but rounded to the nearest hundred million pounds):

Northern	7	N.W. Thames	9	Wessex	6	Mersey	6
Yorkshire	8	N.E. Thames	9	Oxford	4	N. Western	10
Trent	9	S.E. Thames	9	S. Western	7		
E. Anglia	4	S.W. Thames	7	W. Midlands	11		

(a) Calculate the true mean expenditure per RHA.
(b) Calculate the true standard deviation of expenditure per RHA.
(c) There are 14 possible simple random samples of size 13 that could be drawn from the population of RHAs. For each of these 14 samples calculate the sample mean expenditure per RHA.
(d) Calculate the mean of the 14 values obtained in part (c) (this is the average value of the sample means).
(e) Calculate the standard deviation of the 14 values obtained in part (c) (this is the standard error of the sample means).
(f) According to theory the sample mean is an unbiased estimate of the population mean, in which case your answer to part (d) should be equal to your answer to part (a). Check that this is the case.

(g) According to theory the standard error of the sample means (the answer to part (e)) is $S\sqrt{(1-f)/n}$ where S is the true standard deviation (the answer to part (b)). Check that this is the case.

(h) Draw histograms of the raw data and the sample means (from part (c)) on separate graphs. Consider what would happen if you tried to overlay these histograms on the same set of axes. This should convince you that the spread of mean values is less than the spread of original data values.

10.2 The GP data in Appendix 2 are a sample of patients. However, for the sake of an exercise, consider the 32 patients as a population in its own right from which samples can be drawn.

(a) Draw a simple random sample of size 8 from the 32 patients. To be able to check your answers you should number the patients consecutively using Appendix 2 as your sampling frame. Then use Table A.6 starting at row 6 column 1 and reading left to right across.

(b) Estimate the mean systolic blood pressure from your sample.

(c) Estimate the standard error of this mean.

(d) Estimate the proportion of patients with BMI in class 1 from your sample.

(e) Estimate the standard error of this proportion.

(f) Estimate the total number of patients with a BMI in class 1 from your sample.

(g) Estimate the standard error of this total.

10.3 Now select a stratified sample from the GP 'population' using smoking status as the stratification factor.

(a) Draw a proportionately stratified sample of overall size 8. Again to be able to check your answers you should number the patients within each smoking group consecutively using the exercise set as the sampling frame. Then use Table A.6 starting at column 1 in the bottom row and reading from left to right. When you reach the end of the bottom row simply go to the first column of the first row and continue. Draw the random numbers for non-smokers first and then continue from where this selection finishes to draw random numbers for smokers.

(b) Answer Questions 10.2(b)–(g) for non-smokers.

(c) Answer Questions 10.2(b)–(g) for smokers.

(d) Answer Questions 10.2(b)–(g) overall (smokers and non-smokers combined).

10.4 As in Question 10.1, you are in the totally unrealistic situation of having complete knowledge of the GP 'population' of size N = 32. Compare your answers to Questions 10.2 and 10.3 with these 'population' values (the answers to the exercises of earlier chapters give virtually all the required information). How well has your 1 in 4 sample performed?

Notice that the exercises of earlier chapters have shown that smoking status has no significant effect on blood pressure or BMI so the stratified sample should not be expected to perform any better than the simple random sample for the overall estimates.

Questions 10.2 and 10.3 have involved only one sample of each type. To obtain more understanding of the problems when sampling it would be useful to repeat all the estimation with various different samples and then to look at the variation in the answers obtained.

Appendices

Appendix 1: Hospital ward data

The following data are derived from a real-life random sample of one month's discharges from a ward of an acute hospital.

Source of admission	Sex	No. of operations	Diagnosis code	Age	Disposal	Length of stay
5	2	0	2989	84	2	24
5	2	0	486	77	1	3
5	2	1	2500	52	1	13
5	2	0	3592	56	1	3
5	1	0	436	70	1	19
5	1	1	4439	67	2	65
5	1	0	4254	61	1	31
6	2	0	4151	73	6	15
2	2	0	3229	82	1	5
6	2	0	2830	70	2	30
2	2	1	280	93	1	3
6	2	0	*	82	1	11
2	2	0	2918	46	1	8
3	2	0	9651	33	1	1
2	2	0	4511	69	1	24
2	2	0	7802	77	1	1
2	2	0	6827	49	2	4
2	2	0	4939	22	1	4
3	2	0	9691	43	1	2
2	2	0	436	81	6	2
3	2	0	9650	35	1	1
2	2	0	4320	59	1	54
2	2	0	5184	76	1	12
3	2	0	9654	16	1	0
2	2	1	410	67	1	30
2	2	0	410	76	1	4
2	2	0	*	89	1	19
2	2	0	3429	71	2	44
2	2	0	*	84	2	24

Source of admission	Sex	No. of operations	Diagnosis code	Age	Disposal	Length of stay
2	2	0	*	79	1	4
2	2	1	7800	78	2	22
3	2	0	9690	34	3	3
2	2	0	075	18	1	4
3	2	0	9627	36	4	0
2	2	0	410	92	6	0
2	2	0	5198	82	2	24
2	2	0	303	38	4	2
2	2	0	9657	53	1	1
2	2	0	410	80	2	7
2	2	1	1519	91	1	3
2	2	0	7890	58	1	4
3	2	0	9694	31	1	0
2	2	3	4552	39	1	3
3	1	0	8540	1	1	1
2	2	0	4280	72	1	7
5	1	1	5301	62	3	4
2	2	0	7291	52	1	2
2	2	0	436	58	6	1
2	2	0	9916	81	1	30
2	2	0	7802	72	1	4
2	2	1	7847	72	1	3
2	2	0	4511	39	1	6
2	2	0	9429	84	1	24
2	2	0	4151	83	1	18
2	2	0	2500	67	1	5
2	2	0	436	93	6	22

* Denotes a missing value; Sex: 1 = male, 2 = female; Diagnosis codes are ICD 9th revision; Length of stay is measured in days; Age is measured in years.

Source of admission codes

1 = waiting (elective admission) list
2 = emergency
3 = accident
4 = booked readmission
5 = transfer
6 = booked admission
7 = baby born in hospital
8 = other

Disposal codes

1 = home
2 = transfer to hospital
3 = transfer to psychiatric hospital
4 = self-discharge
5 = dead with post mortem
6 = dead without post mortem
7 = transfer to convalescent home
8 = other

Appendix 2: GP data

The following data are derived from a real-life random sample of 32 men aged 45–64 selected from the list of a certain General Practice.

Smoking status	Body mass index	Systolic blood pressure	Diastolic blood pressure
0	1	152	71
0	1	124	77
0	1	105	61
1	1	146	96
1	2	167	120
0	0	156	94
0	0	133	89
0	2	144	81
0	1	186	138
1	0	103	75
0	1	98	67
0	1	131	87
1	1	155	99
0	1	163	90
1	1	136	74
1	1	129	66
0	1	170	112
0	2	160	85
0	1	142	86
1	1	142	82
0	1	115	76
1	1	201	119
0	1	129	83
0	1	158	92
0	2	113	70
1	0	149	84
1	0	157	98
1	2	132	78
0	2	146	88
0	1	175	103
1	1	142	79
0	0	118	68

Smoking status: 0 = non-smoker, 1 = smoker.
Body mass index (BMI) = weight/(height)^2:
using the scale of Garrow (1981) 0 = 20 to less than 25kg/m^2 ('desirable');
1 = 25 to less than 30kg/m^2; 2 = 30 to less than 40kg/m^2 ('clinically obese').
Blood pressures both measured in mmHg.

Appendix 3: Summary of the most commonly used procedures in statistical inference

Note: Here it is assumed that the population is infinite, the sample is drawn by simple random sampling and normal approximations are reasonable. Population standard deviations are taken to be unknown.

Formulae for estimation

Parameter (or function of parameters) to be estimated	Point estimate	Confidence interval
μ	$\bar{x}$	$\bar{x} \pm t_{n-1}\, s/\sqrt{n}$
$\mu_1 - \mu_2$ (independent data) $(\sigma_1 = \sigma_2)$	$\bar{x}_1 - \bar{x}_2$	$\bar{x}_1 - \bar{x}_2 \pm t_{n_1+n_2-2}\, s_p \sqrt{1/n_1 + 1/n_2}$
$\mu_1 - \mu_2$ (independent data) $(\sigma_1 \neq \sigma_2;\ \text{large } n)$	$\bar{x}_1 - \bar{x}_2$	$\bar{x}_1 - \bar{x}_2 \pm z\sqrt{s_1^2/n_1 + s_2^2/n_2}$
$\mu_1 - \mu_2$ (paired data)	$\bar{x}_1 - \bar{x}_2$	$\bar{x}_d \pm t_{n-1}\, s_d/\sqrt{n}$
π	p	$p \pm z\sqrt{p(1-p)/n}$
$\pi_1 - \pi_2$ (independent data)	$p_1 - p_2$	$p_1 - p_2 \pm z \sqrt{\dfrac{p_1(1-p_1)}{n_1} + \dfrac{p_2(1-p_2)}{n_2}}$
$\pi_1 - \pi_2$ (paired data)	$p_1 - p_2$	$p_1 - p_2 \pm \dfrac{z}{n} \sqrt{n_{SF} + n_{FS} - \dfrac{(n_{SF} - n_{FS})^2}{n}}$

Procedures for hypothesis tests

Null hypothesis (H_0)	Test statistic	Distribution of test statistic when H_0 is true
$\mu = \mu_0$	$\dfrac{\bar{x} - \mu_0}{s/\sqrt{n}}$	T_{n-1}
$\mu_1 = \mu_2$ (independent data) $(\sigma_1 = \sigma_2)$	$\dfrac{\bar{x}_1 - \bar{x}_2}{s_p\sqrt{1/n_1 + 1/n_2}}$	$T_{n_1+n_2-2}$
$\mu_1 = \mu_2$ (independent data) $(\sigma_1 \neq \sigma_2;$ large n)	$\dfrac{\bar{x}_1 - \bar{x}_2}{\sqrt{s_1^2/n_1 + s_2^2/n_2}}$	Normal (Z)
$\mu_1 = \mu_2$ (paired data)	$\dfrac{\bar{x}_d}{s_d/\sqrt{n}}$	T_{n-1}
$\sigma_1 = \sigma_2$	s_1^2/s_2^2	F_{n_1-1,n_2-1}
$\pi = \pi_0$	$\dfrac{p - \pi_0}{\sqrt{\pi_0(1 - \pi_0)/n}}$	Normal (Z)
$\pi_1 = \pi_2$ (independent data)	$\dfrac{p_1 - p_2}{\sqrt{p_c(1 - p_c)(1/n_1 + 1/n_2)}}$	Normal (Z)
$\pi_1 = \pi_2$ (paired data)	$\dfrac{n_{SF} - n_{FS}}{\sqrt{n_{SF} + n_{FS}}}$	Normal (Z)

Appendix 4: Tables

Table A.1 The standard normal distribution. Cumulative probabilities (p) for the standard normal (Z) distribution. This table gives the probability (p) of Z taking a value below z, where z ranges from 0 to 3.69 in steps of 0.01. The first two figures of z label the rows and the third labels the columns; for example $P(Z < 1.24) = 0.8925$.

z	0.00	0.01	0.02	0.03	0.04	0.05	0.06	0.07	0.08	0.09
0.0	0.5000	0.5040	0.5080	0.5120	0.5160	0.5199	0.5239	0.5279	0.5319	0.5359
0.1	0.5398	0.5438	0.5478	0.5517	0.5557	0.5596	0.5636	0.5675	0.5714	0.5753
0.2	0.5793	0.5832	0.5871	0.5910	0.5948	0.5987	0.6026	0.6064	0.6103	0.6141
0.3	0.6179	0.6217	0.6255	0.6293	0.6331	0.6368	0.6406	0.6443	0.6480	0.6517
0.4	0.6554	0.6591	0.6628	0.6664	0.6700	0.6736	0.6772	0.6808	0.6844	0.6879
0.5	0.6915	0.6950	0.6985	0.7019	0.7054	0.7088	0.7123	0.7157	0.7190	0.7224
0.6	0.7257	0.7291	0.7324	0.7357	0.7389	0.7422	0.7454	0.7486	0.7517	0.7549
0.7	0.7580	0.7611	0.7642	0.7673	0.7704	0.7734	0.7764	0.7794	0.7823	0.7852
0.8	0.7881	0.7910	0.7939	0.7967	0.7995	0.8023	0.8051	0.8078	0.8106	0.8133
0.9	0.8159	0.8186	0.8212	0.8238	0.8264	0.8289	0.8315	0.8340	0.8365	0.8389
1.0	0.8413	0.8438	0.8461	0.8485	0.8508	0.8531	0.8554	0.8577	0.8599	0.8621
1.1	0.8643	0.8665	0.8686	0.8708	0.8729	0.8749	0.8770	0.8790	0.8810	0.8830
1.2	0.8849	0.8869	0.8888	0.8907	0.8925	0.8944	0.8962	0.8980	0.8997	0.9015
1.3	0.9032	0.9049	0.9066	0.9082	0.9099	0.9115	0.9131	0.9147	0.9162	0.9177
1.4	0.9192	0.9207	0.9222	0.9236	0.9251	0.9265	0.9279	0.9292	0.9306	0.9319
1.5	0.9332	0.9345	0.9357	0.9370	0.9382	0.9394	0.9406	0.9418	0.9429	0.9441
1.6	0.9452	0.9463	0.9474	0.9484	0.9495	0.9505	0.9515	0.9525	0.9535	0.9545
1.7	0.9554	0.9564	0.9573	0.9582	0.9591	0.9599	0.9608	0.9616	0.9625	0.9633
1.8	0.9641	0.9649	0.9656	0.9664	0.9671	0.9678	0.9686	0.9693	0.9699	0.9706
1.9	0.9713	0.9719	0.9726	0.9732	0.9738	0.9744	0.9750	0.9756	0.9761	0.9767
2.0	0.9772	0.9778	0.9783	0.9788	0.9793	0.9798	0.9803	0.9808	0.9812	0.9817
2.1	0.9821	0.9826	0.9830	0.9834	0.9838	0.9842	0.9846	0.9850	0.9854	0.9857
2.2	0.9861	0.9864	0.9868	0.9871	0.9875	0.9878	0.9881	0.9884	0.9887	0.9890
2.3	0.9893	0.9896	0.9898	0.9901	0.9904	0.9906	0.9909	0.9911	0.9913	0.9916
2.4	0.9918	0.9920	0.9922	0.9925	0.9927	0.9929	0.9931	0.9932	0.9934	0.9936
2.5	0.9938	0.9940	0.9941	0.9943	0.9945	0.9946	0.9948	0.9949	0.9951	0.9952

Table A.1 Continued

z	0.00	0.01	0.02	0.03	0.04	0.05	0.06	0.07	0.08	0.09
2.6	0.9953	0.9955	0.9956	0.9957	0.9959	0.9960	0.9961	0.9962	0.9963	0.9964
2.7	0.9965	0.9966	0.9967	0.9968	0.9969	0.9970	0.9971	0.9972	0.9973	0.9974
2.8	0.9974	0.9975	0.9976	0.9977	0.9977	0.9978	0.9979	0.9979	0.9980	0.9981
2.9	0.9981	0.9982	0.9982	0.9983	0.9984	0.9984	0.9985	0.9985	0.9986	0.9986
3.0	0.9987	0.9987	0.9987	0.9988	0.9988	0.9989	0.9989	0.9989	0.9990	0.9990
3.1	0.9990	0.9991	0.9991	0.9991	0.9992	0.9992	0.9992	0.9992	0.9993	0.9993
3.2	0.9993	0.9993	0.9994	0.9994	0.9994	0.9994	0.9994	0.9995	0.9995	0.9995
3.3	0.9995	0.9995	0.9995	0.9996	0.9996	0.9996	0.9996	0.9996	0.9996	0.9997
3.4	0.9997	0.9997	0.9997	0.9997	0.9997	0.9997	0.9997	0.9997	0.9997	0.9998
3.5	0.9998	0.9998	0.9998	0.9998	0.9998	0.9998	0.9998	0.9998	0.9998	0.9998
3.6	0.9998	0.9998	0.9999	0.9999	0.9999	0.9999	0.9999	0.9999	0.9999	0.9999

Table A.2   Student's t distribution. Cumulative probabilities for Student's t distribution. For different values of the cumulative probability (p) and the degrees of freedom (d.f.) this table gives the value t such that the probability of Student's t taking a value below t is p.

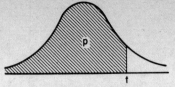

				p			
d.f.	0.90	0.95	0.975	0.99	0.995	0.999	0.9995
1	3.078	6.314	12.706	31.821	63.657	318.309	636.621
2	1.886	2.920	4.303	6.965	9.925	22.327	31.599
3	1.638	2.353	3.182	4.541	5.841	10.215	12.925
4	1.533	2.132	2.776	3.747	4.604	7.173	8.610
5	1.476	2.015	2.571	3.365	4.032	5.893	6.869
6	1.440	1.943	2.447	3.143	3.707	5.208	5.959
7	1.415	1.895	2.365	2.998	3.499	4.785	5.408
8	1.397	1.860	2.306	2.896	3.355	4.501	5.041
9	1.383	1.833	2.262	2.821	3.250	4.297	4.781
10	1.372	1.812	2.228	2.764	3.169	4.144	4.587
11	1.363	1.796	2.201	2.718	3.106	4.025	4.437
12	1.356	1.782	2.179	2.681	3.055	3.930	4.318
13	1.350	1.771	2.160	2.650	3.012	3.852	4.221
14	1.345	1.761	2.145	2.624	2.977	3.787	4.140
15	1.341	1.753	2.131	2.602	2.947	3.733	4.073
16	1.337	1.746	2.120	2.583	2.921	3.686	4.015
17	1.333	1.740	2.110	2.567	2.898	3.646	3.965
18	1.330	1.734	2.101	2.552	2.878	3.610	3.922
19	1.328	1.729	2.093	2.539	2.861	3.579	3.883
20	1.325	1.725	2.086	2.528	2.845	3.552	3.850
21	1.323	1.721	2.080	2.518	2.831	3.527	3.819
22	1.321	1.717	2.074	2.508	2.819	3.505	3.792
23	1.319	1.714	2.069	2.500	2.807	3.485	3.768
24	1.318	1.711	2.064	2.492	2.797	3.467	3.745
25	1.316	1.708	2.060	2.485	2.787	3.450	3.725
26	1.315	1.706	2.056	2.479	2.779	3.435	3.707
27	1.314	1.703	2.052	2.473	2.771	3.421	3.690
28	1.313	1.701	2.048	2.467	2.763	3.408	3.674
29	1.311	1.699	2.045	2.462	2.756	3.396	3.659
30	1.310	1.697	2.042	2.457	2.750	3.385	3.646
40	1.303	1.684	2.021	2.423	2.704	3.307	3.551
60	1.296	1.671	2.000	2.390	2.660	3.232	3.460
80	1.292	1.664	1.990	2.374	2.639	3.195	3.416
100	1.290	1.660	1.984	2.364	2.626	3.174	3.390
120	1.289	1.658	1.980	2.358	2.617	3.160	3.373
∞	1.282	1.645	1.960	2.326	2.576	3.090	3.291

Table A.3 The chi-squared distribution. Cumulative probabilities for the chi-squared distribution. For different values of the cumulative probability (p) and the degrees of freedom (d.f.) this table gives the value x such that the probability of chi-squared taking a value below x is p.

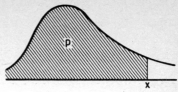

d.f.	0.90	0.95	0.975	p 0.99	0.995	0.999	0.9995
1	2.706	3.841	5.024	6.635	7.880	10.832	12.127
2	4.605	5.991	7.378	9.211	10.598	13.823	15.208
3	6.251	7.815	9.348	11.345	12.839	16.272	17.747
4	7.779	9.488	11.143	13.277	14.862	18.477	20.007
5	9.236	11.071	12.833	15.087	16.752	20.520	22.120
6	10.645	12.592	14.450	16.813	18.550	22.464	24.120
7	12.017	14.067	16.013	18.476	20.280	24.329	26.037
8	13.362	15.507	17.535	20.091	21.957	26.131	27.888
9	14.684	16.919	19.023	21.668	23.591	27.893	29.684
10	15.987	18.307	20.484	23.211	25.190	29.603	31.437
11	17.275	19.675	21.920	24.726	26.758	31.277	33.174
12	18.549	21.026	23.337	26.218	28.303	32.920	34.854
13	19.812	22.362	24.736	27.690	29.822	34.549	36.505
14	21.064	23.685	26.119	29.143	31.321	36.140	38.132
15	22.307	24.996	27.489	30.580	32.805	37.711	39.759
16	23.542	26.296	28.846	32.002	34.270	39.276	41.340
17	24.769	27.588	30.192	33.410	35.723	40.808	42.904
18	25.990	28.870	31.527	34.807	37.160	42.326	44.476
19	27.204	30.144	32.853	36.192	38.587	43.843	46.005
20	28.412	31.411	34.170	37.568	40.001	45.331	47.550
21	29.615	32.671	35.480	38.934	41.407	46.823	49.049
22	30.814	33.925	36.781	40.292	42.800	48.287	50.571
23	32.007	35.173	38.077	41.640	44.187	49.758	52.044
24	33.196	36.416	39.365	42.982	45.563	51.200	53.546
25	34.382	37.653	40.647	44.316	46.934	52.651	54.996
26	35.563	38.886	41.924	45.644	48.294	54.075	56.477
27	36.742	40.114	43.195	46.966	49.651	55.509	57.909
28	37.916	41.338	44.462	48.281	50.998	56.916	59.374
29	39.088	42.558	45.723	49.591	52.342	58.334	60.788
30	40.256	43.773	46.980	50.894	53.676	59.727	62.237
35	46.059	49.803	53.204	57.345	60.280	66.649	69.250
40	51.806	55.759	59.344	63.696	66.776	73.436	76.155
45	57.506	61.657	65.412	69.961	73.176	80.112	82.941
50	63.168	67.506	71.422	76.158	79.499	86.695	89.627
60	74.398	79.083	83.301	88.384	91.962	99.657	102.793
70	85.528	90.533	95.026	100.433	104.225	112.375	115.699
80	96.579	101.881	106.632	112.336	116.339	124.900	128.393
90	107.566	113.147	118.139	124.126	128.314	137.301	140.911
100	118.499	124.344	129.565	135.814	140.189	149.531	153.353
110	129.386	135.483	140.920	147.423	151.964	161.686	165.594
120	140.234	146.570	152.215	158.960	163.666	173.702	177.803

Table A.4 The F distribution. Cumulative probabilities for the F distribution. For different values of the first (d.f. [1]) and second (d.f. [2]) degrees of freedom this table gives the values f_1 and f_2 such that:
(i) the probability of F taking a value below f_1 (the upper figure) is 0.95;
(ii) the probability of F taking a value below f_2 (the lower figure) is 0.99.

d.f. [2]	1	2	3	4	5	6	7	8	9	10	11
1	161.4	199.5	215.7	224.6	230.2	234.0	236.8	238.9	240.5	241.9	243.0
	4052	4999.5	5403	5625	5764	5859	5928	5981	6022	6056	6083
2	18.51	19.00	19.16	19.25	19.30	19.33	19.35	19.37	19.38	19.40	19.40
	98.50	99.00	99.17	99.25	99.30	99.33	99.36	99.37	99.39	99.40	99.41
3	10.13	9.55	9.28	9.12	9.01	8.94	8.89	8.85	8.81	8.79	8.76
	34.12	30.82	29.46	28.71	28.24	27.91	27.67	27.49	27.35	27.23	27.13
4	7.71	6.94	6.59	6.39	6.26	6.16	6.09	6.04	6.00	5.96	5.94
	21.20	18.00	16.69	15.98	15.52	15.21	14.98	14.80	14.66	14.55	14.45
5	6.61	5.79	5.41	5.19	5.05	4.95	4.88	4.82	4.77	4.74	4.70
	16.26	13.27	12.06	11.39	10.97	10.67	10.46	10.29	10.16	10.05	9.96
6	5.99	5.14	4.76	4.53	4.39	4.28	4.21	4.15	4.10	4.06	4.03
	13.75	10.92	9.78	9.15	8.75	8.47	8.26	8.10	7.98	7.87	7.79
7	5.59	4.74	4.35	4.12	3.97	3.87	3.79	3.73	3.68	3.64	3.60
	12.25	9.55	8.45	7.85	7.46	7.19	6.99	6.84	6.72	6.62	6.54
8	5.32	4.46	4.07	3.84	3.69	3.58	3.50	3.44	3.39	3.35	3.31
	11.26	8.65	7.59	7.01	6.63	6.37	6.18	6.03	5.91	5.81	5.73
9	5.12	4.26	3.86	3.63	3.48	3.37	3.29	3.23	3.18	3.14	3.10
	10.56	8.02	6.99	6.42	6.06	5.80	5.61	5.47	5.35	5.26	5.18
10	4.96	4.10	3.71	3.48	3.33	3.22	3.14	3.07	3.02	2.98	2.94
	10.04	7.56	6.55	5.99	5.64	5.39	5.20	5.06	4.94	4.85	4.77
11	4.84	3.98	3.59	3.36	3.20	3.09	3.01	2.95	2.90	2.85	2.82
	9.65	7.21	6.22	5.67	5.32	5.07	4.89	4.74	4.63	4.54	4.46
12	4.75	3.89	3.49	3.26	3.11	3.00	2.91	2.85	2.80	2.75	2.72
	9.33	6.93	5.95	5.41	5.06	4.82	4.64	4.50	4.39	4.30	4.22
13	4.67	3.81	3.41	3.18	3.03	2.92	2.83	2.77	2.71	2.67	2.63
	9.07	6.70	5.74	5.21	4.86	4.62	4.44	4.30	4.19	4.10	4.02
14	4.60	3.74	3.34	3.11	2.96	2.85	2.76	2.70	2.65	2.60	2.57
	8.86	6.51	5.56	5.04	4.69	4.46	4.28	4.14	4.03	3.94	3.86
15	4.54	3.68	3.29	3.06	2.90	2.79	2.71	2.64	2.59	2.54	2.51
	8.68	6.36	5.42	4.89	4.56	4.32	4.14	4.00	3.89	3.80	3.73
16	4.49	3.63	3.24	3.01	2.85	2.74	2.66	2.59	2.54	2.49	2.46
	8.53	6.23	5.29	4.77	4.44	4.20	4.03	3.89	3.78	3.69	3.62
17	4.45	3.59	3.20	2.96	2.81	2.70	2.61	2.55	2.49	2.45	2.41
	8.40	6.11	5.18	4.67	4.34	4.10	3.93	3.79	3.68	3.59	3.52
18	4.41	3.55	3.16	2.93	2.77	2.66	2.58	2.51	2.46	2.41	2.37
	8.29	6.01	5.09	4.58	4.25	4.01	3.84	3.71	3.60	3.51	3.43
19	4.38	3.52	3.13	2.90	2.74	2.63	2.54	2.48	2.42	2.38	2.34
	8.18	5.93	5.01	4.50	4.17	3.94	3.77	3.63	3.52	3.43	3.36
20	4.35	3.49	3.10	2.87	2.71	2.60	2.51	2.45	2.39	2.35	2.31
	8.10	5.85	4.94	4.43	4.10	3.87	3.70	3.56	3.46	3.37	3.29
21	4.32	3.47	3.07	2.84	2.68	2.57	2.49	2.42	2.37	2.32	2.28
	8.02	5.78	4.87	4.37	4.04	3.81	3.64	3.51	3.40	3.31	3.24
22	4.30	3.44	3.05	2.82	2.66	2.55	2.46	2.40	2.34	2.30	2.26
	7.95	5.72	4.82	4.31	3.99	3.76	3.59	3.45	3.35	3.26	3.18
23	4.28	3.42	3.03	2.80	2.64	2.53	2.44	2.37	2.32	2.27	2.24
	7.88	5.66	4.76	4.26	3.94	3.71	3.54	3.41	3.30	3.21	3.14
24	4.26	3.40	3.01	2.78	2.62	2.51	2.42	2.36	2.30	2.25	2.22
	7.82	5.61	4.72	4.22	3.90	3.67	3.50	3.36	3.26	3.17	3.09
25	4.24	3.39	2.99	2.76	2.60	2.49	2.40	2.34	2.28	2.24	2.20
	7.77	5.57	4.68	4.18	3.85	3.63	3.46	3.32	3.22	3.13	3.06
26	4.23	3.37	2.98	2.74	2.59	2.47	2.39	2.32	2.27	2.22	2.18
	7.72	5.53	4.64	4.14	3.82	3.59	3.42	3.29	3.18	3.09	3.02
27	4.21	3.35	2.96	2.73	2.57	2.46	2.37	2.31	2.25	2.20	2.17
	7.68	5.49	4.60	4.11	3.78	3.56	3.39	3.26	3.15	3.06	2.99
28	4.20	3.34	2.95	2.71	2.56	2.45	2.36	2.29	2.24	2.19	2.15
	7.64	5.45	4.57	4.07	3.75	3.53	3.36	3.23	3.12	3.03	2.96
29	4.18	3.33	2.93	2.70	2.55	2.43	2.35	2.28	2.22	2.18	2.14
	7.60	5.42	4.54	4.04	3.73	3.50	3.33	3.20	3.09	3.00	2.93
30	4.17	3.32	2.92	2.69	2.53	2.42	2.33	2.27	2.21	2.16	2.13
	7.56	5.39	4.51	4.02	3.70	3.47	3.30	3.17	3.07	2.98	2.91
35	4.12	3.27	2.87	2.64	2.49	2.37	2.29	2.22	2.16	2.11	2.07
	7.42	5.27	4.40	3.91	3.59	3.37	3.20	3.07	2.96	2.88	2.80
40	4.08	3.23	2.84	2.61	2.45	2.34	2.25	2.18	2.12	2.08	2.04
	7.31	5.18	4.31	3.83	3.51	3.29	3.12	2.99	2.89	2.80	2.73

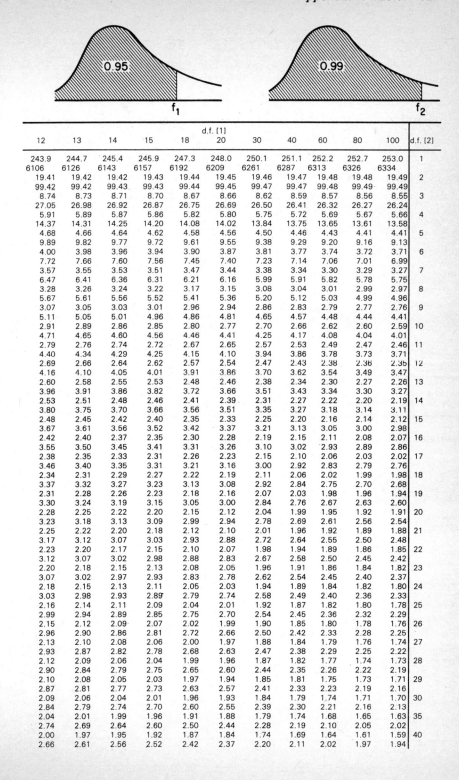

12	13	14	15	18	d.f. [1] 20	30	40	60	80	100	d.f. [2]
243.9	244.7	245.4	245.9	247.3	248.0	250.1	251.1	252.2	252.7	253.0	1
6106	6126	6143	6157	6192	6209	6261	6287	6313	6326	6334	
19.41	19.42	19.42	19.43	19.44	19.45	19.46	19.47	19.48	19.48	19.49	2
99.42	99.42	99.43	99.43	99.44	99.45	99.47	99.47	99.48	99.49	99.49	
8.74	8.73	8.71	8.70	8.67	8.66	8.62	8.59	8.57	8.56	8.55	3
27.05	26.98	26.92	26.87	26.75	26.69	26.50	26.41	26.32	26.27	26.24	
5.91	5.89	5.87	5.86	5.82	5.80	5.75	5.72	5.69	5.67	5.66	4
14.37	14.31	14.25	14.20	14.08	14.02	13.84	13.75	13.65	13.61	13.58	
4.68	4.66	4.64	4.62	4.58	4.56	4.50	4.46	4.43	4.41	4.41	5
9.89	9.82	9.77	9.72	9.61	9.55	9.38	9.29	9.20	9.16	9.13	
4.00	3.98	3.96	3.94	3.90	3.87	3.81	3.77	3.74	3.72	3.71	6
7.72	7.66	7.60	7.56	7.45	7.40	7.23	7.14	7.06	7.01	6.99	
3.57	3.55	3.53	3.51	3.47	3.44	3.38	3.34	3.30	3.29	3.27	7
6.47	6.41	6.36	6.31	6.21	6.16	5.99	5.91	5.82	5.78	5.75	
3.28	3.26	3.24	3.22	3.17	3.15	3.08	3.04	3.01	2.99	2.97	8
5.67	5.61	5.56	5.52	5.41	5.36	5.20	5.12	5.03	4.99	4.96	
3.07	3.05	3.03	3.01	2.96	2.94	2.86	2.83	2.79	2.77	2.76	9
5.11	5.05	5.01	4.96	4.86	4.81	4.65	4.57	4.48	4.44	4.41	
2.91	2.89	2.86	2.85	2.80	2.77	2.70	2.66	2.62	2.60	2.59	10
4.71	4.65	4.60	4.56	4.46	4.41	4.25	4.17	4.08	4.04	4.01	
2.79	2.76	2.74	2.72	2.67	2.65	2.57	2.53	2.49	2.47	2.46	11
4.40	4.34	4.29	4.25	4.15	4.10	3.94	3.86	3.78	3.73	3.71	
2.69	2.66	2.64	2.62	2.57	2.54	2.47	2.43	2.38	2.36	2.35	12
4.16	4.10	4.05	4.01	3.91	3.86	3.70	3.62	3.54	3.49	3.47	
2.60	2.58	2.55	2.53	2.48	2.46	2.38	2.34	2.30	2.27	2.26	13
3.96	3.91	3.86	3.82	3.72	3.66	3.51	3.43	3.34	3.30	3.27	
2.53	2.51	2.48	2.46	2.41	2.39	2.31	2.27	2.22	2.20	2.19	14
3.80	3.75	3.70	3.66	3.56	3.51	3.35	3.27	3.18	3.14	3.11	
2.48	2.45	2.42	2.40	2.35	2.33	2.25	2.20	2.16	2.14	2.12	15
3.67	3.61	3.56	3.52	3.42	3.37	3.21	3.13	3.05	3.00	2.98	
2.42	2.40	2.37	2.35	2.30	2.28	2.19	2.15	2.11	2.08	2.07	16
3.55	3.50	3.45	3.41	3.31	3.26	3.10	3.02	2.93	2.89	2.86	
2.38	2.35	2.33	2.31	2.26	2.23	2.15	2.10	2.06	2.03	2.02	17
3.46	3.40	3.35	3.31	3.21	3.16	3.00	2.92	2.83	2.79	2.76	
2.34	2.31	2.29	2.27	2.22	2.19	2.11	2.06	2.02	1.99	1.98	18
3.37	3.32	3.27	3.23	3.13	3.08	2.92	2.84	2.75	2.70	2.68	
2.31	2.28	2.26	2.23	2.18	2.16	2.07	2.03	1.98	1.96	1.94	19
3.30	3.24	3.19	3.15	3.05	3.00	2.84	2.76	2.67	2.63	2.60	
2.28	2.25	2.22	2.20	2.15	2.12	2.04	1.99	1.95	1.92	1.91	20
3.23	3.18	3.13	3.09	2.99	2.94	2.78	2.69	2.61	2.56	2.54	
2.25	2.22	2.20	2.18	2.12	2.10	2.01	1.96	1.92	1.89	1.88	21
3.17	3.12	3.07	3.03	2.93	2.88	2.72	2.64	2.55	2.50	2.48	
2.23	2.20	2.17	2.15	2.10	2.07	1.98	1.94	1.89	1.86	1.85	22
3.12	3.07	3.02	2.98	2.88	2.83	2.67	2.58	2.50	2.45	2.42	
2.20	2.18	2.15	2.13	2.08	2.05	1.96	1.91	1.86	1.84	1.82	23
3.07	3.02	2.97	2.93	2.83	2.78	2.62	2.54	2.45	2.40	2.37	
2.18	2.15	2.13	2.11	2.05	2.03	1.94	1.89	1.84	1.82	1.80	24
3.03	2.98	2.93	2.89	2.79	2.74	2.58	2.49	2.40	2.36	2.33	
2.16	2.14	2.11	2.09	2.04	2.01	1.92	1.87	1.82	1.80	1.78	25
2.99	2.94	2.89	2.85	2.75	2.70	2.54	2.45	2.36	2.32	2.29	
2.15	2.12	2.09	2.07	2.02	1.99	1.90	1.85	1.80	1.78	1.76	26
2.96	2.90	2.86	2.81	2.72	2.66	2.50	2.42	2.33	2.28	2.25	
2.13	2.10	2.08	2.06	2.00	1.97	1.88	1.84	1.79	1.76	1.74	27
2.93	2.87	2.82	2.78	2.68	2.63	2.47	2.38	2.29	2.25	2.22	
2.12	2.09	2.06	2.04	1.99	1.96	1.87	1.82	1.77	1.74	1.73	28
2.90	2.84	2.79	2.75	2.65	2.60	2.44	2.35	2.26	2.22	2.19	
2.10	2.08	2.05	2.03	1.97	1.94	1.85	1.81	1.75	1.73	1.71	29
2.87	2.81	2.77	2.73	2.63	2.57	2.41	2.33	2.23	2.19	2.16	
2.09	2.06	2.04	2.01	1.96	1.93	1.84	1.79	1.74	1.71	1.70	30
2.84	2.79	2.74	2.70	2.60	2.55	2.39	2.30	2.21	2.16	2.13	
2.04	2.01	1.99	1.96	1.91	1.88	1.79	1.74	1.68	1.65	1.63	35
2.74	2.69	2.64	2.60	2.50	2.44	2.28	2.19	2.10	2.05	2.02	
2.00	1.97	1.95	1.92	1.87	1.84	1.74	1.69	1.64	1.61	1.59	40
2.66	2.61	2.56	2.52	2.42	2.37	2.20	2.11	2.02	1.97	1.94	

Table A.4   Continued

d.f. [2]	1	2	3	4	5	6	7	8	9	10	11
45	4.06	3.20	2.81	2.58	2.42	2.31	2.22	2.15	2.10	2.05	2.01
	7.23	5.11	4.25	3.77	3.45	3.23	3.07	2.94	2.83	2.74	2.67
50	4.03	3.18	2.79	2.56	2.40	2.29	2.20	2.13	2.07	2.03	1.99
	7.17	5.06	4.20	3.72	3.41	3.19	3.02	2.89	2.78	2.70	2.63
60	4.00	3.15	2.76	2.53	2.37	2.25	2.17	2.10	2.04	1.99	1.95
	7.08	4.98	4.13	3.65	3.34	3.12	2.95	2.82	2.72	2.63	2.56
70	3.98	3.13	2.74	2.50	2.35	2.23	2.14	2.07	2.02	1.97	1.93
	7.01	4.92	4.07	3.60	3.29	3.07	2.91	2.78	2.67	2.59	2.51
80	3.96	3.11	2.72	2.49	2.33	2.21	2.13	2.06	2.00	1.95	1.91
	6.96	4.88	4.04	3.56	3.26	3.04	2.87	2.74	2.64	2.55	2.48
90	3.95	3.10	2.71	2.47	2.32	2.20	2.11	2.04	1.99	1.94	1.90
	6.92	4.85	4.01	3.53	3.23	3.01	2.84	2.72	2.61	2.52	2.45
100	3.94	3.09	2.70	2.46	2.31	2.19	2.10	2.03	1.97	1.93	1.89
	6.90	4.82	3.98	3.51	3.21	2.99	2.82	2.69	2.59	2.50	2.43

d.f. [1] spans columns 1–11.

12	13	14	15	18	d.f. [1] 20	30	40	60	80	100	d.f. [2]
1.97	1.94	1.92	1.89	1.84	1.81	1.71	1.66	1.60	1.57	1.55	45
2.61	2.55	2.51	2.46	2.36	2.31	2.14	2.05	1.96	1.91	1.88	
1.95	1.92	1.89	1.87	1.81	1.78	1.69	1.63	1.58	1.54	1.52	50
2.56	2.51	2.46	2.42	2.32	2.27	2.10	2.01	1.91	1.86	1.82	
1.92	1.89	1.86	1.84	1.78	1.75	1.65	1.59	1.53	1.50	1.48	60
2.50	2.44	2.39	2.35	2.25	2.20	2.03	1.94	1.84	1.78	1.75	
1.89	1.86	1.84	1.81	1.75	1.72	1.62	1.57	1.50	1.47	1.45	70
2.45	2.40	2.35	2.31	2.20	2.15	1.98	1.89	1.78	1.73	1.70	
1.88	1.84	1.82	1.79	1.73	1.70	1.60	1.54	1.48	1.45	1.43	80
2.42	2.36	2.31	2.27	2.17	2.12	1.94	1.85	1.75	1.69	1.65	
1.86	1.83	1.80	1.78	1.72	1.69	1.59	1.53	1.46	1.43	1.41	90
2.39	2.33	2.29	2.24	2.14	2.09	1.92	1.82	1.72	1.66	1.62	
1.85	1.82	1.79	1.77	1.71	1.68	1.57	1.52	1.45	1.41	1.39	100
2.37	2.31	2.27	2.22	2.12	2.07	1.89	1.80	1.69	1.63	1.60	

Table A.5 Critical values of the correlation coefficient.

n	d.f.	Significance level		
		5%	1%	0.1%
3	1	0.997	0.999	1.000
4	2	.950	.990	0.999
5	3	.878	.959	.991
6	4	.811	.917	.974
7	5	.754	.875	.951
8	6	0.707	0.834	0.925
9	7	.666	.798	.898
10	8	.632	.765	.872
11	9	.602	.735	.847
12	10	.576	.708	.823
13	11	0.553	0.684	0.801
14	12	.532	.661	.780
15	13	.514	.641	.760
16	14	.497	.623	.742
17	15	.482	.606	.725
18	16	0.468	0.590	0.708
19	17	.456	.575	.693
20	18	.444	.561	.679
21	19	.433	.549	.665
22	20	.423	.537	.652
27	25	0.381	0.487	0.597
32	30	.349	.449	.554
37	35	.325	.418	.519
42	40	.304	.393	.490
47	45	.288	.372	.465
52	50	0.273	0.354	0.443
62	60	.250	.325	.408
72	70	.232	.302	.380
82	80	.217	.283	.357
92	90	.205	.267	.338
102	100	.195	.254	.321

Table A.6 Random numbers.

```
1 0 2 7 5 3 9 6 2 3 7 1 5 0 5 4 3 6 2 3 5 4 5 1 5 0 1 4 2 8 0 2 1 2 2 9 8 8 8 7
8 5 9 0 2 2 5 8 5 2 9 0 2 2 7 6 9 5 7 0 0 2 8 4 7 4 6 9 0 6 1 3 9 8 8 6 0 6 5 0
4 4 3 3 2 9 8 8 9 0 4 9 0 7 5 5 6 9 5 0 2 0 2 7 5 9 5 1 9 7 5 3 5 7 0 4 2 2 2 6
4 7 5 7 2 2 5 2 7 5 7 4 5 3 1 1 7 6 1 1 2 1 1 6 1 2 4 4 3 1 8 9 1 6 9 1 4 7 7 5
0 3 2 0 5 4 2 0 7 0 5 6 7 7 5 9 9 5 6 0 1 9 7 5 2 9 9 4 1 1 2 3 5 9 3 0 1 4 4 7
6 4 1 7 1 8 4 3 9 7 3 7 6 6 5 5 8 6 0 8 7 4 5 0 4 3 4 3 2 3 2 9 1 6 2 4 1 5 6 2
9 1 1 4 6 1 7 1 0 3 4 0 1 5 6 9 4 4 4 6 5 4 6 6 3 5 0 1 8 7 6 1 2 3 7 6 3 6 8 0
2 7 7 1 2 9 9 3 5 2 8 9 6 4 7 8 3 2 9 7 6 5 2 8 9 9 8 2 4 1 1 0 9 7 5 2 4 1 9 1
1 2 9 6 1 7 7 0 7 2 7 6 1 7 9 3 3 8 2 6 7 2 9 6 2 8 7 3 2 7 6 4 7 8 1 6 7 2 8 1
5 4 3 0 6 1 1 3 6 0 5 0 6 1 5 6 4 0 2 0 1 9 2 2 3 0 6 1 4 3 8 9 6 0 0 9 8 2 3 9
8 3 3 2 9 9 2 9 3 0 0 6 1 9 7 1 1 1 3 2 6 9 1 7 8 6 3 4 5 0 7 6 3 7 4 1 7 6 5 4
2 7 1 7 2 5 6 1 9 1 7 6 1 9 5 4 9 9 7 3 9 7 2 1 4 4 8 7 3 9 6 3 2 4 2 2 7 4 3 0
4 0 8 9 2 1 8 8 5 6 8 4 1 1 7 5 7 4 8 8 2 3 5 5 4 8 9 8 1 9 4 8 7 9 8 1 9 2 6 2
5 1 6 6 1 7 4 8 2 9 9 6 0 0 8 3 8 1 2 3 5 8 0 9 2 1 3 9 3 9 2 0 8 3 4 6 3 0 7 5
9 5 2 2 6 3 3 4 5 8 9 1 7 8 2 2 5 0 2 2 7 7 2 1 1 4 1 9 5 8 6 6 4 9 2 5 0 3 5 1
9 3 8 3 7 3 7 0 8 0 8 8 7 1 8 5 6 4 4 4 5 7 5 0 1 9 8 2 6 0 7 7 3 8 9 5 9 3 3 3
4 2 0 2 3 3 1 8 3 3 5 5 9 6 6 6 8 8 3 8 1 6 8 0 7 7 5 1 1 7 9 6 4 9 7 6 9 9 2 8
4 2 4 2 1 3 3 3 6 6 0 0 1 8 3 7 5 8 8 0 5 4 3 2 0 0 9 6 2 5 1 6 1 5 3 7 3 4 1 2
6 6 7 1 6 7 5 4 7 9 2 5 6 4 3 4 8 2 1 5 2 8 9 7 8 8 8 4 8 4 5 1 6 2 9 0 1 7 7 1
7 3 0 5 5 3 8 5 6 3 1 8 0 6 4 7 7 1 0 0 3 2 3 1 5 9 7 2 3 4 2 8 7 0 8 3 1 2 9 0
0 2 8 0 1 2 2 4 3 4 7 8 2 2 5 0 5 7 0 2 0 7 0 1 1 3 0 0 7 8 8 0 9 4 9 3 1 4 5 3
2 2 8 9 8 1 3 2 3 2 7 2 4 8 9 2 9 5 7 5 8 8 5 6 7 5 5 3 7 9 1 7 5 3 8 1 5 4 1 7
9 4 4 5 6 4 8 4 1 7 2 8 0 6 5 7 7 1 9 6 8 1 3 6 3 7 6 5 4 2 6 2 4 3 8 4 4 5 2 3
1 0 3 0 0 5 0 7 2 1 3 4 5 9 1 8 8 5 9 5 2 1 8 7 7 3 1 6 7 8 3 7 1 5 9 8 1 6 6 6
7 3 3 9 2 1 9 4 0 1 8 4 2 8 2 0 5 0 3 5 5 7 8 2 8 8 1 3 5 2 5 3 7 6 7 3 6 8 2 2
4 7 9 1 8 7 3 6 4 5 6 9 0 3 0 1 2 4 2 5 1 3 6 4 4 2 7 4 3 6 6 7 7 7 6 7 0 0 9 2
3 9 2 4 2 6 7 7 6 2 3 7 8 2 4 6 9 3 9 6 8 2 7 5 7 5 1 6 9 5 0 5 3 0 6 8 8 3 0 2
7 7 2 9 0 9 1 2 4 1 7 7 2 9 5 7 3 4 8 9 9 4 9 5 4 5 7 0 5 9 8 5 3 8 0 4 0 4 8 0
0 4 7 8 2 0 0 7 1 7 1 5 6 8 1 2 3 8 2 6 0 1 9 0 6 8 3 0 8 3 8 0 1 9 8 9 9 8 6 5
8 3 8 1 5 3 0 8 0 9 2 3 2 2 6 1 9 9 4 1 2 7 9 0 3 5 4 3 0 7 0 9 6 2 2 6 4 5 8 3
9 7 6 7 7 4 5 4 9 6 1 4 6 3 2 8 9 8 1 1 1 8 3 3 8 2 6 0 9 0 4 1 3 3 1 1 7 7 5 9
5 2 8 0 2 6 8 9 1 3 3 8 7 0 0 8 7 3 2 2 6 4 7 0 8 3 4 4 4 9 2 4 2 0 9 3 1 2 5 9
8 0 6 9 4 3 2 7 3 3 5 6 3 9 8 8 7 3 3 1 2 4 4 4 8 7 3 3 0 8 2 1 4 0 0 6 7 7 9 1
0 0 4 8 2 4 0 8 7 3 9 2 3 7 1 9 6 9 8 7 9 1 7 9 8 6 2 7 4 7 9 1 3 1 7 0 5 3 5 2
1 4 9 1 9 7 3 7 5 3 4 0 4 6 2 6 2 9 2 5 9 6 4 2 5 7 2 2 9 4 3 4 5 9 7 1 2 3 5 9
5 0 6 2 2 8 5 1 9 4 1 0 1 5 1 8 0 6 0 2 3 9 9 4 1 3 9 1 5 4 5 0 6 0 2 7 2 6 6 8
1 7 5 9 5 3 0 8 5 8 0 6 8 0 0 0 7 5 7 1 9 5 1 3 7 6 9 1 2 4 5 5 3 4 0 9 9 7 1 2
7 3 1 7 9 9 4 5 8 5 2 8 6 3 1 7 9 9 3 1 2 4 6 2 7 5 8 2 7 8 8 9 2 7 5 9 1 8 6 2
3 7 9 5 7 4 9 6 2 5 4 4 9 5 6 6 4 2 0 2 3 1 4 8 8 2 2 1 7 6 8 7 8 6 7 5 0 7 9 5
7 6 9 5 1 8 7 6 7 6 2 8 1 8 6 0 4 4 9 2 7 6 0 9 4 6 9 6 3 9 3 7 2 7 1 2 3 0 4 4
```

Table A.7 English female life table 1980–1982.

Age x	l_x	d_x	q_x	T_x	$\overset{\circ}{e}_x$	Age x	l_x	d_x	q_x	T_x	$\overset{\circ}{e}_x$
0	100 000	984	0.00984	7 700 187	77.002	55	93 034	581	0.00624	2 328 129	25.025
1	99 016	71	0.00072	7 601 014	76.766	56	92 453	634	0.00686	2 235 381	24.178
2	98 945	45	0.00045	7 502 036	75.820	57	91 819	690	0.00752	2 143 241	23.342
3	98 900	31	0.00031	7 403 116	74.855	58	91 129	750	0.00824	2 051 782	22.515
4	98 869	25	0.00025	7 304 232	73.878	59	90 379	815	0.00901	1 961 003	21.698
5	98 844	22	0.00022	7 205 376	72.896	60	89 564	883	0.00986	1 871 026	20.890
6	98 822	20	0.00020	7 106 543	71.913	61	88 681	955	0.01077	1 781 898	20.093
7	98 802	19	0.00019	7 007 731	70.927	62	87 726	1031	0.01176	1 693 688	19.307
8	98 783	19	0.00019	6 908 939	69.941	63	86 695	1113	0.01284	1 606 471	18.530
9	98 764	18	0.00018	6 810 165	68.954	64	85 582	1198	0.01400	1 520 326	17.765
10	98 746	18	0.00018	6 711 410	67.966	65	84 384	1289	0.01528	1 435 336	17.010
11	98 728	18	0.00018	6 612 673	66.979	66	83 095	1387	0.01669	1 351 590	16.266
12	98 710	17	0.00018	6 513 953	65.991	67	81 708	1494	0.01828	1 269 180	15.533
13	98 693	18	0.00019	6 415 252	65.002	68	80 214	1611	0.02008	1 188 211	14.813
14	98 675	22	0.00022	6 316 569	64.014	69	78 603	1739	0.02212	1 108 793	14.106
15	98 653	25	0.00026	6 217 903	63.028	70	76 864	1877	0.02443	1 031 048	13.414
16	98 628	30	0.00030	6 119 262	62.044	71	74 987	2028	0.02704	955 111	12.737
17	98 598	32	0.00033	6 020 649	61.062	72	72 959	2187	0.02998	881 126	12.077
18	98 566	35	0.00035	5 922 066	60.082	73	70 772	2356	0.03329	809 247	11.435
19	98 531	34	0.00035	5 823 518	59.103	74	68 416	2530	0.03698	739 640	10.811
20	98 497	35	0.00035	5 725 004	58.124	75	65 886	2708	0.04110	672 474	10.207
21	98 462	35	0.00036	5 626 524	57.144	76	63 178	2884	0.04566	607 927	9.622
22	98 427	35	0.00036	5 528 079	56.164	77	60 294	3058	0.05072	546 176	9.059
23	98 392	36	0.00037	5 429 670	55.184	78	57 236	3226	0.05637	487 397	8.516
24	98 356	38	0.00038	5 331 296	54.204	79	54 010	3387	0.06271	431 760	7.994
25	98 318	38	0.00039	5 232 959	53.225	80	50 623	3534	0.06982	379 431	7.495
26	98 280	41	0.00041	5 134 660	52.245	81	47 089	3663	0.07779	330 563	7.020
27	98 239	42	0.00043	5 036 400	51.267	82	43 426	3765	0.08669	285 295	6.570

Table A.7 Continued

Age x	l_x	d_x	q_x	T_x	$\overset{\circ}{e}_x$
28	98 197	44	0.00045	4 938 182	50.288
29	98 153	48	0.00048	4 840 007	49.311
30	98 105	51	0.00052	4 741 877	48.335
31	98 054	54	0.00056	4 643 797	47.359
32	98 000	59	0.00060	4 545 770	46.385
33	97 941	64	0.00065	4 447 799	45.413
34	97 877	70	0.00071	4 349 890	44.442
35	97 807	75	0.00078	4 252 047	43.474
36	97 732	83	0.00085	4 154 277	42.5C7
37	97 649	92	0.00093	4 056 586	41.543
38	97 557	100	0.00103	3 958 983	40.581
39	97 457	111	0.00114	3 861 475	39.622
40	97 346	123	0.00127	3 764 073	38.667
41	97 223	137	0.00141	3 666 787	37.715
42	97 086	153	0.00157	3 569 632	36.768
43	96 933	170	0.00176	3 472 622	35.825
44	96 763	190	0.00196	3 375 772	34.887
45	96 573	212	0.00219	3 279 103	33.955
46	96 361	236	0.00245	3 182 634	33.028
47	96 125	263	0.00274	3 086 390	32.1C8
48	95 862	293	0.00305	2 990 394	31.195
49	95 569	325	0.00340	2 894 676	30.289
50	95 244	360	0.00378	2 799 267	29.3S0
51	94 884	398	0.00419	2 704 200	28.5C0
52	94 486	439	0.00465	2 609 512	27.618
53	94 047	483	0.00514	2 515 241	26.744
54	93 564	530	0.00567	2 421 432	25.880

Age x	l_x	d_x	q_x	T_x	$\overset{\circ}{e}_x$
83	39 661	3832	0.09661	243 743	6.146
84	35 829	3851	0.10750	205 992	5.749
85	31 978	3813	0.11922	172 087	5.381
86	28 165	3706	0.13160	142 019	5.042
87	24 459	3534	0.14448	115 716	4.731
88	20 925	3300	0.15772	93 038	4.446
89	17 625	3017	0.17116	73 783	4.186
90	14 608	2698	0.18468	57 690	3.949
91	11 910	2360	0.19814	44 457	3.733
92	9 550	2019	0.21143	33 755	3.534
93	7 531	1690	0.22442	25 243	3.352
94	5 841	1384	0.23703	18 584	3.182
95	4 457	1111	0.24914	13 461	3.020
96	3 346	873	0.26096	9 582	2.864
97	2 473	676	0.27331	6 693	2.706
98	1 797	516	0.28715	4 574	2.545
99	1 281	388	0.30330	3 048	2.380
100	893	288	0.32252	1 972	2.210
101	605	209	0.34538	1 232	2.038
102	396	148	0.37231	739	1.866
103	248	100	0.40349	422	1.698
104	148	65	0.43881	227	1.535
105	83	40	0.47780	115	1.380
106	43	22	0.51960	54	1.234
107	21	12	0.56277	23	1.100
108	9	5	0.60521	9	0.976
109	4	3	0.64382	3	0.862
110	1	1	0.67391	1	0.755

Tables A.1–A.4 were produced on the statistical computer package INSTAT available from the Statistical Services Centre, Reading University, P.O. Box 217, Whiteknights, Reading RG6 2AN.

Table A.5 is reproduced from *A Basic Course in Statistics* by G.M. Clarke and D. Cooke (2nd Edition, Edward Arnold (1983)) with permission of the authors.

Table A.6 is reproduced from *Statistical Methods in Agriculture and Experimental Biology* by R. Mead and R.N. Curnow (Chapman and Hall (1983)) with permission of the publishers and authors.

Table A.7 has been extracted from *English Life Tables No. 14* (Office of Population Censuses and Surveys, Series DS No. 7 (1987)) with permission of OPCS.

Appendix 5: Glossary of abbreviations

BMI	body mass index
CI	confidence interval
CV	coefficient of variation
d.f.	degrees of freedom
DGM	District General Manager
DHA	district health authority
DHSS	Department of Health and Social Security
DIS	District Information System
FPC	Family Practitioner Committee
GP	general practitioner
IACC	Inter-Authority Comparisons and Consultancy
ICD	International Classification of Diseases
NHS	National Health Service
OPCS	Office of Population Censuses and Surveys
PI	Performance Indicator
RAWP	Resource Allocation Working Party
RCGP	Royal College of General Practitioners
RHA	regional health authority
RIS	Regional Information System
RGSS	regression sum of squares
RMS	residual mean square
RSS	residual sum of squares
s.r.s	simple random sample
s.d.	standard deviation
SFR	standardized fertility ratio
SMR	standardized mortality ratio
SS	sum of squares
VDU	visual display unit
WHO	World Health Organization

Appendix 6: Solutions to exercises

(where appropriate)

3.1 (a)
90 to less than 110		3
110	"	130	6
130	"	140	4
140	"	150	7
150	"	160	5
160	"	170	3
170	"	190	3
190	"	210	1

(b)
less than	110	9%
"	"	130	28%
"	"	140	41%
"	"	150	63%
"	"	160	78%
"	"	170	87%
"	"	190	97%
"	"	210	100%

3.3 (b)

	0	1	2	Total
No	3	13	4	20
Yes	3	7	2	12
Total	6	20	6	32

   (c) 133   142   145
        149   142   149.5

4.1 (a) 143.03; (b) 143; (c) 142; (d) 103; (e) 14.25; (f) 23.86;
    (g) 113; (h) 186.

4.2 (a) 143.28; (b) 144.3; (c) 24.45; (d) 15.67.
    (Using the solution to 3.1 (a) and (b).)

4.3 (b) Diastolic: 2.62; systolic: 0.03.

4.4 (a) 0.375; (b) 1:1.67; (c) 18.75%.

4.5 (c) Smokers: 146.58; non-smokers: 140.90.
    (d) Smokers: 23.62; non-smokers; 24.35.

4.6 (b) 3.19; (c) 3; (d) 2.45; (e) 1.25.

4.7 (a) 3295.1; (b) 534.2

4.8 (a)
100	100	100	100
120	113	123	115
140	139	152	138
185	183	198	177
309	299	342	227
467	443	507	339
646	610	697	474
882	811	903	647

   (b) Population size.

4.9 (b) 165.51, 166.94, 167.08, 166.24, 165.68, 165.35, 165.44, 165.31, 163.44, 162.61, 162.20, 161.25.
(c) 1.130, 0.971, 0.885, 1.014.

4.10 (b) 16.21; (c) 1.62; (d) 81.0%; (e) 6.96; (f) 42.5;
(g) 1.63.

5.1 (a) 59.78 per thousand; (b) 27.6, 95.4, 126.2, 73.6, 23.6, 4.5, 0.4;
(c) 1.76 per woman; (d) 12.8 per thousand.

5.2 (a) 1:0.98; (b) 8.7 per thousand.

(c)

males	7.9	0.1	0.3	0.6	0.6	1.4	5.0	13.2	62.9
females	8.1	0.6	0.3	0.2	0.3	1.0	3.7	7.7	46.7

(d) 10.1 per thousand; (e) 0.86; (f) 10.1 per thousand.

5.3 (a) 0.16%; (b) 5263.6 thousand; (c) 0.10%.

5.4 (a)

x	l_x	d_x	p_x	q_x
30	90 000	121	.9986575	.0013425
31	89 879	129	.9985651	.0014349
32	89 750	124	.9986152	.0013848
33	89 626	131	.9985337	.0014663
34	89 495	116	.9987074	.0012926

5.6 (a) 0.0176; (b) 0.0144.

5.7 (a) 1991; (b) 5889.

6.1 (a) Ace (clubs), 2 (clubs), . . ., Queen (clubs), King (clubs)

       .   .

       .   .

       .   .

Ace (spades), 2 (spades), . . ., Queen (spades), King (spades);
(b) yes; (c) 0.077.

6.2 (a) 2, 3, 4, 5, 6, 7, 8, 9, 10, 11, 12; (b) no; (c) 0.139; (d) 0.5.

6.3 0.0014.

6.4 (a) 91 390; (b) 52 360; (c) 0.573; (d) 0.1.

6.5 (a) 0.091; (b) 0.111.

6.6 (a) (i) no, (ii) no; (b) (i) yes, (ii) no; (c) (i) no, (ii) yes;
(d) (i) no, (ii) no; (e) (i) no, (ii) yes.

6.7 (a) 0.1; (b) 0.1; (c) 0.9.

6.8 (a) 0.498; (b) 0.00001.

6.9 (a) 0.016; (b) 10.95 (using the binomial).

6.10 0.147 (using the geometric).

6.11 (a) 0.865; (b) 0.982.

6.12 (a) 0.841; (b) 0.997; (c) 9.2.

7.1 (a) 143.03 ± 8.60; (b) 0.375; (c) 0.375 ± 0.220; (d) no;
(e) 5.68 ± 17.96; (f) 0.10 ± 0.29; (g) fail to reject the assertion;
(h) test statistic $= -2.58, 0.005 > P > 0.0025$; (i) test statistic $=$
$-1.41, P = 0.0793$; (j) test statistic $= 0.65, P > 0.1$; (k) test
statistic $= -0.23, P = 0.4090$.

7.2 test statistic $= 21.34, P < 0.0001$ (using unequal variances).

7.3 57.8 ± 13.3.

7.4 323.

7.5 (a) not significant; (b) 11.

8.1 (b) 0.84; (c) $y = 41.32 + 1.17x$;

(d) regression	12 567	1	12 567
residual	5 082	30	169
total	17 649	31	

(e) 0.71; (j) 41.32 ± 24.54; (k) 1.17 ± 0.28;
(l) 158.32; (m) 158.32 ± 36.62.

8.2 (a) x_2, bread; (b) x_1 and x_3, salt and high sodium; (c) model with
x_1 and x_3: $y = -656 - 75.9x_1 + 81.4x_3$;
(d) $y = 22\ 043 - 10.9(\text{year})$; $R^2 = 0.947$, RMS $= 272$.

8.3 (b) $y = -10.4 + 8.99x$; (c) $\log y = 1.68 + 0.322x$; (d) linear r^2
$= 0.89$, exponential $r^2 = 0.98$.

9.1 $X^2 = 1.30$ (or 1.08 with a continuity correction), $P > 0.1$.

9.2 $P = 0.044$ (by Fisher's exact test).

9.3 (a) $X^2 = 3.6, 0.1 > P > 0.05$; (b) There could be an order effect (it
is better to vary the order).

9.4 (a) $P = 0.12$; (b) $P = 0.667$.

9.5 (a) 0.0017; (b) 0.0006368; (c) 2.67; (d) 2.67.

9.6   (a) 2.47;   (b) 0.93 to 6.57.

9.7   (a) $X^2 = 148.2, P < 0.0005$.

9.8   (a) $X^2 = 2.88, P > 0.1$;   (b) no ($X^2_{(t)} = 0.07$).

9.9   Model 7: proportion of day cases differs in the two districts, and this is consistent across age groups. No difference in proportions of day cases in children and adults in the two districts.

9.10 $X^2 = 11.1$ (pooling 5 to 12), $0.025 > P > 0.01$.

10.1   (a) 7.57;   (b) 2.10;   (d) 7.57;   (e) 0.16.

10.2   (a) Random numbers selected: 17, 18, 8, 23, 29, 16, 24, 15;
      (b) 146.5;   (c) 4.62;   (d) 0.625;   (e) 0.158;   (f) 20;   (g) 5.06.

10.3   (a) Random numbers selected: 18, 9, 12, 10, 14 (non-smokers), 2, 12, 6 (smokers);   (b) 143.0, 7.81, 0.6, 0.212, 12, 4.24;   (c) 146.0, 9.66, 0.667, 0.289, 8, 3.46;   (d) 144.125, 6.08, 0.625, 0.171, 20, 5.48.

10.4

True values	Non-smokers	Smokers	Overall
Mean	140.90	146.58	143.03
Proportion	0.65	0.583	0.625
Total	13	7	20

References

Alderson, M. (1983). *An Introduction to Epidemiology*, 2nd edition. MacMillan, London.

Anscombe, F.J. (1973). Graphs in statistical analysis. *The American Statistician*, **27**, 17–21.

Armitage, P. and Berry, G. (1987). *Statistical Methods in Medical Research*, 2nd edition. Blackwell Scientific Publications, Oxford.

Bailey, N.T.J. (1962). Calculating the scale of inpatient accommodation. In *Towards a Measure of Medical Care*. Nuffield Provincial Hospitals Trust, Oxford University Press, Oxford.

Barber, B. and Johnson, D. (January 1973). The presentation of acute hospital in-patient statistics. *The Hospital and Health Services Review*, 11–14.

Barker, D.J.P. and Rose, G. (1984). *Epidemiology in Medical Practice*, 3rd edition. Churchill Livingstone, London.

Barnett, V. (1974). *Elements of Sampling Theory*. Hodder and Stoughton, London.

Barr, A. and Logan, R.F. (1977). Policy alternatives for resource allocation. *Lancet*, **1**, 994–7.

Bartholomew, D.J. and Forbes, A.F. (1979). *Statistical Techniques for Manpower Planning*. John Wiley, Chichester.

Bauman, K.E. (1980). *Research Methods for Community Health and Welfare*. Oxford University Press, New York.

Bishop, Y.M.M., Fienberg, S.E. and Holland, F.W. (1975). *Discrete Multivariate Analysis*. Massachusetts Institute of Technology Press.

Breslow, N.E. and Day, N.E. (1980). *Statistical Methods in Cancer Research. Volume 1 – The Analysis of Case-Control Studies*. International Agency for Research in Cancer, Lyon.

Central Statistical Office (annually). *Annual Abstract of Statistics*. HMSO, London.

Central Statistical Office (annually). *Key Data*. HMSO, London.

Central Statistical Office (monthly). *Monthly Digest of Statistics*. HMSO, London.

Central Statistical Office (annually). *Regional Trends*. HMSO, London.

Central Statistical Office (annually). *Social Trends*. HMSO, London.

Central Statistical Office (periodically). *Guide to Official Statistics*. HMSO, London.

Chatfield, C. (1980). *The Analysis of Time Series: An Introduction*. Chapman and Hall, London.

Clarke, G.M. and Cooke, D. (1983). *A Basic Course in Statistics*, 2nd edition. Edward Arnold, London.

Cochran, W.G. (1954). Some methods for strengthening the common χ^2 tests. *Biometrics*, **10**, 417–51.

Cochran, W.G. (1977). *Sampling Techniques*, 3rd edition. John Wiley, New York.

Conover, W.J. (1980). *Practical Nonparametric Statistics*, 2nd edition. John Wiley, New York.

Cook, G.A., Sellwood, J., Francis, L.M.A. and Court, S. (1987). Uptake of rubella vaccine among susceptible adults. *Community Medicine*, **9**, 254–9.

Coulter, A. (1987). Measuring morbidity. *British Medical Journal*, **294**, 263–4.

Cowie, A. (1986). Medical statistical information - a guide to sources. *Health Libraries Review*, **3**, 203–21.

Cummins, R.O. (1983). Recent changes in salt use and stroke mortality in England and Wales. Any help for the salt-hypertension debate? *Journal of Epidemiology and Community Health*, **37**, 25–8.

Department of Health and Social Security, Office of Population Censuses and Surveys and Welsh Office (1974). *Hospital In-patient Enquiry, Main Tables*. HMSO, London

Department of Health and Social Security (1976). Sharing Resources for Health Care in England. *Report of the Resource Allocation Working Party*. HMSO, London.

Department of Health and Social Security (1985). *Performance Indicators for the NHS: Guidance for Users*. DHSS, London.

Department of Health and Social Security (1986). *A Guide to Health and Social Services Statistics*. HMSO, London.

Department of Health and Social Security and the Welsh Office (quarterly). *Health Trends*. HMSO, London.

Draper, N.R. and Smith, H. (1981). *Applied Regression Analysis*, 2nd edition. John Wiley, New York.

Doll, R. and Hill, A.B. (1950). Smoking and carcinoma of the lung. *British Medical Journal*, **2**, 739–48.

Doll, R. and Hill, A.B. (1964). Mortality in relation to smoking: ten years' observations of British doctors. *British Medical Journal*, **1**, 1399–410.

Doll, R. and Peto, R. (1976). Mortality in relation to smoking: 20 years' observations on male British doctors. *British Medical Journal*, **2**, 1525–36.

Edwards, J.D. and Wilkins, R.G. (1987). Atrial fibrillation precipitated by acute hypovolaemia. *British Medical Journal*, **294**, 283–4.

Ellis, I.O., Hinton, C.P., MacNay, J., Elston, C.W., Robins, A., Owainati, A.A.R.S., Blamey, R.W., Baldwin, R.W. and Ferry, B. (1985). Immunocytochemical staining of breast carcinoma with the monoclonal antibody NCRC 11: a new prognostic indicator. *British Medical Journal*, **290**, 881–3.

Everitt, B.S. (1977). *The Analysis of Contingency Tables*. Chapman and Hall, London.

Fanshel, S. and Bush, J.W. (1970). A health-status index and its application to health-services outcomes. *Operations Research*, **18**, 1021–66.

Field, D.J., Milner, A.D., Hopkin, I.E. and Madeley, R.J. (1985). Changing overall workload in neonatal units. *British Medical Journal*, **290**, 1539–42.

Fienberg, S.E. (1977). *The Analysis of Cross-Classified Categorical Data*. Massachusetts Institute of Technology Press.

Fisher, R.A. and Yates, F. (1957). *Statistical Tables for Biological, Agricultural and Medical Research*, 5th edition. Oliver and Boyd, Edinburgh.

Folks, J.L. (1981). *Ideas of Statistics*. John Wiley, New York.

Forbes, A.F. (1971). Non-parametric methods of estimating the survivor function. *The Statistician*, **20**, 27–52.

Forster, D.P. (1977). Mortality, morbidity and resource allocation. *Lancet*, **1**, 997–8.

Gardner, M.J. and Altman, D.G. (1986). Confidence intervals rather than P values: estimation rather than hypothesis testing. *British Medical Journal*, **292**, 746–50.

Garrow, J.S. (1981). *Treat Obesity Seriously. A Clinical Manual*. Churchill Livingstone, London.

Gart, J.J. (1969). An exact test for comparing matched proportions in cross-over designs. *Biometrika*, **56**, 75–80.

Gilchrist, W.J., Lee, Y.C., Tan, H.C., MacDonald, J.B. and Williams, B.O. (1987). Prospective study of drug reporting by general practitioners for an elderly population referred to a geriatric service. *British Medical Journal*, **294**, 289–90.

Gore, S.M. and Altman, D.G. (1982). *Statistics in Practice*. British Medical Association, London.

Government Statistical Service (annually). *Government Statistics: a Brief Guide to Sources*. HMSO, London.

Huff, D. (1954). *How to Lie with Statistics*. Pelican, Harmondsworth.

Ilersic, A.R. and Pluck, R.A. (1977). *Statistics*, 14th edition. HFL, London.

Institute of Statisticians (1984). Data Protection Act. *The Professional Statistician*, **3**, part 8, 1–2.

Jarman (1983). Identification of underpriviliged areas. *British Medical Journal*, **286**, 1705–9.

Jarman (1984). Underpriviliged areas: validation and distribution of scores. *British Medical Journal*, **289**, 1587–91.

Kendall, M.G. and Stuart, A. (1973). *The Advanced Theory of Statistics, Volume 2*, 3rd edition. Griffin, London.

Körner, E. (1984a). A report on the collection and use of information about hospital clinical activity in the National Health Service. *Steering Group on Health Information*. HMSO, London.

Körner, E. (1984b). A report on the collection and use of information about patient transport services in the National Health Service.

Steering Group on Health Information. HMSO, London.

Körner, E. (1984c). A report on the collection and use of information about manpower in the National Health Service. *Steering Group on Health Information*. HMSO, London.

Körner, E. (1984d). A further report on information about activity in hospitals and the community in the National Health Service. *Steering Group on Health Information*. HMSO, London.

Körner, E. (1984e). Supplement to the first and fourth reports to the Secretary of State. *Steering Group on Health Information*. HMSO, London.

Körner, E. (1984f). A report on the collection and use of information about services for and in the community in the National Health Service. *Steering Group on Health Information*. HMSO, London.

Körner, E. (1984g). A report on the collection and use of financial information in the National Health Service. *Steering Group on Health Information*. IIMSO, London.

Levin, J. (1985). *Trends in Perinatal Mortality in the West Midlands 1974–1981*, MSc thesis. Department of Applied Statistics, University of Reading.

Lilienfield, A.M. and Lilienfield, D.E. (1980). *Foundations of Epidemiology*. Oxford University Press, New York.

Macfarlane, A., Chalmers, I. and Adelstein, A.M. (1980). The role of standardization in the interpretation of perinatal mortality rates. *Health Trends*, **12**, 45–50.

McCullagh, P. and Nelder, J.A. (1983). *Generalised Linear Models*. Chapman and Hall, London.

Mead, R. and Curnow, R.N. (1983). *Statistical Methods in Agriculture and Experimental Biology*. Chapman and Hall, London.

Mednick, S.A. and Baert, A.E. (Eds) (1981). *Prospective Longitudinal Research. An Empiricial Basis for the Primary Prevention of Psychosocial Disorders*. Oxford University Press, New York.

Montgomery, D.C. and Peck, E.A. (1982). *Introduction to Linear Regression Analysis*. John Wiley, New York.

Morgan, M. and Chinn, S. (1983). ACORN group, social class and child health. *Journal of Epidemiology and Community Health*, **37**, 196–203.

Moser, C.A. and Kalton, G. (1971). *Survey Methods in Social Investigation*, 2nd edition. Heinemann, London.

Neal, D.G. and Tate, M.J. (1972). *Study of Maternity Services in Reading and Berkshire*. Operational Research (Health Services) Unit, Department of Applied Statistics, University of Reading.

Office of Health Economics (periodically). *Compendium of Health Statistics*. OHE, London.

Office of Population Censuses and Surveys, Social Survey Division (annually). *General Household Survey*. Series GHS, HMSO, London.

Office of Population Censuses and Surveys, HIPE Section (1975). *Classification of Surgical Operations*. OPCS, Titchfield.

Office of Population Censuses and Surveys (1981). *Census 1981 Preliminary Report, England and Wales*. HMSO, London.

Office of Population Censuses and Surveys (1987). *English Life Tables No. 14*. HMSO, London.

Oxford Community Health Care Project (1978). *OXMIS Problem Codes*. OXMIS Publications, Oxford.

Oxford Regional Health Authority (1985). *Hospital Statistics*. Oxford.

Palmer, S.R., West, P.A., and Dodd, P. (1980). Randomness in the RAWP formula: the reliability of mortality data in the allocation of National Health Service revenue. *Journal of Epidemiology and Community Health*, **34**, 212–16.

Pocock, S.J. (1983). *Clinical Trials: A Practical Approach*. John Wiley, Chichester.

Pollard. A.H., Yusuf, F. and Pollard, G.N. (1981). *Demographic Techniques*, 2nd edition. Pergamon, Sydney.

Puska, P. *et al*. (1983). Change in risk factors for coronary heart disease during 10 years of a community intervention programme. North Karelia project. *British Medical Journal*, **287**, 1840–44.

Roberts, A. (1986). Performance indicators for the National Health Service. *Statistical News*, **75**, 75.23–26.

Royal College of General Practitioners (1984). *Occasional paper 26: Classification of diseases, problems and procedures*. RCGP, Edinburgh.

Royal College of General Practitioners, Office of Population Censuses and Surveys and Department of Health and Social Security (1958). *Morbidity Statistics from General Practice*. HMSO, London.

Royal College of General Practitioners, Office of Population Censuses and Surveys and Department of Health and Social Security (1982). *Morbidity Statistics from General Practice 1970–71: Socio-economic analyses*. HMSO, London.

Royal College of General Practitioners, Office of Population Censuses and Surveys and Department of Health and Social Security (1986). *Morbidity Statistics from General Practice 1981–82*. HMSO, London.

Ryan, B.F., Joiner, B.L. and Ryan, T.A. (1985). *Minitab Handbook*, 2nd edition. Duxbury, Boston.

Schlesselman, J.J. (1974). Sample size requirements in cohort and case-control studies of disease. *American Journal of Epidemiology*, **33**, 381–4.

Schlesselman, J.J. (1982). *Case-control Studies: Design, Conduct and Analysis*. Oxford University Press, New York.

Senn, S.J. and Samson, W.B. (1982). Estimating hospital catchment populations. *The Statistician*, **31**, 81–96.

Shaper, A.G., Pocock, S.J., Phillips, A.N. and Walker, M. (1986). Identifiying men at high risk of heart attacks: strategy for use in general practice. *British Medical Journal*, **293**, 474–9.

Shryock, H.S. and Siegel, J.S. (and Stockwell, E.G., Ed.) (1976). *The Methods and Materials of Demography*, Condensed edition. Academic Press, New York.

Siegel, S. (1956). *Nonparametric Statistics for the Behavioural Sciences*. McGraw-Hill, New York.

Snedecor, G.W. and Cochran, W.G. (1980). *Statistical Methods*, 7th edition. Iowa State University Press.

Thirkettle, G.L. (1981). *Weldon's Business Statistics and Statistical Method*, 9th edition. MacDonald and Evans, Plymouth.

Thunhurst, C. (1985). The analysis of small area statistics and planning for health. *The Statistician*, **34**, 93–106.

Tufte, E.R. (1983). *The Visual Display of Quantitative Information*. Graphics Press, Cheshire, Connecticut.

United Nations (annually). *Demographic Yearbook*, UN, New York.

Wilson, S.R. (1985). Analysing case-control data in GLIM. *GLIM Newsletter*, **9**, 22–6.

Woodward, M. (1982). On projecting the number of long-stay mentally handicapped patients in West Berkshire hospitals. *Community Medicine*, **4**, 217–30.

World Health Organization (1977). *International Classification of Diseases, Injuries and Causes of Death*, 9th revision. HMSO, London.

Yates, F. (1934). Contingency tables involving small numbers and the chi-square test. *Journal of the Royal Statistical Society Supplement*, **1**, 217–35.

Yeomans, K.A. (1968). *Statistics for the Social Scientist. Volume I: Introductory Statistics, Volume II: Applied Statistics*. Penguin, Harmondsworth.

Index